Advanced Mine Ventilation

Woodhead Publishing Series in Energy

Advanced Mine Ventilation

Respirable Coal Dust, Combustible Gas and Mine Fire Control

Pramod Thakur, Ph.D.
President, Expert Solutions for Mine safety, LLC
Morgantown, West Virginia, USA

ELSEVIER
WP
WOODHEAD
PUBLISHING
An imprint of Elsevier

Woodhead Publishing is an imprint of Elsevier
The Officers' Mess Business Centre, Royston Road, Duxford, CB22 4QH, United Kingdom
50 Hampshire Street, 5th Floor, Cambridge, MA 02139, United States
The Boulevard, Langford Lane, Kidlington, OX5 1GB, United Kingdom

Notices

Knowledge and best practice in this field are constantly changing. As new research and experience broaden our understanding, changes in research methods, professional practices, or medical treatment may become necessary.

Practitioners and researchers must always rely on their own experience and knowledge in evaluating and using any information, methods, compounds, or experiments described herein. In using such information or methods they should be mindful of their own safety and the safety of others, including parties for whom they have a professional responsibility.

To the fullest extent of the law, neither the Publisher nor the authors, contributors, or editors, assume any liability for any injury and/or damage to persons or property as a matter of products liability, negligence or otherwise, or from any use or operation of any methods, products, instructions, or ideas contained in the material herein.

Library of Congress Cataloging-in-Publication Data
A catalog record for this book is available from the Library of Congress

British Library Cataloguing-in-Publication Data
A catalogue record for this book is available from the British Library

ISBN: 978-0-08-100457-9

For information on all Woodhead Publishing publications visit our website at https://www.elsevier.com/books-and-journals

Publisher: Joe Hayton
Acquisition Editor: Maria Convey
Editorial Project Manager: Charlotte Kent
Production Project Manager: Swapna Srinivasan
Cover Designer: Victoria Pearson

Typeset by TNQ Technologies

This book is dedicated to my family:
Parents: Late Radha and Jagdish Thakur
Wife: Meera, and
Our sons: Drs. Gautam and Anand Thakur
For their love and support.
May it keep all coal miners safe around the world.

Contents

Preface

Maintaining a healthy and safe air environment in an underground coal mine is a sine qua non for efficient coal production. There are three main hazards associated with coal mining air environment. First, coal mining process creates many respirable dusts, such as, coal and silica dust, and diesel particulate matter (DPM) which can be hazardous to human health in high concentrations. Some of these dust clouds can also be explosive.

Secondly, coal seams inherently contain many combustible gases, such as, methane and ethane that can become explosive when mixed with insufficient volumes of air. Several thousand fatalities have occurred because of methane and dust explosions in the coal mines of the world since the beginning of coal mining some 200 years ago. The above two problems are minimized by mine ventilation. Large volumes of air, often 20 tons of air for each ton of coal mined, are circulated through the mine workings to dilute the gas and dust concentrations to safe levels. Proper design of mine ventilation is thus crucial to mine safety.

Thirdly, all coal seams have a tendency for spontaneous combustion (slow oxidation of coal) when it comes in contact with air. If no corrective actions are taken, the coal can catch fire leading to the loss of the mining section and sometimes, the entire mine.

The purpose of this book is to address the four related issues in detail to make coal mining a safe and profitable business. In modern, highly productive coal mines, ventilation alone cannot provide adequate control of pollutants. Special techniques for minimizing the risks of respirable dust, combustible gases and mine fires are needed.

Most of the existing knowledge on mine ventilation has been derived from the following three books published 50–60 years ago:

1. Mine Ventilation: ed. A Roberts (1960).
2. Mine Ventilation and Air Conditioning: ed H Hartmann (1961). The book was reprinted in 1982 and 1997 with only minor changes.
3. Mine Ventilation: A Skochinsky and V. Kamarov (1969)

These books provide good basic knowledge of mine ventilation (and air conditioning in deep metal mines) but they could not cover the new developments in all four critical areas mentioned above. A tremendous amount of research and innovation over the past 50 years have resulted in:

1. Digital (computerized) design of mine ventilation able to compute air flows, pressure losses and the concentrations of many pollutant in all junctions and airways of the mine.

2. Adequate control of respirable dust such that coal workers' pneumoconiosis (CWP) has been eliminated in some countries and minimized in most countries.
3. Efficient drainage of methane pre-mining and post-mining such that mine explosions are minimized, if not, eliminated. Each large coal mine has become a potential field for gas production with added profit.
4. Design of coal mines such that it minimizes the risk of spontaneous combustion of coal and consequently reduces the chances of mine fire.

Advanced Mine Ventilation discusses these four topics in four sections named:

1. Coal Mine Ventilation
2. Respirable dust control
3. Combustible gas control
4. Mine fire control

The author had the unique opportunity to research and work on all four topics over his 50 years' service to the coal industry.

The coal mine ventilation section has six chapters comprising, (1) Underground coal mine atmosphere, (2) Air flow in mine airways, (3) Turbulent dispersion of pollutants in mine airways, (4) Estimation of ventilation air quantity, (5) Ventilation network analysis, and (6) Mechanical and natural ventilation.

The respirable dust control section has six chapters comprising, (7) Health hazards of respirable dusts, (8) Characteristics of respirable coal dust particles, (9) Generation of respirable coal dust, (10) Respirable dust control, (11) Diesel Exhaust control, and (12) Respirable dust sampling and measurements.

The combustible gas control section has eight chapters comprising, (13) Origin of gases in coal mines, (14) Reservoir properties of coal seams, (15) Pre-mining degasification of coal seams, (16) Post-Mining degasification of coal mines, (17) Floor gas emissions and gas outburst, (18) Gas transport in underground coal mines, (19) Measurement and monitoring of mine gases, and (20) Economics of coal mine degasification.

The mine fire management section has four chapters comprising, (21) Spontaneous combustion of coal, (22) Prevention of frictional ignitions, (23) Gas and dust explosions, and (24) Mine sealing and recovery.

Advance mine ventilation is a unique book that is not only an excellent reference book for the practicing mine engineer but also a great text book for two graduate level courses in Mining Engineering program. I recommend teaching mine ventilation and respirable dust control in one semester , and combustible gas control and mine fire management in the second semester.

I owe thanks and gratitude to a number of people for helping me to learn the contents of this book and actually practicing it in the industry for the past 50 years.

Coal mine ventilation: to late Dr. H Hartmann of the University of Alabama who invited me to write chapters in his books on Mine ventilation and air conditioning.

Respirable dust control: to Dr. AK Sinha and late Dr. R Stefanko under whose guidance I did my MS (Characterization of coal dust particles) and Ph.D. theses (Computer-aided analysis of diesel exhaust dispersion in mine airways) respectively.

Combustible gas control: to late William Poundstone and late Eustace Frederick of CONSOL Energy for providing support in discovering and applying various methane control techniques in the mines of CONSOL Energy.

Mine fire management: to late L David Hughes of Andrew Yule and Co and late Donald Mitchell, a friendly consultant, who taught me the elements of mine fire fighting enabling me to fight and control many fires on my own.

My thanks are also due to Joycc Conn for typing the manuscript and Sowmya Devraja for drafting all figures. Finally my thanks are due to Natasha Welford and Charlotte Kent of Elsevier for their help and encouragement in completing this text.

Pramod Thakur, Ph. D.

Section One

Mine Ventilation

Underground Coal Mine Atmosphere

1

Chapter Outline

Advanced Mine Ventilation. https://doi.org/10.1016/B978-0-08-100457-9.00001-8

1.1 Introduction to Coal Mining

Coal is the most abundant and the cheapest fossil fuel in the world today. Over the past 200 years, it has played a vital role in the growth and stability of world economy. The current world human population of about 7300 million consumes 5×10^{20} J of energy per year. It is likely to increase to 7.5×10^{20} J/year in the next 20 years. Fossil fuels at present provide 87% of all energy consumed. Nuclear and hydro power provide 12%. Solar, wind, and geothermal energy barely provide 1% [1] as shown in Table 1.1.

Barring a breakthrough in nuclear fusion, fossil fuels will remain the main source of energy in the foreseeable future, as they have been in the past 200 years. Ninety percent of all fossil fuel energy in the world is in coal seams. It is, therefore, natural to anticipate that coal's share in the energy mix will increase. At present, coal provides 26% of global energy demand and generates 41% of the world's electricity. Coal deposits are widespread in 70 countries of the world. Coal is a very affordable and reliable source of energy. The total proved, mineable reserve of coal exceeds 1 T tons to a depth of about 3300 feet. Indicated reserves (mostly nonmineable) to a depth of 10,000 feet range from 17 to 30 T tons [2]. Current (2015) world coal production is about 8000 million ton/year. Coal production from top 10 countries are shown in Table 1.2 [3].

Total tonnage mined in these 10 countries comprises nearly 90% of global production. Coal production may continue to increase if they start converting coal into synthetic gases and liquid fuels, such as diesel and aviation fuels.

Coal mining is done in two ways: surface mining and underground mining. Most thick and shallow deposits of coal are mined by surface mining methods. The depth of surface mining is generally less than 200–300 feet. The overlying soil and rocks are removed to expose coal before it is mined out. Nearly 50% of the global production of coal is obtained by surface mining. The mine is open to the atmosphere, hence no ventilation is needed.

Table 1.1 World Energy Reserve and Consumption

Fuel Type	Energy Consumed (EJ/year)[a]	Proved Reserve (ZJ)[b]
Coal	120	290
Gas	110	15.7
Oil	180	18.4
Nuclear	30	2–17[c]
Hydro	30	N.A.
All others	4	Uncertain

[a] $E = 10^{18}$.
[b] $Z = 10^{21}$.
[c] reprocessing not considered. 1000 J = 0.948 BTU.

Table 1.2 Global Coal Production

Country	Annual Production (Metric)[a] t (2013)
China	3561
United States	904
India	613
Indonesia	489
Australia	459
Russia	347
South Africa	256
Germany	191
Poland	143
Kazakhstan	120

[a]1 metric ton = 1.1 short tons.
Adapted from World Coal Statistics.

However, most of the coal deposits are deeper than 300 feet and thus are mined by underground mining methods. Mine shafts or inclines are dug to access the coal seam. A series of tunnels are driven to create a large block of coal, called a longwall panel, that is mined by machines. Coal is transported out of the mine by conveyor belts and hoists (Fig. 1.1).

1.2 Underground Mine Atmosphere

The underground mine atmosphere has many pollutants. They are mostly solids (respirable dust) and gases (such as methane, carbon dioxide, etc.). Liquid pollutants, such as mists, are not an issue in the mining industry.

The provision of an adequate air environment to promote health, safety, and comfort of mine workers has always been and will continue to be a prime requisite for successful coal mining operations. Although the definition of an adequate environment varies from country to county, it generally means the provision of sufficient circulating air, often at specified velocities, to maintain at least 19.5% oxygen in the working areas; concentrations of solids (respirable dust) and gaseous pollutants, such as methane, carbon dioxide, etc., below specified limits; and heat and humidity below specified limits. Because most coal mines are shallow (less than 3000 ft deep), temperature and humidity control is not warranted. It is a concern only in deep metal mines (up to 15,000 ft) for copper, silver, and gold.

Table 1.3 shows the maximum allowable concentrations of these atmospheric pollutants in US underground coal mines [4].

Figure 1.1 A typical underground coal mine layout.

1.3 Properties of Air

Because large quantities of air, sometimes 20 tons of air for each ton of coal mined, is circulated through the mine airways, it is important to know its properties. The chemical composition of air is as follows:

Component Gas	Percent (Volume)	Percent (Weight)
Nitrogen	78.09	75.55
Oxygen	20.95	23.13
Carbon dioxide	0.03	0.05
Other gases (argon, water vapor, etc)	0.93	1.27

Air is a physical mixture of these gases with a specific gravity of 1.00. It is a colorless, odorless, and tasteless gas that supports life and combustion via its oxygen content. Some important properties of air are listed in Table 1.4.

Table 1.3 Permissible Exposure Limits for Coal Mine Air Contaminants

Contaminant	8-hr Time Weighted Limit (ppm)	Ceiling Limit (ppm)
Methane (fresh air)	–	1%–1.25%
Methane (return air)	–	2%
Carbon dioxide	–	0.5%
Carbon monoxide	50	77
Nitrogen oxide	25	37.5
Nitrogen dioxide	3	5
Sulfur Dioxide	5	10
Radon (uranium mines)	1.0 working level	–
Respirable coal dust (United States)	1.5 mg/m^3	–
Respirable silica dust (United States)	0.1 mg/m^3[a]	–
Diesel particulate matter (coal)	120 $\mu g/m^3$[b]	–
Diesel particulate matter (metal mines)	160 $\mu g/m^3$	–

[a]Coal dust concentration is limited by silica concentration in the respirable dust by the following formula: respirable coal dust = 10/silica concentration in %, mg/m^3.
[b]Applicable only in Pennsylvania and West Virginia in the United States. It is the most stringent standard in the world for diesel particulate matter.

Table 1.4 Properties of Air

Molecular Weight	28.97
Specific gravity	1.00
Density (at STP[a])	0.075 lb/ft^3
Atmospheric air pressure	14.7 psi
Specific heat at constant pressure	0.240 Btu/lb°F
Specific heat at constant volume	0.171 Btu/lb°F
Critical temperature and pressure	Nitrogen: −147.1°C at 492.5 psi Oxygen: −118.8°C at 730.6 psi

[a]STP Standard temperature of 70°F and pressure 29.92 inches of Hg.

1.3.1 Gas Laws Related to Air

Air just like all other gases follows many laws of physics that are essential to understand its behavior. Only the most pertinent laws will be discussed here.

1.3.1.1 Boyle's Law

The volume of air, V is inversely proportional to pressure, P at constant temperature, T. For a given volume of gas changing from volume V_1 and pressure P_1 or to volume V_2 and pressure P_2,

$$P_1V_1 = P_2V_2 \tag{1.1}$$

1.3.1.2 Charles' Law

The volume of air (gas) is directly proportional to the absolute temperature, T at constant pressure.

Mathematically,

$$\frac{V_1}{V_2} = \frac{T_1}{T_2} \tag{1.2}$$

Combining Eq. (1.1) and Eq. (1.2) we get the generalized gas law:

$$\frac{P_1V_1}{T_1} = \frac{P_2V_2}{T_2} = R \tag{1.3}$$

where R is the gas constant with a value of 53.35 ft-lb/lb mass °R.

1.3.1.3 Dalton's Law of Partial Pressure

It states that the pressure exerted by a mixture of gases is equal to the sum of separate pressures that each gas would exert if it alone occupied the whole volume.

Mathematically,

$$P\,V = V\,(P_1 + P_2 + P_3,\ \text{etc}) \tag{1.4}$$

where P is the pressure and P_1, P_2, P_3, etc are partial pressures. In normal air, there are only two gases, dry air and water vapor.

$$P = P_a + P_v$$

where P_a , P_v are partial pressures of air and vapor.

An example:

Let us assume P (barometric pressure) is 30 inches of Hg
Partial pressure of vapor is 0.5 inch Hg

Calculate the density of dry air if the temperature was 70°F
$P_a = P - P_v = 30–0.5 = 29.5$ inch Hg
Or $P_a = 2085.8$ lb/ft^2

$$\text{Va (specific volume of air)} = \frac{53.35(460 + 70)}{2085.8} = 13.556 \text{ ft}^3/\text{lb}$$

Hence density of dry air; $= \rho_a = \frac{1}{V_a} = 0.0738$ lb/ft^3

Density of the moist air is also calculated by using another equation:

$$\rho_a = \frac{D(P - 0.378 P_v)}{T} \quad (1.5)$$

where D is 1.3258 if pressure is measured in inches of mercury; P_1 = barometric pressure; P_v = vapor pressure; T = dry-bulb temperature in Rankine.

This yields the density of dry air in the above example as 0.0743 lb/ft^3, which is quite close to the previous volume of 0.0738 lb/ft^3.

1.3.1.4 Graham's Law of Diffusion

It states that the rate of diffusion of a gas is adversely proportional to the square root of the ratio of the densities (specific gravity) of the gas, ρ_g and air, ρ_a.

$$\text{Diffusion rate is proportional to } \sqrt{\frac{\rho_a}{\rho_g}} \quad (1.6)$$

In other words, a gas lighter than air will diffuse faster than one heavier than air.

For example, methane has a specific gravity of 0.55 and carbon dioxide has a specific gravity of 1.5 compared with air, hence methane will diffuse 1.65 times faster than carbon dioxide.

1.3.1.5 Air Density at Higher Altitude

The air density is normally measured at sea level, and it decreases as the altitude increases. The temperature also normally goes down as the altitude increases. Madison [5] provides a mathematical relationship as follows:

$$\frac{W_2}{W_1} = \left(\frac{288 - 0.00198\,H}{288}\right)^{4.526} \quad (1.7)$$

where W_2 is the density at height, H over the sea level; W_1 is the density at the sea level.

An example:

Calculate the density of air, W_2 on top of Mt. Everest at 29,000 ft where

$$W_1 = 0.075\ \mathrm{lb/ft^3}\ \text{at the sea level.}$$

$$W_2 = 0.075\left(\frac{288 - 0.00198 \times 29,000}{288}\right)^{4.526} = 0.060\ \mathrm{lb/ft^3}$$

1.3.1.6 *Pressure Versus Fluid Head*

In coal mining practices, ventilation pressures are small in magnitude and hence they are measured by inches of water or mercury. Water and other liquid pressures are measured in pounds per square inch.

The equation for conversion is shown in Eq. (1.8).

$$p = W_1H_1 = W_2H_2 \tag{1.8}$$

where p is the pressure in lbs/ft^2; W_1, W_2 are the density of the fluid in lb/ft^3; H_1, H_2 are height of the fluid column in feet.

An example:

Atmospheric pressure of air is measured in inches of mercury. Typically it is 30 inches Hg. Convert it into inches of water.
Density of mercury is 13.6; water is 1.00.

Using Eq. (1.8), atmospheric pressure $= 30 \times 13.6 = H_2 \bullet 1$.
Hence, $H_2 = 408$ inches or 34 feet of water.
Example 1:
A mine fan is running at 10 inches of water. Convert it into lb/ft^2.

$p = W_2H_2 = 62.4\ \mathrm{lb/ft^3} \times \frac{10}{12} = 52\ \mathrm{lb/ft^2}$
It is important to remember that:
1 inch of water = 5.2 lb/ft^2
1 inch of Hg = 13.6 inches of water
1 psi = 2.036 inches of Hg = 27.7 inches of water.

1.4 Pollutant Control Strategy

In modern, highly mechanized and productive coal mines, it is not possible to dilute the respirable dust and gases generated by the mining process to safe levels by ventilation alone. Engineering control of a pollutant requires the following strategy:

1. Minimize the generation of dust or gas at the source.
2. Suppress the dust/gas at the source.
3. Collect or contain the pollutant at the source.
4. Dilute the remaining pollutant by ventilation to safe levels.

Thus in the case of respirable dust, the generation of dust can be minimized by pretreating the coal with water and using sharp cutting bits to cut coal. Next, well-designed water sprays can be used to suppress the dust at the site. The dust that gets airborne can be next collected by a scrubber on site. Finally, the remaining airborne dust can be diluted to a safe level of less than 1.5 mg/m^3 by adequate ventilation air.

Similarly for methane in coal, the emission of gas can be minimized at the source by drainage of gas ahead of mining. Water sprays on cutting machines create a good mixing of gas and air. Finally, enough air is circulated to dilute the gas to less than 1% by volume and render it safe.

For diesel exhaust, the control strategy is as follows:

1. Select engines and fuel (low sulfur) that have a low specific diesel particulate matter emission, preferably less than 5 gm/bhp-hr.
2. Make the exhaust go over a catalytic convertor, where harmful gases, such as CO are converted to CO_2.
3. Next, let the exhaust go through a filtration system where most particulates are collected with an efficiency of 90%–95%.
4. Finally, dilute the exhaust with enough air to render it harmless.

1.5 Enforcement of Ventilation Standards

The final goal of a safe coal mining air environment is achieved by the 3 E's.

Engineering: The purpose of this book is to provide the engineering control principles.
Education: It is the domain of academic institutions. Technology transfer must be a part of the education process.
Enforcement: The basic requirement for engineering control must be enforced.

An agency of the US federal government does the inspection and enforcement in coal mines. It is called Mine Safety and Health Agency or MSHA in the United States.

1.5.1 Mine Ventilation Regulations

Each coal mining country has its own ventilation regulations, but they are very similar. Large countries, such as the United States, have not only federal regulations that apply to all states but each state also has its own "state regulations" for local situations that may not be adequately covered by the federal regulations. State regulations generally are stricter than federal regulations.

1.5.1.1 US Federal Regulations

Federal regulations for all mining activity are specified in the "Code of Federal Regulations (CFR)". It is divided into 50 titles. Title 29 of the CFR regulates tunnel construction work. Title 30 is devoted to mineral resources and is the most pertinent for coal mining. CFR 30, part 70 deals with health standards, whereas CFR 30, part 75 deals with safety standards in coal mines [6].

All ventilation standards (as shown in Table 1.1) are established by the CFR Title 30 or the threshold limits established by the American Conference of Governmental Industrial Hygienists [7]. Some excerpts from these vast documents are provided here to illustrate the safety measures. Reference should be made to the original documents for details.

1.5.2 Maximum Concentration of Explosive Gases in Coal Mine Air

The legal maximum concentration of several gases in coal mines are limited by MSHA as shown below:

Gases	Maximum Limits (%)
Methane	1 (intake), 2 (return)
Acetylene	0.4
Propane	0.4
Hydrogen	0.8
Carbon monoxide	2.5
Hydrogen sulfide	0.4

These limits are much lower than the minimum explosive limits of these gases in air. It will be discussed in detail later in the book.

1.5.3 Some Highlights of Code of Federal Regulations 30, Parts 70 and 75

1.5.3.1 Respirable Dust Measurement

Respirable dust is theoretically defined as particles smaller than 5 μm, but it is actually measured by the amount collected by the approved sampling instrument on a filter. The standard for respirable dust requires that the average concentration (of five consecutive shifts) to which a miner is exposed be at or below 1.5 mg/m^3. In addition, the respirable dust concentration in the intake air to the same working section should be below 1 mg/m^3.

When the respirable coal dust contains more than 5% quartz, the dust standard is lowered by the following formula:

$$\text{Respirable Dust Standard} = \frac{10}{\text{Percent Silica}}\,\text{mg}/\text{m}^3$$

Thus if silica concentration is 10%, the respirable dust standard is reduced to 1 mg/m^3. Additional details on dust measurement will be provided later in the book.

Both the mechanized mining units and the designated areas are sampled for an effective control of respirable coal dust. Respiratory equipment, either a filter type respirator or a supplied-air type device, is provided to a miner required to inspect dusty areas with more than 1.5 mg/m^3 of dust concentration.

1.5.3.2 Methane Measurements

30 CFR 75 requires frequent checks on methane levels. Each working area is examined within 3 h of start of work for methane concentrations. A certified person uses an approved device to measure and record all methane readings in a working area.

The second examination is on-shift measurement of both methane and oxygen. At the working face, a reading should be taken every 20 minutes. In addition, each mining machine is fitted with a methane detector. At 1.5% the machine gives a visible (yellow light) and audible alarm. At 2% a red light comes on and electric power to the machine is cut. The machine is not restarted until excess methane is cleared and safe levels of methane are obtained.

1.5.3.3 Minimum Air Requirements

Minimum air quantities, often at specified velocities, are required to make sure all contaminants are adequately diluted.

Federal government requires a minimum of 6000 CFM for the development sections and 9000 CFM for longwall sections. These air quantities may be grossly inadequate if the coal seam is moderately or highly gassy [8]. State regulations of West Virginia require larger air quantities because mines are gassier.

1.5.3.4 Permissible Electrical/Diesel Equipment

To further secure the safety of coal mines, all electrical or diesel equipment working inbye of last open cross-cut must be permissible. Such equipment will not ignite an explosive mixture of methane and air. This is verified in laboratory before a 'permissible' certificate is issued. The 30 CFR is a large document and goes in detail to secure mine safety. Reference can be made to it for additional information.

1.6 Definition of Air (Gas) Properties [9]

For a clear understanding, various properties of air and gases are defined in the following section.

1.6.1 Atomic Weight

Atomic weight is the relative weight of the atom on the basis of oxygen as 16. For a pure isotope, the atomic weight rounded off to the nearest integer gives the total number of neutrons and protons making up the atomic nucleus. These weights expressed in grams are called gram atomic weights.

1.6.2 Avogadro's Number

Avogadro's law says that equal volumes of different gases at the same pressure and temperature contain the same number of molecules. The number of molecules in 1 g-molecular weight of a substance is 6.02×10^{23} ($\pm 1\%$).

1.6.3 British Thermal Unit

It is the quantity of heat required to raise the temperature of one pound of water by 1° of Fahrenheit at, or near, its point of maximum density (39.1°F).

1.6.4 Critical Temperature and Pressure

Critical temperature is that temperature above which a gas cannot be liquefied by pressure alone. The pressure under which a substance may exist as gas in equilibrium with liquid at the critical temperature is the critical pressure.

1.6.5 Density

It is the concentration of matter, measured by mass per unit volume. It is expressed as lb/ft^3.

1.6.6 Dew Point

It is the temperature at which condensation of vapor in the air takes place.

1.6.7 Enthalpy or Heat Content

It is a thermodynamic quantity equal to the sum of internal energy in a system plus the product of the pressure–volume work done on the system.

Thus:

$$H = E + pv \text{ (Btu/lb)}$$

Where

H = enthalpy or heat content
E = internal energy of the system
p = pressure
v = volume

1.6.8 Entropy

It is the capacity factor for isothermally unavailable energy. The increase in the entropy of a body, ds, during an infinitesimal stage of a reversible process is equal to the infinitesimal amount of heat, Q, absorbed divided by the absolute temperature of the body, T.

Thus for a reversible process:

$$ds = \frac{Q}{T}\ \text{Btu/lb/}^{\circ}\text{F}$$

1.6.9 Mole Volume

The volume occupied by a mole or a gram molecular weight of any gas at standard conditions is 22.414 L.

1.6.10 Relative Humidity

It is the ratio of the quantity of water vapor present in the atmosphere to the quantity which would saturate it at the existing temperature. It is also the ratio of the pressure of water vapor present to the pressure of saturated water vapor at the same temperature.

1.6.11 Specific Heat

Heat required to raise the temperature of unit weight of a gas by 1°F at constant pressure (C_p) or volume (C_v) and is measured in Btu/lb°F.

1.6.12 Viscosity

It is the drag or shear resistance of air to motion. It is measured in lbs/ft^2. This is also called absolute (μ) viscosity. Absolute viscosity divided by mass density is called kinematic viscosity (υ).

Problems

1.1 Calculate the height of a column of dry air equivalent to 1 psi pressure. Assume standard conditions for atmospheric pressure and temperature.

1.2 Repeat the above calculation for water.

1.3 Calculate the density of air at 10,000 ft. Assume air density at sea level is equal to 0.075 lb/ft^3.

1.4 Calculate the maximum allowable respirable dust concentration in a coal mine if the respirable dust contains (a) 10%, (b) 15%, and (c) 20% silica.

1.5 Calculate the specific gravity of methane, ethane, propane, hydrogen, carbon dioxide, nitrogen, and oxygen if the specific gravity of air is 1.00. (Hint: specific gravity is proportional to the molecular weight).

References

[1] World Energy Reserves and Consumptions, http://en.wikipedia.org/wiki/world_energy_consumption/; 2013.
[2] Landis ER, Weaver JW. Global coal occurrences: hydrocarbons from coal. AAPG Study of Coal Geology 1993;38:1−12.
[3] World Coal Statistics. World coal association. 2013. http://worldcoal.org/.
[4] Thakur PC. Gas and dust control. In: Darling P, editor. Chapter 15.4 SME mining engineering handbook; 2011. p. 1595−609.
[5] Madison RD. Fan engineering. Buffalo, New York: Buffalo Forge Company; 1999.
[6] Anon. Code of federal regulations. Title 30. Washington, DC: Mineral Resources, US Government Printing Office; 1980.
[7] Anon. TLVs threshold limit value for chemical substances and physical agents in workroom environments. Cincinnati, Ohio: American Conference of Governmental Industrial Hygienists; 1979.
[8] Thakur PC, Zachwieja J. Methane control and ventilation for 1000-ft wide longwall faces. In: Proceedings of Conference on Longwall USA, Pittsburgh, PA, USA; 2001. p. 167−79.
[9] Hodgman CD, et al. Handbook of chemistry and physics. Cleveland, Ohio, USA: The Chemical Rubber Publishing Company; 1962. p. 3125−93.

Air Flow in Mine Airways

2

Chapter Outline

2.1 Introduction

The chapter deals with the fundamentals of fluid flow in pipes and mine airways. Basic pressure loss equation is derived and various ways to estimate the friction factor, λ, are presented. In fully turbulent flow only the degree of roughness determines the friction factor. However, mine airways are very different from conventional pipes. Atkinson's equation is developed further, and a large collection of friction factors from the US and British coal mines are presented. Mine airways are typically rectangular and also have a lot of obstructions that cannot be theoretically analyzed. Actual data with experience alone can yield reliable results. Airflow in ventilation ducts made of steel or fiberglass are discussed. Shock losses owing to obstructions and changes in airway directions are also discussed. Total resistance of a mine to airflow, R, is used to create a "characteristic curve" for the mine. It is useful in determining the correct fan size for the mine (to be discussed later in the book). The concept of "equivalent orifice" is mathematically analyzed and used to determine the area of a regulator in a mine airway to restrict the air flow to a predetermined value. Mine airways change the cross section many times

Advanced Mine Ventilation. https://doi.org/10.1016/B978-0-08-100457-9.00002-X

owing to geology or ground control conditions. Such airways are treated as "airways in series." Similarly, often two to five airways are needed in parallel to carry a given volume of air. Both "series" and "parallel" airways are mathematically analyzed for air flow distribution and pressure losses. Finally, an equation is provided to estimate the horsepower needed to run a fan that can provide a prescribed ventilation air volume at a required pressure differential.

A coal mine is basically a network of roadways that are mostly rectangular in section. The roadways/airways have many bends, support pillars, and other obstructions that cause shock losses. The air flow is mainly created by mechanical fans and at times assisted by natural ventilation (to be discussed later in the book). In view of large sizes of roadways (20 × 6 ft, for example) and high velocities required by law, the flow is generally turbulent. It is also a steady-state flow. The flowing air is a real viscous fluid that creates friction. Even though air is definitely compressible, the pressure differential to create the flow in mine airways is low (less than ½ psi), and hence the air is treated as a noncompressible fluid. It often takes a number of fans with 1000–3000 horsepower drivers to keep a typical coal mine well ventilated. A large coal mine producing 5 to 7 million tons per year may circulate 3 to 5 million cubic feet of air per minute at 5–15 in. of water gauge (W.G.).

2.2 Derivation of Basic Fluid Flow Equation

Solutions of mine air flow problems are derived from energy principles, the equation of continuity, and equation of fluid resistance [1]. Resistance to flow in mine airways is offered not only by frictional resistance but also by roof support structures in the airways and bends in the airways, which create turbulence and additional dissipation of energy.

Early experiments (by Darcy, c. 1850) on the flow of water in pipes indicated that the pressure loss was directly proportional to the length of pipeline and the velocity head, $\left(\frac{v^2}{2g}\right)$, and inversely proportional to the diameter of the pipe, d. Mathematically, it can be expressed as:

$$h = \frac{\lambda\, l v^2}{2\, g d} \tag{2.1}$$

where h is the pressure loss over a distance; l, in feet of the fluid; v is the velocity in ft/s; l is the length of the pipe in feet; d is the diameter of the pipe, in feet; g is the acceleration due to gravity (32 ft/s^2); λ is a coefficient of proportionality, commonly called the friction factor. It is dimensionless.

2.2.1 *Determination of λ in Eq. (2.1)*

To calculate the head (pressure) loss in Eq. (2.1), the only thing not known is the friction factor, λ.

The theory of dimensional analysis will be used to determine the variables that can be used to predict λ. When a viscous fluid flows in a pipe, the frictional stress, τ_o, is dependent on the following variables only:

v, the velocity of the fluid, ft/s (L/T)
d, the pipe diameter, ft (L)
ρ, the density of the fluid, lb/ft^3 $\left(\frac{M}{L^3}\right)$
μ, the viscosity of the fluid, poise $\left(\frac{M}{LT}\right)$

and e, the pipe roughness, in ft (L)

Mathematically,

$$\tau_o = F\,(v,\ d,\ \rho,\ \mu,\ e) \tag{2.2}$$

Dimensional analysis converts Eq. (2.2) into

$$\frac{M}{T^2L} = \left(\frac{L}{T}\right)^a (L)^b \left(\frac{M}{L^3}\right)^c \left(\frac{M}{LT}\right)^d (L)^e$$

Comparing the power of mass (M), length (L), and time (T) on both sides for

M: $1 = c + d$
L: $-1 = a + b - 3c - d + e$
T: $-2 = -a - d$ or $2 = a + d$

We can eliminate three unknowns by converting a, b, and c into d and e.

$a = 2 - d;\ b = -(d + e)$ and $c = 1 - d$.

Thus Eq. (2.2) can be rewritten as:

$$\tau_o = c v^{2-d}\, d^{-(d+e)} \rho^{1-d} \mu^d e^e$$

$$\text{or } \tau_o = c\left(\frac{\mu}{vd\,\rho}\right)^d \left(\frac{e}{d}\right)^e \rho\cdot v^2 \tag{2.3}$$

where c is the constant of proportionality.

Eq. (2.3) shows that frictional losses in a pipe are basically a function of two variables: $\left(\frac{v\,d\,\rho}{\mu}\right)$ is defined as the Reynold's number, R, and $\left(\frac{e}{d}\right)$ is the roughness factor.

Stanton [2] and Nikuradse [3] have carried out extensive research on measuring λ for Reynold's number ranging from 10^3 to 10^6 and $\left(\frac{e}{d}\right)$ ranging from 1×10^{-4} to 1×10^{-2}. References can be made to these works for details.

The roughness, e, for various commercial pipes are shown in Table 2.1.

Colebrook [4] studied roughness of many pipes and came up with a single equation that can be used very conveniently.

Table 2.1 Roughness for Various Pipes

Type of Pipe	e, in.
Wrought iron pipe	0.0017
Well tubing/line pipe	0.0007
Cast iron	0.0050
Galvanized iron	0.0060
Uncoated cast iron	0.0100
Wood pipe	0.007 to 0.036
Concrete	0.012 to 0.12
Riveted steel	0.035 to 0.35

$$\frac{1}{\sqrt{\lambda}} = 2\log\frac{d}{e} + 1.14 - 2\log\left[1 + \frac{9.28}{R\left(\frac{e}{d}\right)\sqrt{\lambda}}\right] \quad (2.4)$$

The value of λ is obtained by several iterations. However, Eq. (2.4) has been made user-friendly by the Moody [5] diagram that shows the value of λ against varying Reynold's number and different $\left(\frac{e}{d}\right)$ ratios ranging from 10^{-5} to 10^{-1}. Mostly, the λ values range from 0.01 to 0.09, representing very smooth to wholly rough pipes. For fully turbulent flow in smooth pipes, Vennard [1] provides another equation for λ that may be easier to use.

$$\frac{1}{\sqrt{\lambda}} = -0.80 + 2.0\log R\sqrt{\lambda} \quad (2.5)$$

When the flow becomes completely turbulent—that is beyond the transition zone—the frictional coefficient is no longer a function of Reynold's number but becomes a function of e/d only. The friction factor in this region of flow is completely independent of the physical properties of the flowing fluid. For fully turbulent flow, the λ factor is expressed by an equation obtained experimentally by Nikuradse [6].

$$\frac{1}{\sqrt{\lambda}} = 2\log\frac{d}{e} + 1.14 \quad (2.6)$$

Thus for a 6″ diameter cast iron pipe with roughness of 0.005 in:

$$\frac{d}{e} = \frac{6}{0.005} = 1200$$

Hence,

$$\frac{1}{\sqrt{\lambda}} = 2\log 1200 + 1.14 = 7.3$$

or

$$\lambda = \left(\frac{1}{7.3}\right)^2 = 0.0188$$

An example:

For a flow rate of 75 gpm water, calculate the head (pressure) loss for 3000 feet of 3″ diameter drill pipe. The roughness of pipe, e is 0.006.

Step 1. Calculate the fluid velocity.

$$\text{Q, fluid flow rate} = 75 \text{ gpm} = 10 \text{ ft}^3/\text{min} = 0.167 \text{ ft}^3/\text{s}$$

$$\text{A, cross} - \text{section of pipe} = \frac{\pi}{4}\left(\frac{3}{12}\right)^2 = 0.049 \text{ ft}^2$$

$$\text{Hence velocity, V} = \frac{\text{Q}}{\text{A}} = \frac{0.167}{0.049} = 3.41 \text{ ft/s}$$

Step 2. Calculate the Reynold's number, R.

$$\text{R} = \frac{\text{Vd}}{(\mu/\rho)} = \frac{3.41 \times 0.25}{1.217 \times 10^{-5}} = 70,000$$

(μ/ρ is the kinematic viscosity of water $= 1.217 \times 10^{-5}$ ft^2/s)

Hence the flow is fully turbulent.

Step 3. Calculate λ from Eq. (2.6).

$$\frac{1}{\sqrt{\lambda}} = 2 \log\left(\frac{3}{0.006}\right) + 1.114$$

This gives $\lambda = 0.0234$.

Step 4. Calculate head loss using Eq. (2.1).

$$\text{h} = \frac{0.0234 \times 3000(3.41)^2}{2 \times 32 \times 0.25} \text{ feet of water}$$

$$= 51 \text{ ft} = 22.2 \text{ psi}$$

2.3 Traditional Equations for Pressure Loss Calculation in Mines

Eq. (2.1) works well for circular pipes and ventilation tubing made of steel or fiberglass, but it is not directly usable for mine airways that are actually rectangular in shape and have uncertain degree of roughness.

A new term, hydraulic radius, R_h, is introduced as the ratio of area to perimeter.

$$R_h = \frac{A}{P} \tag{2.7}$$

where A is the area of the airway and P is the perimeter.

For a circular pipe of diameter, D:

$$R_h = \frac{A}{P} = \frac{\pi D^2}{4\pi D} = \frac{D}{4} \tag{2.8}$$

Substituting Eq. (2.8) in Eq. (2.1),

$$h = \frac{\lambda\, l\, v^2}{4R_h\, 2\, g} \tag{2.9}$$

Atkinson [7] used mining engineering units to further modify Eq. (2.9) such that:

$$H = \frac{KPLQ^2}{5.2\, A^3} \tag{2.10}$$

where H is expressed in inches of water. L is in feet. V was replaced by $\frac{Q}{A}$ where Q was the air quantity in cubic feet/minute and A is the cross-sectional area in ft^2. g was replaced by 32 ft/s^2.

Density of air was 0.075 lb/ft^3 at standard temperature and pressure.

Where

$$K = \lambda \left(810 \times 10^{-10}\right) \frac{lb\ min^2}{ft^4}$$

Because K is a very small number, it is multiplied by 10^{10} to get whole numbers such as 10, 20, or 100.

Thus:

$$K = 810\, \lambda \frac{lb\ min^2}{ft^4} \tag{2.11}$$

It should be noted that K must be corrected if the air density is different from 0.075 lb/ft^3 by using Eq. (2.12).

$$\text{Corrected } K = K\left(\frac{W}{0.075}\right) \tag{2.12}$$

where W is the actual density of air in lb/ft^3.

Eq. (2.10) is commonly used in mining engineering and is known as Atkinson's equation [7].

2.4 Determination of Mine Airway Friction Factor, K

Because the size and shape of mine airways are greatly variable, they cannot be treated as pipelines of uniform size and measureable roughness. The friction factor, K, is hence best obtained by direct measurements.

Many authors have measured the friction factor using models or actual mine airways. These values are listed in different units. They are all modified and expressed as K in Eq. (2.11) in this book.

2.4.1 Historical Data on Friction Factors for Mine Airways

In 1923–24, McElroy [8] found friction factors for metal and some coal mines. Richardson [9] added additional data and jointly published them. Greenwadd [10] created additional data a few years later. The data are summarized in Table 2.2.

If the roadways are sinuous or curved, the friction factor should be multiplied by a factor of 1.4 in each category.

Table 2.3 shows friction factor values for British coal mines. It is a summary of vast data published by Pursall [11] in 1960.

The author used the data published by the National Coal Board of the United Kingdom [12] to plan ventilation for many coal mines successfully. Their data are shown in Table 2.4.

McPherson [13] provides some data on K for the longwall faces that are very heavily obstructed with shearers, conveyors, and shield supports. The value of K increases

Table 2.2 Values of K for Straight Airways (Actual K Multiplied by 10^{10})

Type of Airway	Degree of Roughness	Clean	Moderately Obstructed
Smooth-bored	Minimum	10	25
	Average	15	30
	Maximum	20	35
Sedimentary rock	Minimum	30	45
	Average	55	70
	Maximum	70	85
Timbered at 5 ft center	Minimum	80	95
	Average	95	110
	Maximum	105	120
Igneous rock	Minimum	90	105
	Average	145	160
	Maximum	195	210

Table 2.3 Friction Factors in British Coal Mines

Type of Airway	Condition of Airway	Friction Factor, K (Ave)
Shafts	Brick lined	29
Arched	Straight to slightly waved	28
Smooth lined circular/semi circular arch	Concrete-lined	24
Steel-arch girdered	Timber-lined	62
Unlined airways	Uniform to irregular cross section	20–55
Timbered airways	Uniform to irregular	48–81

Table 2.4 Values of K for Rectangular Airways (Coal)

Condition of Airway	Value of K
Smooth, concrete-lined	20
Steel girders on brick walls	50
Unlined, straight airways	65
Unlined, rough airways	85
Wood beams on timber legs	100

as expected to 200–350. The corresponding values of λ for K values of 20 and 200 are 0.0246–0.246, respectively, making the mine airways extremely rough pipes.

A ventilation engineer must use experience as a guide to select proper K values when planning the total pressure requirement for mines. There is a tendency to generally underestimate the pressure requirements.

An example:

Two vertical shafts of 1000 ft depth and 20 ft diameter are connected by a 3000 ft long rectangular tunnel of a cross section of 20 × 8 ft. Calculate the pressure loss if 200,000 CFM air is flowing from one shaft to the other.

Assume K for shaft as 20 and K for the tunnel as 30.

1. For shaft:

   $L = 1000$ ft

   $P = \Pi D = 3.14 \times 20 = 62.8$ ft

   $Q = 200{,}000\ \text{ft}^3/\text{min}$

   $A = \dfrac{\Pi D^2}{4} = 314\ \text{ft}^2$

   Now calculate $H = \dfrac{20 \times 10^{-10} \times 62.8 \times 1000(200{,}000)^2}{5.2 \times (314)^3} = 0.312$ in.

2. For the tunnel:

$$H = \frac{30 \times 10^{-10} \times 56 \times 3000 \times (200,000)^2}{5.2 \times (160)^3} = 0.948 \text{ in.}$$

3. Total pressure loss = 2(0.312) + 0.948 = 1.572 in.

2.5 Air Flow in Ventilation Duct/Pipes

In the development section of a coal mine, an auxiliary fan with ventilation ducts are used to improve the air supply to the working face. It dilutes gases better and gets dusty air out of the working faces. A small fan is used to create the air flow. Table 2.5 shows typical friction factors for different types of ducts.

The length of tubing in use generally does not exceed 100 feet in the mines, but it can be longer for tunnels.

An example:

Calculate the pressure loss in a 500-ft-long steel duct of 2 ft. diameter required to deliver 15,000 CFM air at the working face.

Assume

$K = 20$
$\text{Area} = \frac{\Pi d^2}{4} = 3.14 \text{ ft}^2$
$\text{Perimeter} = \Pi\, d = 3.14 \times 2 = 6.28 \text{ ft}$
$Q = 15{,}000 \text{ CFM}$
$L = 500 \text{ ft.}$

Using Eq. (2.5)

$$H(W.G.) = \frac{20 \times 10^{-10} \times 6.28 \times 500 \times (15,000)^2}{5.2 \times (3.14)^3} = 8.78 \text{ inches of water.}$$

2.6 Shock Losses in Mine Airways

As the air current changes direction, it creates a shock loss that increases the total pressure loss for the ventilation network. The following are three ways to determine and compensate for this loss:

Table 2.5 Typical Friction Factors for Different Types of Ducts

Type of Duct	Friction Factor, K (New) (Average for Used Ducts	
Steel or fiberglass	15	21
Canvass or plastic	20	25
Spiral canvass	23	28

2.6.1 Direct Calculation of Shock Loss

The shock loss, H_S, is a direct function of the velocity head, $H_V = \frac{V^2}{2g}$; the shock loss. H_S can be expressed as

$$H_S = K_L\frac{V^2}{2g} \tag{2.13}$$

The velocity head, H_V, can be again expressed in familiar units. H_V is in inches of water, and velocity in ft/min.

$$H_V = \frac{WV^2}{5.2(64.4)(60)^2}$$

Where V = velocity in ft/min. W = density of air in lbs/ft^3

Hence,

$$H_S = K_L\frac{WV^2}{(1098)^2} \tag{2.14}$$

where K_L is a dimensionless, shock loss factor. The value of K_L can be mathematically calculated.

An example:

Fig. 2.1 shows a rounded bend in an airway of a cross section b × d.

Assume:

Deflection angle = θ
Bend radius = r
Airway width = b

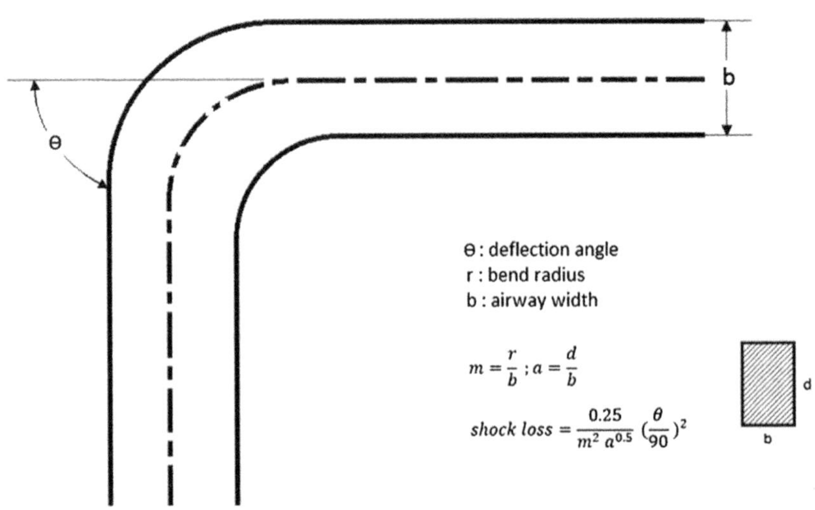

Figure 2.1 Rounded bend in cross-sectional airway.

Airway height = d
Radius ratio = m = r/b
Aspect ratio = a = d/b

McElroy [8] gives an estimate for K_L as

$$K_L = \frac{0.25}{m^2\, a^{0.5}} \left(\frac{\theta}{90}\right)^2 \tag{2.15}$$

For a sharp bend, where r = b/2

$$K_L = \frac{0.60}{m a^{0.5}} \left(\frac{\theta}{90}\right)^2 \tag{2.16}$$

Appendix A shows other equations for different situations.

2.6.2 Shock Losses by Increasing Friction Factor

McElroy [8] advocates compensating for shock loss by increasing the friction factor for the airway. Only limited data are available in this area. It is, therefore, not commonly used.

2.6.3 Equivalent Length For Shock Losses

The most popular method for compensating for shock losses is to increase the length of an airway by a calculated amount.

Pressure loss due to shock, $H_S = \frac{K_L V^2}{2g}$

In mining units:

$$H_S = K_L \left(\frac{WV^2}{(1,098)^2}\right)$$

Equating H_S to an equivalent length of L_e we can write:

$$K_L \frac{WV^2}{(1,098)^2} = \frac{KL_e V^2}{5.2\, R_H}$$

or

$$L_e = \frac{3,235\, R_H K_L}{10^{10}\, K} \tag{2.17}$$

Some calculated equivalent lengths are shown in Table 2.6.

Table 2.6 Equivalent Lengths for Shock Losses[a]

Source	Shock Loss in Feet
Acute bend, sharp	150
Acute bend, round	3
Right angle bend, round	1
Right angle bend, sharp	70
Obtuse angle bend, round	1
Obtuse angle bend, sharp	15
Doorway	70
Overcast	65
Inlet to a duct	20
Discharge from a duct	65
90° split in airway	200
Mine car (blocking 20% of airway)	100
Mine car (blocking 45% of airway)	500

[a]Adapted from Mine Ventilation by Hartman et al. [14].

2.7 Mine Characteristics Curve

When the pressure and shock losses for an entire mine network is determined and ventilation surveys (to be discussed later in the book) determine the air quantity needed to properly ventilate the mine, it is necessary to create a "mine characteristic" to select a matching fan. A plot of pressure requirements in the y-axis for different ventilation quantities on the x-axis is known as "mine characteristic" (curve).

For derivation of this curve, we again use a modified Eq. (2.18). It can be rewritten as

$$H = RQ^2 \tag{2.18}$$

where

$$R = \left(\frac{KPL}{5.2\ A^3}\right).$$

R is the sum total of all resistances to flow if we can use the analogy of the voltage needed to make a current of Q to flow through a conductor of resistance, R. It is usually the resistance of the split with the highest resistance called "free split." Other splits may be regulated.

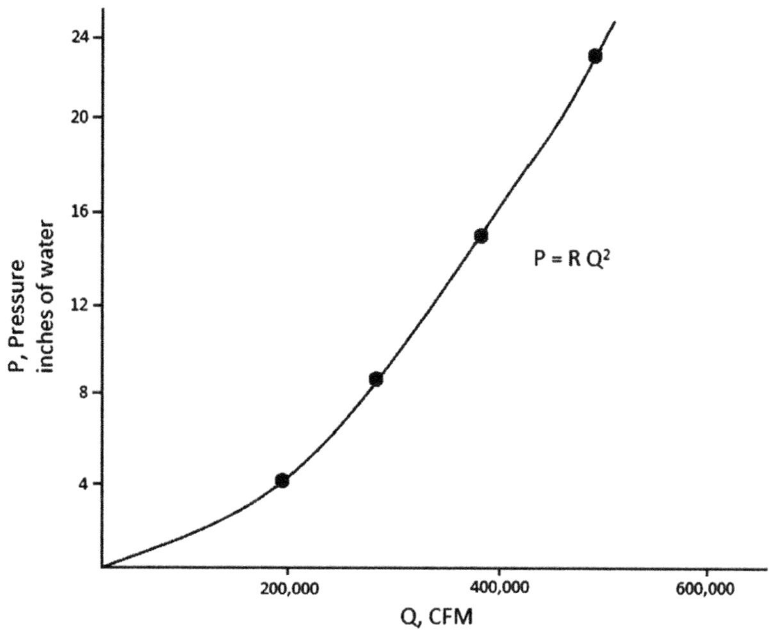

Figure 2.2 Mine characteristic curve for example problem.

An example: A single fan mine needs a ventilation rate of 250,000 CFM air at 4 in. of W.G. Calculate and plot the mine characteristic.

Using Eq. (2.18), a set of data is created and plotted in Fig. 2.2.

Mine Q (CFM)	Total Pressure (W.G.)
200,000	4
300,000	9
400,000	16
500,000	25

Eq. (2.18) leads to a new term called equivalent orifice (EO). It compares the total mine resistance to a circular opening in a thin plate that creates the same resistance if the same ventilation quantity was flowing through it.

McElroy [8] has developed an equation for it.

$$EO = \frac{3.9 \times 10^{-4}\ Q}{\sqrt{H}} \tag{2.19}$$

where EO is in ft^2, Q is in CFM, and H is in inches of water.

Thus the EO for the mine in the previous problem:

$$\mathrm{EO} = \frac{3.9 \times 10^{-4} \times 250{,}000}{\sqrt{4}} = 48.75\ \mathrm{ft}^2$$

Eq. (2.19) with slight modification can also be used to calculate the size of a regulator in a mine airway. The regulator is needed to restrict the airflow in a given split to deliver a fixed quantity of air.

The EO of the regulator that can dissipate a head of H is first determined by Eq. (2.19), and then a correction is made for contraction factor, K_2. An example will illustrate it better.

Given Q = 100,000 CFM, A = 80 ft^2, find the size of a square regulator to drop two inches of W.G.

Hence

$$\mathrm{EO} = \frac{3.9 \times 10^{-4}\ (100{,}000)}{\sqrt{2}\sqrt{K_2}}$$

$$\mathrm{EO} = 17.5\ \mathrm{ft}^2$$

where K_2 is equal to 2.5. Contraction factors, K_2, for a number of edges of a regulator are listed below [8].

Type of Edge	K_2
Rounded	1.5
Smooth	2.0
Square	2.5
Sharp	3.8

2.8 Ventilation Airways in Series/Parallel

In a ventilation system, two basic combinations of airways often arise: in series or in parallel. These can be analyzed mathematically, but the entire network analysis will need a computer. Computer simulation of a mine ventilation network will be discussed later in the text.

Fig. 2.3A shows three airways in series, and Fig. 2.3B shows three airways in parallel.

2.8.1 *Airways in Series*

Fig. 2.3A shows that the same air quantity flows through all three roadways. The total head loss in the three airways is the sum of head losses in each roadway. If the

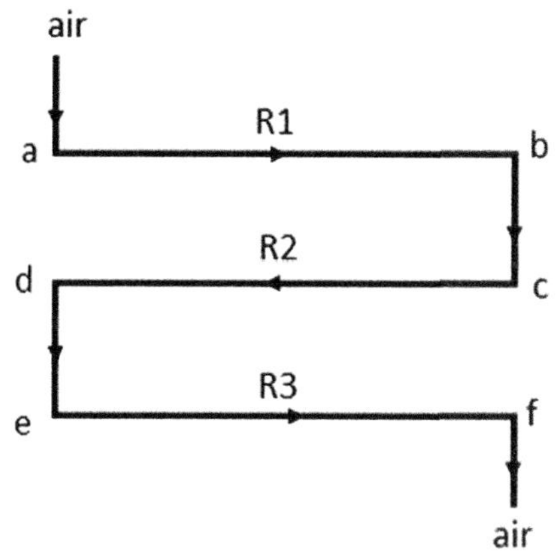

Figure 2.3A Three airways in series.

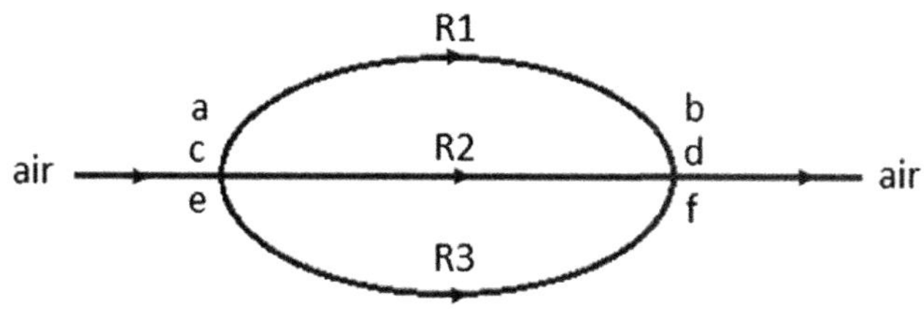

Figure 2.3B Three airways in parallel.

individual resistance of these roadways is R_1, R_2, and R_3, we have total head loss $= H = R_1Q^2 + R_2Q^2 + R_3Q^2$ or

$$H = (R_1 + R_2 + R_3)Q^2 \tag{2.20}$$

2.8.2 Airways in Parallel

Fig. 2.3B shows that the total quantity of air is now divided into three branches or splits.

Here $Q = Q_1 + Q_2 + Q_3$

Where Q_1, Q_2, and Q_3 are air quantities in splits 1, 2, and 3, respectively.

The head loss, however, for each split is the same.

Modifying Eq. (2.18)

$$H_1 = R_1{Q_1}^2 \quad \text{or} \quad Q_1 = \left(\frac{H_1}{R_1}\right)^{\frac{1}{2}}$$

$$H_2 = R_2{Q_2}^2 \quad \text{or} \quad Q_2 = \left(\frac{H_2}{R_2}\right)^{\frac{1}{2}}$$

$$H_3 = R_3 Q_3^{\,2} \quad \text{or} \quad Q_3 = \left(\frac{H_3}{R_3}\right)^{\frac{1}{2}}$$

Hence $Q = Q_1 + Q_2 + Q_3 = \left(\frac{H_1}{R_1}\right)^{\frac{1}{2}} + \left(\frac{H_2}{R_2}\right)^{\frac{1}{2}} + \left(\frac{H_3}{R_3}\right)^{\frac{1}{2}}$
Hence $H_1 = H_2 = H_3 = H$.
If R_{eqv} is the equivalent resistance for network of three airways,

$$Q = \frac{\sqrt{H}}{\sqrt{R_{eqv}}} = \sqrt{H}\left[\frac{1}{R_1^{\,0.5}} + \frac{1}{R_2^{\,0.5}} + \frac{1}{R_3^{\,0.5}}\right]$$

Hence

$$\frac{1}{R_{eqv}^{\,0.5}} = \frac{1}{R_1^{\,0.5}} + \frac{1}{R_2^{\,0.5}} + \frac{1}{R_3^{\,0.5}} \tag{2.21}$$

An example:
Three airways in (Fig. 2.3B) are in parallel with a total air quantity of 90,000 CFM. Their resistances are

$$R_1 = 10\left(\frac{\text{in.} - \text{min}^2 \times 10^{10}}{\text{ft}^6}\right)$$

$$R_2 = 15\left(\frac{\text{in.} - \text{min}^2 \times 10^{10}}{\text{ft}^6}\right)$$

$$R_3 = 20\left(\frac{\text{in.} - \text{min}^2 \times 10^{10}}{\text{ft}^6}\right).$$

Calculate the quantity of air in each split and the head loss.

$$R_{eqv} = \left[\frac{1}{\frac{1}{\sqrt{10}} + \frac{1}{\sqrt{15}} + \frac{1}{\sqrt{20}}}\right]^2 \times 10^{-10}$$
$$= 1.574 \times 10^{-10} \text{ inch-min}^2/\text{ft}^6$$

Hence

$$H = R_{eqv} \times Q^2 = 1.574 \times 10^{-10}(90,000)^2$$
$$= 1.574 \times 81 \times 10^{-2} = 1.275 \text{ in.}$$

Hence

$$Q_1 = \sqrt{\frac{1.275}{10 \times 10^{-10}}} = 35,704 \text{ CFM}$$

$$Q_2 = \sqrt{\frac{1.275}{15 \times 10^{-10}}} = 29,154 \text{ CFM}$$

$$Q_3 = \sqrt{\frac{1.275}{20 \times 10^{-10}}} = 25,249 \text{ CFM}$$

$Q_1 + Q_2 + Q_3$ slightly exceed 90,000 owing to computational rounding.

2.9 Calculation of Air Horsepower

When the total ventilation rate, Q, and pressure required to circulate this air through mine airways are known, it is required to determine the size of the prime mover, typically an electric motor.

If Q is expressed in CFM and P is expressed in inches of water,

$$\text{Air horsepower} = \frac{5.2\,PQ}{33,000} \tag{2.22}$$

Brake horsepower (bhp) of the motor is equal to air horsepower/η, where η is the efficiency of the electric motor. It can range from 0.8 for an induction motor to almost 1.00 for a synchronous induction motor.

An example:

Ventilation planning of a coal mine shows that 500,000 CFM of air is needed at 14 in. of W.G. Calculate the size of the electric motor to drive the fan. Assume motor efficiency equals 0.8.

$$\text{Brake horsepower} = \frac{500,000 \times 14 \times 5.2}{(0.8)33,000} = 1,378.8$$

A 1500 horsepower drive would be a good choice.

Problems

2.1 Calculate the pressure loss in a mine tunnel 3000 ft long and a cross section of 20 × 8 ft if 150,000 CFM of air is needed to keep methane level in compliance. Assume a K of 100.

2.2 Calculate the friction factor, λ, for mine airway with K values of 50, 100, and 200. How does the mine airway compare to a steel pipe?

2.3 Three airways in parallel have the following characteristics:

Airway	Size	L	K
1	20 × 6	900	50
2	18 × 6	1000	60
3	15 × 8	1100	70

The total air quantity in all three airways is 120,000 CFM. Calculate the flow in each airway. Plot the characteristic curve ($H = R\ Q_2$) for each airway and the combined curve for all three airways.

2.4 Repeat the problem (2.3) by arranging airways 1, 2, and 3 in series and plot their individual and combined characteristics.

References

[1] Vennard J. Elementary fluid dynamics. John Wiley and Sons; 1961. p. 570.

[2] Stanton TE. Similarity of motion in relation to the surface friction of fluids. Transactions of the Royal Society London (A) 1914;214.

[3] Nikuradse J. Stromungs gesetze in Rauhen Rohren. VDI-Forschungsheft 1933:361 [in German].

[4] Colebrook CF, White CM. The reduction of carrying capacity of pipes with age. Journal Institute of Civil Engineers, London 1937;7:99.

[5] Moody LF. Friction factors for pipe flow. Transcript of ASME 1944;66:671.

[6] Nikuradse J. VDI-Forschungsheft 1938;(356) [in German].

[7] Atkinson J. On the theory of the ventilation of mines. Transactions of NE Institute of Mining Engineer 1854;3:118.

[8] McElroy GE. Ventilation of mines. U.S. Bureau Mine Bulletin; 1935. p. 385.

[9] McElroy GE, Richardson AS. Resistance in metal mine airway. U.S. Bureau Mine Bulletin; 1927. p. 261.

[10] Greenwadd HP, McElroy GE. Coal mine ventilation factors. U.S. Bureau Mine Bulletin; 1929. p. 285.

[11] Pursall BR. In: Roberts A, editor. Ventilation planning. London: Cleaver-Hume Press; 1960. p. 264–7.

[12] Planning the ventilation of new and reorganized collieries. NCB Information Bulletin; 1955, 55/153.

[13] McPherson MJ. The resistance of airflow on a longwall face. In: Mutmansky JM, editor. Proceedings 3rd U.S. Mine Ventilation Symposium. Littleton, Colorado: SME; 1987. p. 465–77.

[14] Hartmann HL, et al. Mine ventilation and air conditioning. John Wiley and Sons; 1982. p. 156–7.

Turbulent Dispersion of Pollutants in Mine Airways

3

Chapter Outline

3.1 Mine Ventilation Systems

Mine ventilation systems consist essentially of three components, namely, the mine geometry, fluids circulating through the mine (mainly air contaminated with pollutant masses), and mine fans. Because of the velocity of the air and the size of airways, the flow is generally turbulent. Efficient design of mine ventilation systems depends heavily on a thorough knowledge of how pollutant masses originate, disperse, and flow in the mine airways. Mathematical analysis of such turbulent mass transfer processes is, therefore, an important area of study.

Depending on the geometry of the mine sections, these transfer processes can be classified into three parts, namely:

1. Flow through conduits: A fully developed laminar or turbulent transfer of air, gas, etc. occurs in conduits with well-defined geometry and surface properties, for example, flow of air and toxic gases in mine airways.
2. Flow through broken workings: Flow through broken workings can be laminar or turbulent depending on the degree of consolidation of the broken strata and the pressure difference across the area. In general, it is regarded as being laminar, for example, flow of methane and air through gobs (mined out areas).
3. Flow through porous media: An example of this kind of flow is the seepage of methane from coal beds and adjacent strata in situ. This subject is discussed in detail by the author in a recent publication [1].

Most of these flow problems can be solved for varying initial and boundary conditions, which provide a very fertile and important area of research. Precedents for their

Advanced Mine Ventilation. https://doi.org/10.1016/B978-0-08-100457-9.00003-1

solution are generally available in other fields such as heat transfer. However, it is necessary to understand the "physics" of the mining problems and to give them correct mathematical representations. Similarly, great care needs to be exercised in the selection of boundary and initial conditions. Above all, experimental verification of such models and determination of system parameters should be an essential part of such endeavors. In this chapter, an attempt has been made to model a few examples of mass transfer in mines by means of partial differential equations, and solutions have been given for typical initial and boundary conditions.

3.2 Generalized Mass Transfer Model

Modeling of mass transfer processes in mine airways is based largely on extensive studies of turbulent dispersions in wind tunnels and the lower atmosphere. Two basic approaches, namely, the gradient transport theory [2] and the statistical theory [3], have been used, but the former is considered to be more suitable for underground mines, as discussed by the author elsewhere [4].

In general, mine airways can be represented by a rectangular parallelepiped, but cylindrical and spherical geometry can be used with advantage for special situations. A generalized mass transfer model for a rectangular parallelepiped is

$$\begin{aligned}\frac{\partial c}{\partial t} &= \frac{\partial}{\partial x}\left(\varepsilon_x \frac{\partial c}{\partial x}\right) + \frac{\partial}{\partial y}\left(\varepsilon_y \frac{\partial c}{\partial y}\right) + \frac{\partial}{\partial z}\left(\varepsilon_z \frac{\partial c}{\partial z}\right) \\ &\quad - u(x)\frac{\partial c}{\partial x} - v(y)\frac{\partial c}{\partial y} - w(z)\frac{\partial c}{\partial z} - \lambda(x,y,z,c) + f(x,y,z,t)\end{aligned} \tag{3.1}$$

Where c is the time-averaged concentration of pollutants in mine air; x, y, z are the three coordinate directions; and t is the time variable; ε_x, ε_y, ε_z are turbulent dispersion coefficients in the x, y, and z directions, respectively; $u(x)$, $v(y)$, and $w(z)$ are components of air velocity in the three coordinate directions, respectively; λ is a generalized decay coefficient for the pollutant, which could be a constant or a function of the concentration and space variables; and $f(x, y, z, t)$ is a source term for the pollutant mass in the airway.

Eq. (3.1) is a nonhomogeneous partial differential equation. Depending on the nature of λ, ε_x, ε_y, and ε_z, it could be linear or nonlinear. In the latter case, an analytical solution is generally not possible and a solution has to be obtained with digital computers using well-known numerical techniques [5]. For linear cases, solutions are obtained in a manner analogous to that used to solve heat transfer problems [6]. Fortunately, the most important cases of mass transfer in mines are simpler than Eq. (3.1), and their solutions can often be obtained in closed forms with resultant ease of numerical computation.

Sources of pollutants in mines can be classified in various ways depending on the nature of the source (e.g., instantaneous or continuous, moving or stationary), the nature of the pollutant (e.g., gaseous or particulate), and the geometry of the source (e.g., point, linear, or areal). Different combinations of these occur in practice, and some of these will be studied here.

3.3 Instantaneous Stationary Point Source

This is a very common situation in the experimental determination of the longitudinal coefficient of turbulent dispersion. A known quantity, Q, of a tracer such as sulfur hexafluoride, SF_6, is released instantaneously at the origin and its concentration measured downstream at several points. A mathematical relation between concentration and space and time variables is needed to determine the coefficient of turbulent dispersion. Eq. (3.1) is modified in this case using the following assumptions:

1. There is no sink or additional source for the tracer in the roadway.
2. Coefficients of turbulent dispersion are constants and homogeneous.
3. Velocity components in the y and z coordinates are negligible and that in the x coordinate is a constant.

Using polar coordination, the mathematical model for this case is given by:

$$\frac{\partial c}{\partial t} = \varepsilon_r \left[\frac{1}{r} \frac{\partial}{\partial r} \left(r \frac{\partial c}{\partial r} \right) + \frac{\partial^2 c}{\partial x^2} \right] - u \frac{\partial c}{\partial x} \tag{3.2}$$

Typical boundary and initial conditions are as follows:

1. $\int_\upsilon c dv = Q$ for $t > 0$, that is, the tracer mass is conserved
2. $\left.\frac{\partial c}{\partial r}\right| = 0, \; at \; r = a$

where a is the radius of the roadway. This assumes that there is no material transfer at walls of the airway.

3. $c = 0$ at $t = 0$ for all x except at the origin

According to Seager and Fitzpatrick [7], a complete solution of Eq. (3.2) with the above boundary and initial condition is

$$c(x,r,t) = \frac{Q}{\pi a^2} \frac{1}{(4\pi\varepsilon_r t)}^{0.5} \left[1 + \sum_{m=1}^{\infty} \frac{Jo(\beta_m R)}{[Jo(\beta_m)]^2} \exp - \frac{\varepsilon r \beta_m^2 t}{a^2} \right] \cdot \exp \left(- \frac{(x-ut)^2}{4\varepsilon_r t} \right) \tag{3.3}$$

where $R = r/a$, Jo is the Bessel function of order zero, and β_m is the m-th zero of $J_1(R)$.

In practice, however, the tracer gets mixed with mine air very intimately across any given cross section because of obstructions in and the roughness of mine airways. In these circumstances, a one-dimensional model of dispersion is considered quite accurate. The concentration is then a function of x and t alone as given below.

$$c(x,t) = \frac{Q}{2A(\pi\varepsilon_x t)}0.5\ \exp\left[-\frac{(x-ut)^2}{4\varepsilon_x t}\right] \tag{3.4}$$

where A is the cross-sectional area of the roadway. Calculations of the dispersion coefficients from experimental data can be done very simply as discussed in great detail by Airey [8] and Klebanov [9].

3.4 Continuous Stationary Point Source

This case is represented by a continuous source in the roadway often obtained in practice. It is also used for the experimental determination of the radial coefficient of turbulent dispersion. The mathematical model is obtained from Eq. (3.2) by simply dropping the $\frac{\partial c}{\partial t}$ term, that is,

$$u\frac{\partial c}{\partial x} = \varepsilon_r\left[\frac{1}{r}\frac{\partial}{\partial r}\left(r\frac{\partial c}{\partial r}\right) + \frac{\partial^2 c}{\partial x^2}\right] \tag{3.5}$$

The boundary conditions for a source at the origin are

1. $C \to 0$ as $s \to \infty$
   where $s = (r^2 + x^2)^{0.5}$
2. $c = \infty$ as $s \to 0$

A third condition results from the conservation of mass, that is,

$$-4\pi s^2\varepsilon_r\left(\frac{\partial c}{\partial s}\right)_x = Q \text{ at } s \to 0$$

A solution of Eq. (3.5) with the above boundary conditions has been derived by Roberts [10] as given below:

$$c = \frac{Q}{4\pi\varepsilon_r s}\exp\left(-\frac{u(s-x)}{2\varepsilon_r}\right) \tag{3.6}$$

Taking logarithms on both sides of Eq. (3.6) and rearranging, the following relationship is obtained:

$$\ln\frac{Q}{4\pi sc} = \ln\varepsilon_r + \frac{u}{2\varepsilon_r}(s-x) \tag{3.7}$$

By plotting the experimental data on semilog paper with $\frac{Q}{4\pi sc}$ as the ordinate and $(s-x)$ as the abscissa, a straight line with gradient $u/2\varepsilon_r$ and intercept $l_n\varepsilon_r$ is obtained. In addition to the determination of the radial coefficient of turbulent dispersion, the model described by Eq. (3.5) can be used to predict concentrations around and at other strategic locations downstream from these continuous sources.

3.5 Dispersion of Respirable Dust From a Heading

A problem of considerable importance in mines is the dispersion of respirable dust from a heading being driven with continuous miners. Respirable dust particles are generally smaller than 10 μm in diameter and behave as a gas. In order that concentrations of respirable dust do not exceed the specified health standards, adequate ventilation must be provided. Eq. (3.1) is modified in this case with the following assumptions:

1. Cutting or loading machines at the heading virtually occupy the entire width. Consequently, symmetry in the direction of width can be assumed.
2. Because of obstructions and the narrow height of headings, a symmetry in the direction of height is also obtained.
3. The concentration of respirable dust is so small that there is no change in air density.
4. The velocity in the direction of the length of the heading is constant.
5. The decay coefficient for respirable dust is a constant. Generally it is a factor of velocity of air and friction factor of the roadway. Mathematical derivations of this constant are given by Beal [11].

On the basis of these assumptions, the model for the dispersion of respirable dust particles is as follows:

$$\frac{\partial c}{\partial t} = \varepsilon_x \frac{\partial^2 c}{\partial x^2} - u\frac{\partial c}{\partial x} - \lambda c \tag{3.8}$$

Typical boundary and initial conditions for this case are

$$\left.\begin{array}{ll} 1.\ c = c_o & \text{at} \quad x = 0 \\ 2.\ c = 0 & \text{at} \quad x = \infty \end{array}\right\} t > 0$$

$$3.\ c = 0 \quad \text{at} \quad t = 0 \text{ for all } x \text{ except at } x = 0$$

Eq. (3.8) is readily solved using the Laplace transform as given below:

$$c(x,t) = \frac{c_o}{2} e^{ux/2\varepsilon_x} \left[\begin{array}{l} e^{-x(\alpha/\varepsilon_x)^{0.5}} erfc\left(\dfrac{x}{2\sqrt{\varepsilon_x t}} - \sqrt{\alpha t}\right) \\ + e^{-x(\alpha/\varepsilon_x)^{0.5}} erfc\left(\dfrac{x}{2\sqrt{\varepsilon_x t}} + \sqrt{\alpha t}\right) \end{array} \right] \tag{3.9}$$

where $\alpha = \left(u^2 + 4\varepsilon_x\lambda\right)/4\varepsilon_x$ and *erfc* is the complimentary error function. By proper substitution in Eq. (3.9), the value of u and consequently the quantity of ventilating air required to meet a given health standard can readily be obtained.

3.6 Dispersion in a Leaky Roadway

In practice, there are air leaks in most of the airways in mines, and the air velocity is, therefore, nonuniform. The leakage factor is a function of the pressure gradient, mechanical design of stoppings, doors, air crossings, etc., and the fracture system in the strata. In many instances, the nonuniform velocity, $u(x)$, can be approximated by $u_o\exp(\pm\psi x)$, where u_o is the velocity at the initial cross section, and ψ is the leakage factor.

The one-dimensional model for a leaky roadway for the dispersion of a tracer is

$$\frac{\partial c}{\partial t} = \varepsilon_x\frac{\partial^2 c}{\partial x^2} - u_o\exp(\pm\psi x)\frac{\partial c}{\partial x} \tag{3.10}$$

The boundary condition at $x = 0$ depends on the mode of release of the tracer. In reality, it is a function of time, that is, $c|_{x=o} = f(t) \rightarrow$ o for large values of t.

Assuming that $c|_{x=\infty} = 0$ and $c = 0$ at $t = 0$, a solution of Eq. (3.10) is obtained after Grekov [12] as follows:

$$c(x,t) = \frac{x}{2\sqrt{\pi\varepsilon_x}}\int_o^t\frac{f(t-\tau)}{\tau^{1,5}}\exp\left(\frac{x^2}{4\,\varepsilon_x\tau}\right)d\tau$$

$$\pm\left(1-e^{\pm\psi x}\right)\frac{u_o}{4\,\psi\,e_x\sqrt{\pi\,\varepsilon_x}}\int_o^t\frac{f(t-\tau)}{\tau^{1,5}}$$

$$\left\{\exp\frac{x^2}{4\,\varepsilon_x\tau}\pm\frac{\psi\sqrt{\pi\,\varepsilon_x}}{2}\exp\left[\pm\frac{\psi x}{2}+\frac{\psi^2\varepsilon_x}{4}\tau\right]erfc\left[\frac{x}{2\sqrt{\varepsilon_x\tau}}\pm\frac{\psi\sqrt{\varepsilon_x\tau}}{2}\right]\right\}d\tau \tag{3.11}$$

In practice, values of ψ are of the order of 10^{-3} to 10^{-4} m^{-1} and hence $u(x)$ can be approximated by $u_o(1 \pm \psi x)$. Under some mining conditions, a steady-state situation may develop and Eq. (3.10) then simplifies to:

$$\varepsilon x\frac{\partial^2 c}{\partial x^2} - u_o(1\pm\psi x)\frac{\partial c}{\partial x} = 0 \tag{3.12}$$

An approximate solution of Eq. (3.12) for very small values of ψ is obtained as:

$$c(x) = A\,\mathrm{Exp}\left(\frac{u_o(1\pm\psi x)}{\varepsilon_x}\right) + B\,\mathrm{Exp}\left(-\frac{u_o(1\pm\psi x)}{\varepsilon_x}\right) \tag{3.13}$$

where A and B are constants and have to be determined from boundary conditions.

3.7 Concentration Growth in a Roadway With Uniformly Distributed Source

When a heading is being driven in a coal mine or any other gas-bearing strata, it is important to know the maximum concentration that can build up at the face. In coal mines, an exhaust-type ventilation is usually employed to keep dust and gas concentrations low. Because of this, gases given out by the walls of the heading are carried to the face, and there is a gradual rise in gas concentration from the outbye end to the inbye end of the heading. A mathematical model for this situation is

$$\varepsilon_x \frac{\partial^2 c}{\partial x^2} + u\frac{\partial c}{\partial x} - q = 0 \tag{3.14}$$

where q is the uniformly distributed source for gases from walls of the roadway.

Assuming the boundary conditions as:

$$\frac{\partial c}{\partial x} = 0 \text{ at } x = 0$$

and

$$\varepsilon_x \frac{\partial c}{\partial x}\bigg|_{x=L} + uc|_{x=L} = qL$$

where L is the length of the roadway, a solution of Eq. (3.14) is obtained as below:

$$c(x) = \frac{q}{u}\left[x - \frac{\varepsilon_x}{u}\left(1 - e^{-ux/\varepsilon_x}\right)\right] \tag{3.15}$$

substituting for c, the maximum permissible value, the quantity of diluent air can be obtained from Eq. (3.15).The derivation of ex is discussed in detail in Chapter 11 of the book.

Problems

3.1 Using Eq. (3.4), calculate ε_x
If $Q = 1$ ft^3 of SF_6 (sulfur hexachloride)
$A = 100$ ft^2
$u = 200$ ft/min
$t = 20$ min
$x = 10{,}000$ ft
$c(x,t) = 5$ ppm by volume (5 parts of 1 million).

3.2 Assume the above experiment is done in a tunnel of 10 feet diameter, and the source is continuous at 0.1 ft^3/min. Use Eq. (3.7) to determine ε_x.
Airflow remains the same at 20,000 CFM.

3.3 Using the ε_x value from Problem (3.1), calculate $c(x,t)$ at 10,000 ft downstream after 6 hours if all conditions are the same and λ is zero.

Use Eq. (3.9).

References

[1] Thakur PC. Advanced reservoir and production engineering for coalbed methane. Elsevier; 2016. p. 210.

[2] Fick A. Uber diffusion. Annual Review of Physical Chemistry 1855;2:59–86.

[3] Taylor Sir GI. Diffusion by continuous movements. Proceedings of the London Mathematical Society 1921;2:196–202.

[4] Thakur PC. Mathematical modelling of tunnel air pollution. In: Proc. Rapid Excavation and Tunnelling Conference, San Francisco, U.S.A.; 1974. p. 283–94.

[5] Ames WF. Numerical methods for partial differential equations. Barnes and Nobel, Inc.; 1969. p. 291.

[6] Carslaw H, Jaeger JC. Conduction of heat in solids. London: Oxford; 1959. p. 495.

[7] Seager JS, Fitzpatrick RD. A theoretical treatment of dispersion into a turbulent stream in a pipe. Safety in Mines Research Establishment, R.R. 1967;245:25.

[8] Airey EM. Diffusion of firedamp in mine airways. Minerals Engineering January 1969: 207–16.

[9] Klebanov FS, Martynyuk GK. A method for experimental determination of the coefficient of longitudinal turbulent diffusion in ventilation currents in mine workings. Fiziko-technicheskie, Problemy Razrabotki Poleznykh 1973;4:71–5.

[10] Roberts OFT. The theoretical scattering of smoke in a turbulent atmosphere. Proceedings of the Royal Society of London. Series A 1915;215:646–54.

[11] Beal SK. Deposition of particles in turbulent flow on channels or pipe walls. Nuclear Science and Engineering 1970;40:1–11.

[12] Grekov SP, Kalyusski AE. Drift of impurities in a turbulent current of variable velocity with a flow rate which varies along the mine working. Fiziko-Technicheskie, Problemy Razrabotki Poleznykh 1972;3:85–9.

Estimation of Ventilation Air Quantity

4

Chapter Outline

Advanced Mine Ventilation. https://doi.org/10.1016/B978-0-08-100457-9.00004-3

4.1 The Modern Mine Layout

Underground coal mining today is done by both (a) room and pillar and (b) longwall mining. Generally, shallow coal seams are mined by the room and pillar technique, but deeper seams are mined by longwall methods. The latter method is gaining ground as coal seams are getting deeper. In the United States, slightly more than 50% of coal is mined by longwall faces. The trend in longwall mining is for larger panels, bigger longwall equipment, and higher production capacity and productivity. Many longwall panels today are 1000 to 1500 ft wide and 10,000 to 12,000 ft long containing 2.5 to 4 million tons of coal.

The main benefits are

1. Improved safety and reduced injury rates because of improved longwall/development coal ratios and fewer longwall moves
2. Improved recovery of in situ coal
3. Improved productivity and reduced cost per ton

On the other hand, the main disadvantages are

1. Ventilation and methane control on the face and in the gob (mined out areas)
2. Respirable dust control
3. Ground control
4. Unknown geologic anomalies, such as faults, wash outs, sand channels, etc
5. Escape from longwall face in case of an emergency, such as fire

Figure 4.1 shows a typical layout of a modern coal mine. A set of mains are driven from the ventilation shafts that are sunk first of all. Branching from the mains, submains are driven on either side. Typically from a submain (of five–seven headings), two sets of gate roads, (three headings in 6–7 ft thick seams) and (four headings in 5 ft or less thick coal seams), are driven. A typical mine will have one longwall panel and three continuous miner sections. One drives the submains, and the other two drive the gate roads.

High emissions of methane can be encountered in development headings, at the longwall face and in the longwall gob areas. Although the gas emitted in development headings and the longwall face comes from the coal seam being mined, the longwall gob acts as a pressure sink and draws gas from several coal seams overlying and underlying the coal seam being mined. Beside the width of the longwall panel, the other factor that has great influence on gas emissions in the gob is the rate of face advance. Daily methane production from the gob is linearly proportional to the rate of face advance. Hence, the gob emission/acre of mining must be determined for a given width and the rate of advance of the longwall face for proper placement of gob wells on the panel and estimation of bleeder air quantities. All mining sections, development or longwall, produce gas. Some premining and postmining methane drainage is generally needed because ventilation alone cannot meet the legal requirements. It will be discussed in detail in Section 4.3 of the chapter.

A mine like this can produce 5 to 7 million tons per year of clean coal. Very large mines may have two such smaller mines in one producing 10 to 12 million tons per

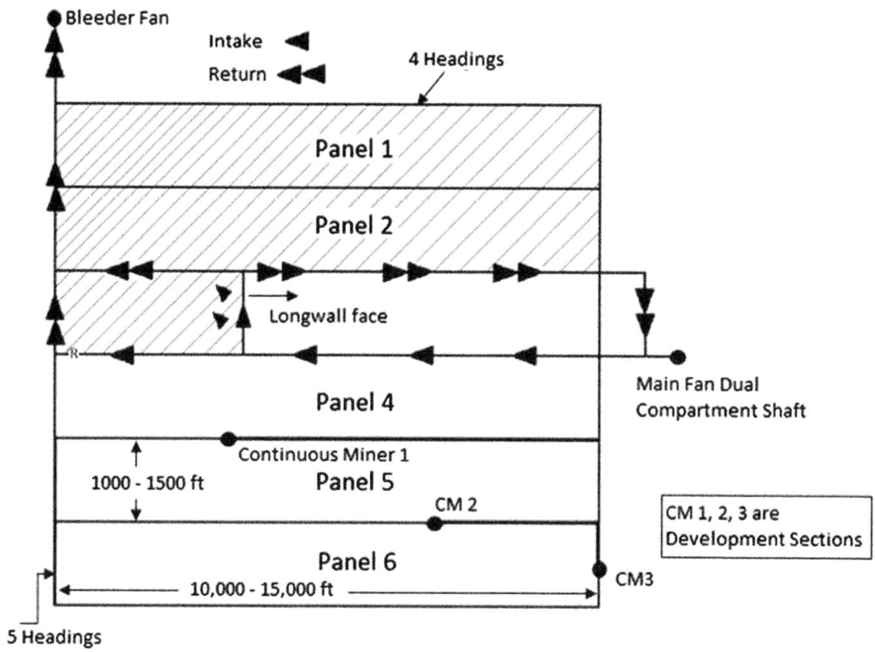

Figure 4.1 A typical modern coal mine layout.

year. Economy of size drives the cost of production further down, but the capital cost goes up. A very large coal mine may cost \$500 million to open up today and produce 12 to 15 million tons of coal per year.

4.2 Methane Emissions

It is safe to assume that all coal seams are gassy because coal and methane in coal are syngenetic in origin; that is, they are derived from the same plant material. The gas content of coal seams varies depending on the rank of coal and the depth of coal seam. Globally, coal seams can be divided into three categories depending on the depth and their gas contents in ft^3/ton as shown in Table 4.1.

Table 4.1 Gassiness of Coal Seams [1]

Category	Depth (ft)	Gas Content (ft^3/ton)
Mildly gassy	500 or less	100 or less
Moderately gassy	500–1500	100–300
Highly gassy	1500–3000	300–700

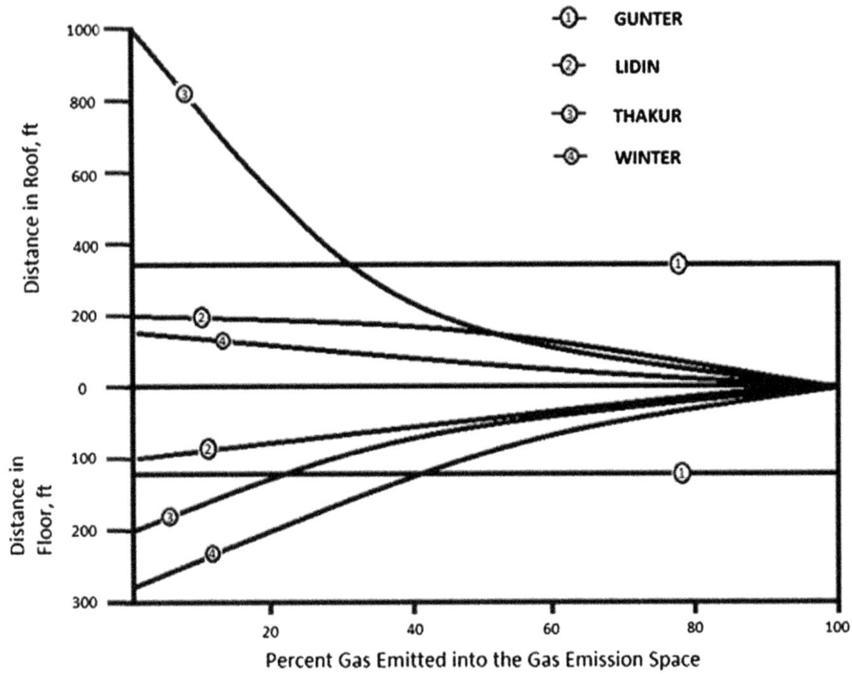

Figure 4.2 Vertical extensions of gas emission space surrounding a longwall gob.

Fig. 4.2 [2] shows the vertical extent of the gas emission space created by longwall mining and the percentage of gas content released by various coal seams contained in the gas emission space as a function of its distance from the mined coal seam. The vertical dimension of the gas emission space is highly dependent on the width of the longwall face. In general, the wider the longwall face, the greater the vertical dimension of the gas emission space and, consequently, the higher the specific gob methane emission (ft^3 of gas emitted per acre/day). In one study, the specific methane emission increased by 50% when the face width increased from 630 to 700 ft [3]. Very wide (1000 + ft) longwall faces also exacerbate methane and respirable dust concentrations at the tailgate and require a larger quantity of air at the intake end of the face to stay in compliance with statutory requirements.

To properly plan for methane control, one must know the amount of ventilation air that is available in strategic areas, e.g., longwall face and bleeders. Table 4.2 shows

Table 4.2 Desirable Ventilation Quantities for Longwall Faces

Category	Face Intake (cfm)	Tailgate (cfm)	Bleeders (cfm)
Mildly gassy	30,000	25,000	100,000–150,000
Moderately gassy	50,000	40,000	150,000–250,000
Very gassy	80,000	60,000	250,000–350,000

desirable ventilation quantities for different categories of coal seams of the northern and southern Appalachian Basin, United States. With proper ventilation planning, it appears possible to deliver these quantities of air to the longwall face and bleeders. It is also assumed that 50% of the original gas content has been drained before mining.

It should also be noted that for 15,000- to 20,000-foot long panels, it is imperative to provide an internal return as shown in Fig. 4.3–4.5. Longwall ventilation must be

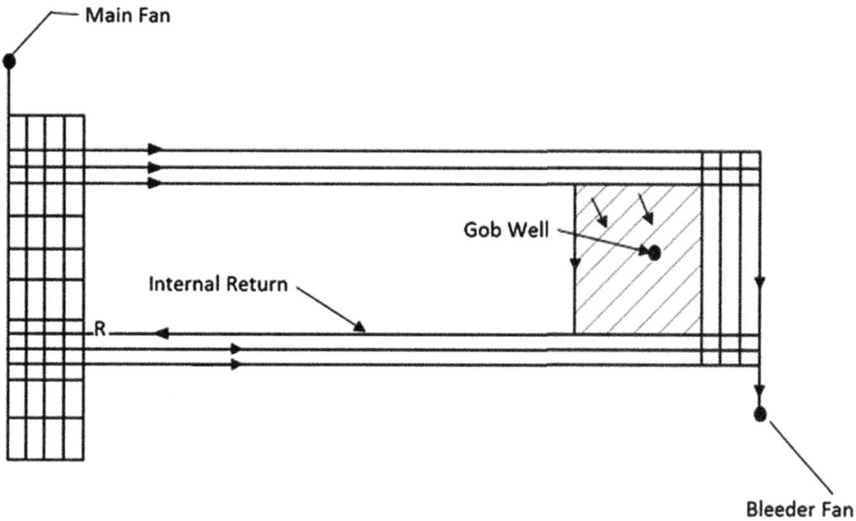

Figure 4.3 Ventilation layout for a longwall panel in mildly gassy coal seams.

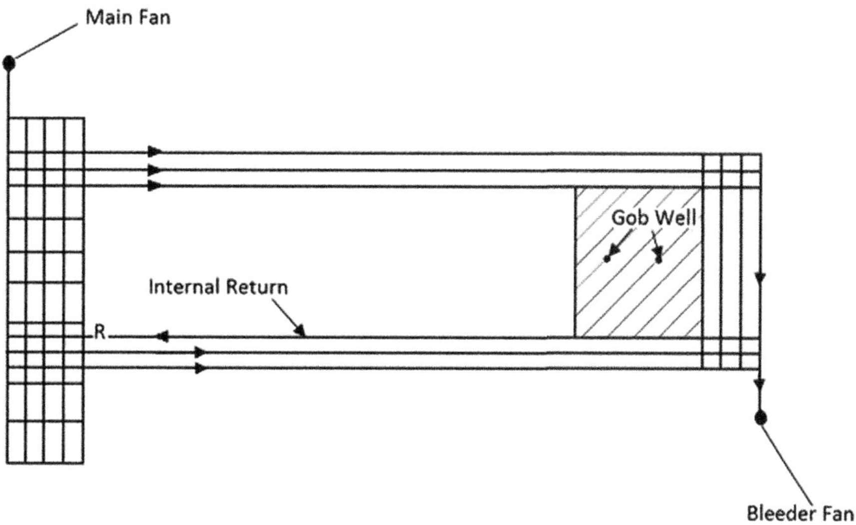

Figure 4.4 Ventilation layout for a longwall panel in moderately gassy coal seams.

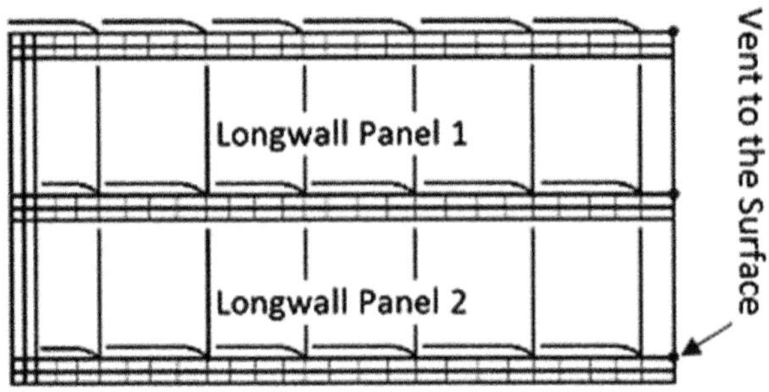

Figure 4.5 Degasification plan for a longwall panel in moderately gassy coal seams.

provided by the combination of a main fan and a bleeder fan. Very high pressure bleeder fans are not desirable for coal mines because all coal seams are liable to spontaneous combustion. High pressure differential promotes spontaneous combustion (to be discussed in Section 4.4).

4.3 Mildly Gassy Coal Seams

A typical layout for a longwall face operating in a mildly gassy coal seam is shown in Fig. 4.3.

4.3.1 Premining Degasification

In coal seams with gas contents less than 100 ft^3/t, there is generally no need for premining degasification. However, it may be desirable to drill the longwall panel horizontally at 1000-foot intervals for exploration and respirable dust control.

4.3.2 Postmining Degasification

At least one gob well within 1000 ft from the setup entry is recommended because longwall gobs always produce some gas. Normally, there exists a very good longitudinal communication over the longwall panels, and a single gob well can have a drainage radius of 3000 to 5000 ft. A second gob well may be needed to cover a 15,000- to 20,000-foot long panel. The gob well diameter can range from 4 to 8 in., and gas production can be enhanced with blowers.

4.3.3 Ventilation Layout and Quantities

A preferred ventilation layout for a very large longwall panel is shown in Fig. 4.3. The main fan shaft provides all the intake air. The bleeder fan assists the main return fan. The internal return may not always be necessary unless the ventilation air on the face declines below the quantities shown in Table 4.1.

4.4 Moderately Gassy Coal Seams

A typical layout for a longwall face operating in moderately gassy coal seams is shown in Fig. 4.4.

4.4.1 *Premining Degasification*

Moderately gassy coal seams generally need to be degassed in advance of mining. Fig. 4.5 shows a degasification scheme with in-mine horizontal boreholes. Boreholes drilled parallel to the development headings degas them, and cross-panel boreholes drilled into the longwall panels degas the longwall panels. If these boreholes are drilled promptly and produced efficiently, nearly 50% of the in situ gas can be drained before mining. The outbye boreholes can be 1000 ft apart, but inbye boreholes should be spaced closer.

4.4.2 *Postmining Degasification*

A gob well must be installed within 500 ft from the setup entry and, based on local experience, additional gob wells should be installed at 30- to 60-acre spacing. Usually the first gob well is the best producer, but other gob wells also help in controlling methane emissions. With proper planning and blowers on the gob wells to assist in methane drainage, 50 to 60% of total gob methane emissions can be captured. The gob well diameter ranges from 6 to 9 in., and gas production is always assisted with well-designed blowers.

4.4.3 *Ventilation Layout and Quantities*

A preferred ventilation layout for longwall panel mining moderately gassy coal seams is shown in Fig. 4.4. An internal return, in this case, is a necessary requirement; otherwise, ventilation air quantities shown in Table 4.2 may be difficult to provide.

4.5 Very Gassy Coal Seams

A typical layout for a longwall face operating in a very gassy coal seam is shown in Fig. 4.6.

4.5.1 *Premining Degasification*

Very gassy coal seams must be drained several years ahead of mining with vertical frac wells. These frac wells can be put at about 20-acre spacing. They can also drain methane from overlying coal seams and reduce gob emissions. Frac wells drilled about 5 years ahead of mining can drain nearly 50% of gas contained in coal, but this may not be enough to sustain a high rate of extraction. Additional degasification is done with

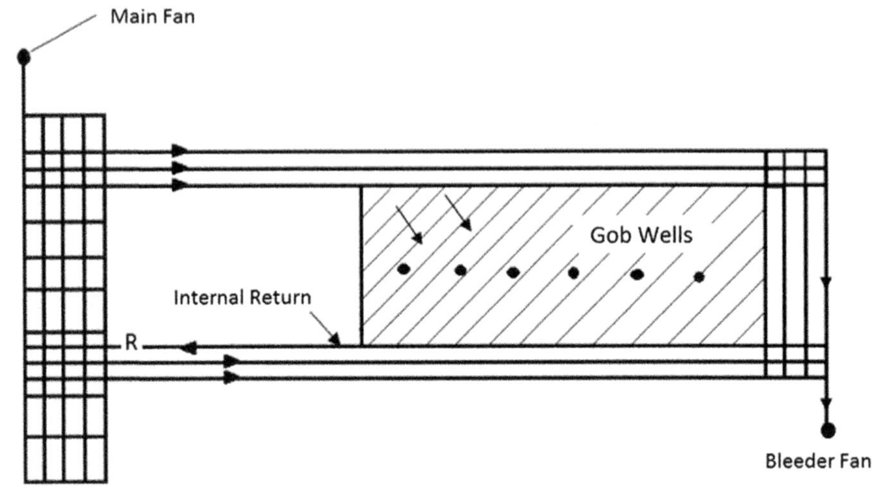

Figure 4.6 Ventilation layout for a longwall panel in very gassy coal seams.

in-mine horizontal drilling to remove nearly 70% of in-place gas before mining as shown in Fig. 4.7. Horizontal boreholes are drilled at 200- to 300-foot intervals and are extended beyond the longwall panels to intersect and degas the next development section.

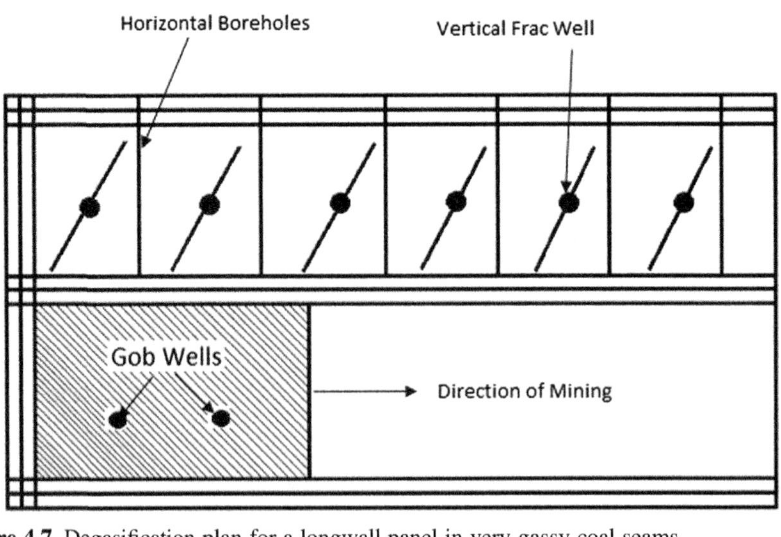

Figure 4.7 Degasification plan for a longwall panel in very gassy coal seams.

4.5.2 Postmining Degasification

Because of very high gas emissions in the gob area, the first gob well must be installed within 50–100 ft from the setup entry. Subsequent gob wells may be drilled at 6–15 acre spacing, depending on the rate of mining and gas emissions experienced. In Virginia and Alabama, the gob wells are generally 9–12 in. in diameter. Powerful blowers with 75–125 Hp motors creating a suction of 5–10 in. of mercury are needed to capture up to 70% of gas emissions.

4.5.3 Ventilation Layout and Quantities

Ventilation layout for a longwall panel in very gassy mines, as shown in Fig. 4.6, is similar to the layout for moderately gassy coal seams, but ventilation air quantities are higher, as shown in Table 4.2. The bleeder shafts are bigger in diameter to handle ventilation air quantities of 250,000 to 350,000 cfm. A well-placed bleeder shaft can serve four to six longwall panels depending on the size of the panels.

4.6 Limitations on the Longwall Face Width Owing to Face Methane Emissions

Ventilation air quantities reaching the tail end of the longwall face is dependent on the ventilation air quantities (and therefore, the air velocity) at the head gate end of the longwall face and the air leak-off on the face. The wider the longwall panel, the higher is the leak-off factor. Fig. 4.8 shows a plot of some actually observed data in coal seams with a thickness of 5–6 ft.

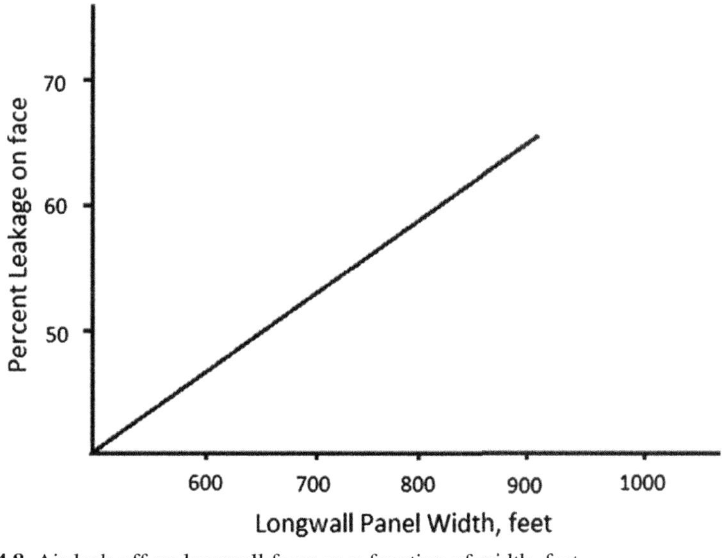

Figure 4.8 Air leak-off on longwall faces as a function of width, feet.

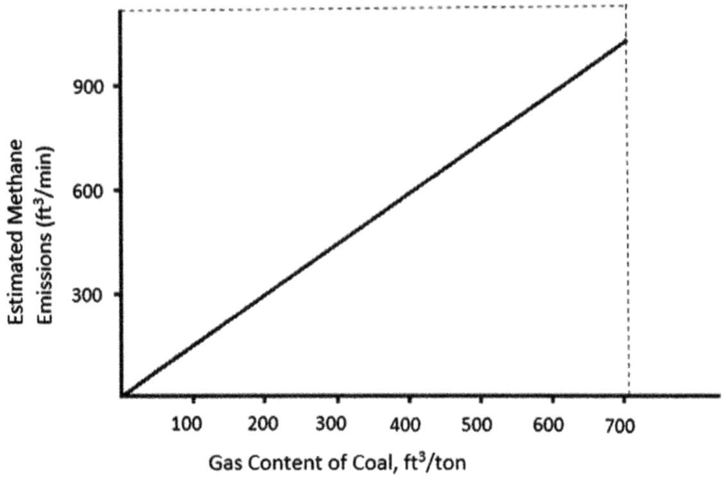

Figure 4.9 Methane emissions on longwall faces as a function of gas content of coal ft^3/t.

It is clear from this observation that the wider the longwall panel, the smaller are the ventilation air quantities reaching the tail end of the face. It is also true, as discussed later in the book, that the wider the longwall panel, the higher is the methane emissions at the tail end of the longwall face. It is, therefore, easy to conclude that for a given set of conditions, a limit on the longwall face width will be reached when it will not be possible to dilute the methane emissions enough to meet the statutory requirements.

4.6.1 Gas Emissions on Longwall Faces

Gas emissions on the longwall face is dependent on (a) the residual gas content of degassed coal seam or the degree of degasification, (b) the rate of coal extraction, and (c) the diffusivity of coal [4]. The latter is a measure of the rate of gas emission from mined coal. Higher rank coal, e.g., low volatile bituminous coals, have a much higher diffusivity than high volatile bituminous coals [5], and as such, they release a higher fraction of their original gas content on mining.

An approximate estimate of methane emissions at the tail end of a longwall face can be derived from the following equation:

$$Q = Q_0 + W(A - B) - C(x) \tag{4.1}$$

where: Q = total methane emissions at the tail end of longwall, ft^3/min; Q_0 = total methane emitted when no mining is done, ft^3/min; W = average rate of mining in t/min; A and B = gas contents of coal before and after mining, respectively; $C(x)$ = methane lost to gob areas by air leak-off and is a function of the distance from the head gate, m^3/min.

Assuming a 1000-ft-wide longwall face has a daily advance of 50 ft and a ventilation leak-off of 50–70%, average methane emissions from highly gassy coal seams at the tail end of the longwall face as a function of degree of degasification is shown in Fig. 4.9. For high coal productivity, highly gassy coal seams should be degassed before mining to reduce the gas content of coal to at least below 100 ft^3/t. It is also prudent to assume a peak emission rate that is 25% higher than the average emission rate. The minimum ventilation air quantity should be able to dilute the peak methane emissions to below statutory requirements, i.e., 1% in the United States.

4.6.2 Gas Layering on Longwall Faces

The second criterion for the adequacy of ventilation on longwall faces is the prevention of gas layering near the roof or floor. Gas layering in any mining roadway including the longwall face is governed by (a) the methane emission rate, (b) ventilation air velocity, and (c) the effective width of the airway. The gas layering number (GLN) is mathematically expressed as follows [6]:

$$\mathrm{GLN} = \frac{V}{41}\left(\frac{D}{Q}\right)^{\frac{1}{3}} \tag{4.2}$$

where Q is the methane emission rate in cfm; V is the air velocity in ft/min; D is the effective width of the airway (longwall face), ft.

A minimum value of 5 for GLN is considered necessary to prevent layering. The higher the value of GLN, the less likely it is that gas layering will occur.

A typical calculation is shown here.

Assume, Q = 500 cfm; D = 10 ft
And a GLN of 6 for safety

The necessary air velocity to prevent gas layering from Eq. (4.2) is 906 ft/min. Assuming a mining height of 6 ft, the minimum ventilation air required at the tail end of a longwall face is 54,317 cfm to prevent gas layering.

4.6.3 Mathematical Modeling of Methane Flow

The flow of methane and air on longwall faces can be modeled mathematically. Fundamental basis of such models are already developed [7]. Main assumptions made are as follows:

1. Symmetry in the directions (y-z) perpendicular to the longwall face (x direction).
2. The density effect of a lighter gas such as methane is neglected.
3. When mining is in progress, a steady-state situation is likely to prevail, i.e., time dependence of methane concentration is discarded.

Taking a mass balance over a small element of the longwall face and applying the above assumptions, the turbulent dispersion of methane on longwall faces can be represented by the following mathematical model:

$$\mathrm{Ex}\frac{d^2c}{d^2x} - u(x)\frac{dc}{dx} + q(x) - p(x) = 0 \tag{4.3}$$

where Ex is the longitudinal coefficient of turbulent dispersion; $u(x) = u_o \exp(-ax)$, where $u(x)$ is the air velocity at any point on the face; u_o is the velocity of air at head gate; a is a leakage coefficient that is experimentally determined; $q(x)$ is the methane source and includes both the steady and transient methane emissions; $p(x)$ is the loss of methane in the gob at any point on the face due to air leakage; c is the concentration of methane at a "small element" of longwall face, defined as Q/V where Q is the total methane emission; V is the volume of air at the same location.

Boundary conditions are

At $x = 0$ (i.e., the head gate); $dc/dx = 0$, i.e., methane concentration is a constant.
At $x = L$ (i.e., the tail gate) $c = 0.01$ (or any other statutory limit).

Eq. (4.3) is a second-order, nonhomogenous differential equation, and no analytical solution can be obtained in a closed form. However, solutions can be obtained using finite difference or finite element techniques and computers. The second challenge here is to accurately measure values of Ex; $q(x)$ and $p(x)$.

The minimum quantity of air thus calculated must be increased by at least 25% to provide some cushion for peak emissions. If this quantity of air cannot be delivered to the tail gate of a longwall face owing to airway size and number (usually three) and the fan size, the width of the longwall panel must be reduced. In moderately gassy mines of the northern Appalachian Basin of the United States, gas emissions on the longwall face limit the width of the panel from 1200 to 1300 ft [8].

4.7 Limitations on the Longwall Face Width Owing to Gob Methane Emissions

In each basin, the coal seams have two emission characteristics. They are as follows:

1. Specific emission of the coal seam
2. Specific gob emissions for the coal seam

4.7.1 Specific Emission of a Coal Seam

It is the amount of gas produced by a section of 100 ft of a horizontal borehole drilled into the coal seam. It ranges from 3 to 25 MCFD/100 ft depending on the rank, permeability, and depth of the coal seam [1]. More on it will be discussed later in Section 4.3.

4.7.2 Specific Gob Emissions for the Coal Seam

As shown in Fig. 4.2, when a coal seam is mined out by a longwall face, the strata above and below get fractured and begin to produce gas. For each basin, it is a characteristic of the coal seam. For example, the Pocahontas #3 coal seam in the Central Appalachian Basin, United States, produces 30 MMCF/acres of the gob.

Table 4.3 Specific Gob Emissions on US Longwall Panels

Width of Longwall Face (ft)	Specific Gob Emissions (MMCF/acre)
450	25
600	30
750	33
900	36
1050	40

Just as excessive methane emissions on a longwall face can limit the width of the panel, very high specific gob emissions can also limit the panel width. The specific gob emission increases with increasing width of longwall panels. Table 4.3 shows some observed data [9] for the Central Appalachian Basin, United States.

The optimum width of a longwall panel is the width where gob wells are most efficient in draining the gob gases, and the total number of gob wells for the panel is the minimum. The most efficient gob drainage is reached when the spacing between the two adjacent gob wells(s) is equal to half the width (w) of the longwall panel. To illustrate this point, spacing of gob wells for longwalls with different widths but the same tonnage of extraction is calculated. The following assumptions are made for this calculation:

1. The rate of extraction is the same for all longwall panels, 1 acre/day.
2. The longwall face is 1000 ft long and needs two gob wells near the setup entry.
3. A gob gas capture ratio of 70–80% will be achieved.
4. Specific gob emission is 30 MMCF/acre.

Table 4.4 shows pertinent data for various widths of longwall panels.

Fig. 4.10 shows a plot of s/w against longwall face widths. The most efficient capture of gob gas is obtained when s/w = 0.5 or the longwall width is 700 ft. Because the gob wells are offset from the center line by 100 ft, the optimum width where gob gas drainage is most efficient is in the range 700–800 ft. A width of 750 ft is a good average. Optimum width for a coal seam with a different specific job emission can be similarly calculated.

Table 4.4 Number of Gob Wells Versus Longwall Width

Width of Face, (ft)	Number of Gob Wells	Spacing/Width, (s/w)
400	24 + 2	0.93
600	26 + 2	0.64
800	28 + 2[a]	0.48
900	32 + 2[a]	0.35
1000	36 + 2[a]	0.26

[a]Total number of gob wells could be slightly higher because of declining capture efficiency.

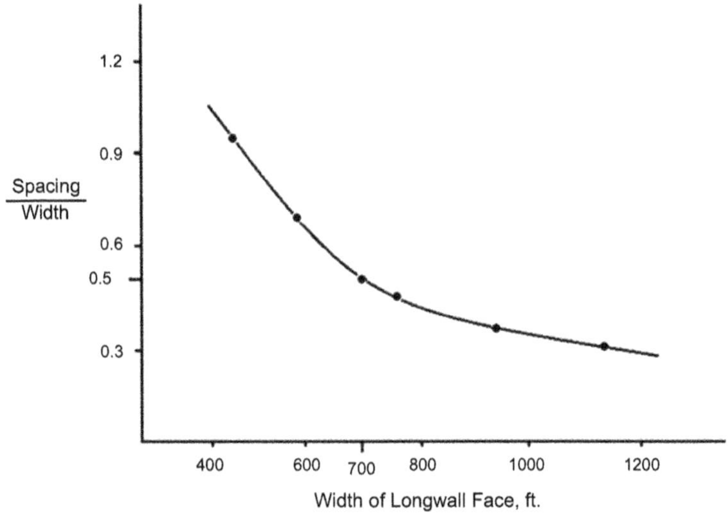

Figure 4.10 Optimal spacing of gob wells on a longwall face.

4.8 Air Quantity Requirements for Development Headings

It is generally not possible to achieve a good rate of advance (such as 100 ft/day) in a three-heading development without advance degasification. Horizontal boreholes drilled in advance (to be discussed in Section 4.3) generally capture 50% of total methane emissions. For moderately gassy mines, the development section still generates 250–400 cfm of methane. To dilute it properly, 50,000 cfm is generally needed at the last open cut.

For highly gassy mines, the air quantities are not much higher than 50,000 cfm/section because 70 to 80% of gas in coal is predrained by vertical, hydraulically fractured wells supplemented by in-mine horizontal drilling.

The air quantities estimated here are higher than those mines where a conveyor is not used for coal transport. US mining laws require part of the air to be used for belt entry ventilation as a separate split.

4.9 Estimation of Total Required Ventilation Air

Once the quantity of air required for each longwall face and development sections has been determined, the total required volume of air to be handled by the mine fan(s) needs to be estimated. This is done by adding to the air quantity needed at the working faces all leakages and air quantities needed in nonworking locations, such as diesel

Table 4.5 Average Leakage of Air in Coal Mines and Typical Air Requirements for Nonworking Areas are Shown in Table 4.6 [10]

Location	Leakage Air, cfm
Separation doors	3000
Air crossings (not explosion proof)	3000
Fan drifts:	
W.G = 5 in.	25,000
10 in.	35,000
20 in.	50,000

Adapted from Roberts A. Mine Ventilation:254–55.

equipment sheds, battery charging stations, etc. This is done for all intakes and returns until the ventilation shafts are reached. An additional amount should be added for the expansion of air in the return shaft and finally adding the surface leakage into the fan house/drift depending on the pressure gauge of the fan. The resultant air quantity is an estimate of total ventilation air for the mine.

Table 4.5 lists observed leakage in typical British coal mines [10].

4.9.1 Expansion of Air in Return Shafts

The barometric pressure in a deep shaft increases by one inch of Hg for every 1000 ft depth. Thus in a 3000 ft deep shaft, the surface air pressure of 30 in. of Hg will increase by 3 in. of Hg at the shaft bottom. Because volume of air is directly proportional to pressure (refer to Chapter 1), the air will expand by 10% in the return shaft of 3000 ft depth. A good rule is to allow an expansion in volume by 1% for every 300 ft of depth.

4.9.2 Air Velocities in Various Branches of a Coal Mine

As discussed earlier, the minimum air velocities on the longwall face (especially at the tail gate) is dictated by the gas layering index. Air velocities in other areas are determined by the most economic size of the airways. Very high velocities are not desirable because it increases power cost and kicks up dust from the floor and sidewalls. To reduce air velocity, the only solution would be to increase the airway size. In coal mines, the height and width are generally fixed, so it is done by increasing the number of airways in parallel. But this increases the construction costs and ultimately the maintenance costs. For each location, there will be a most economic size (to be discussed later in the text), which will give a minimum number of airways at the lowest cost. And this size will give the most economic air velocity. A guideline for optimum air velocities is shown in Table 4.7.

Table 4.6 Air Quantities for Nonworking Areas

Stone Drifts	25 $ft^3/min/ft^2$ of the Cross Section Allow for Leakage in Ducts
Pump houses	5000 cfm
Battery charging station	10,000 cfm
[a]Diesel locomotive	100–140 ft^3/min/bhp

[a]US requirements are more restrictive based on diesel particulate emissions. They will be discussed in Section 4.2.

Table 4.7 Optimum Air Velocities in Coal Mines

Location	Optimum Air Velocity (ft/min)
Working faces	150–500 (use gas layering index to guide)
Conveyors/loading points	Less than 600
Main airways	Less than 1000
Shafts	1500–2000
Fan drifts	Less than 2500

4.10 Standards of Volumetric (Ventilation) Efficiency

A final check on the estimation of ventilation air quantities is done by a well-established criterion, the volumetric/ventilation efficiency. It is defined as the ratio of air quantities required on the working faces/total air quantity at the fan(s). For well-designed ventilation systems, it ranges from 44 to 55% for workings laid out in the coal seam. For horizon mining, it can be even higher at 55 to 65%.

An example:

Let us suppose the mine layout in Fig. 4.1 is for a moderately gassy mine.

Air requirement in each continuous mining section at:

The last open crosscut = 50,000 cfm.
Total for 3 sections = 150,000 cfm.

Air requirement for the longwall:

At the tail gate = 40,000 cfm (a separate split)
At the bleeder shaft = 250,000 cfm
Total air for longwall face = 290,000 cfm

Hence total air at all working faces = 440,000 cfm.

Assuming a ventilation efficiency of 50%, the total capacity of the mine fan(s) should be approximately 880,000 cfm.

Problems

4.1 If a longwall face is 8 ft wide and emits 300 cfm methane, calculate the ventilation air necessary to avoid gas layering. The seam is 6 ft high. Check if the methane concentration will be below 1%.

4.2 Estimate the total ventilation air needed for a mine twice the size of the mine in Fig. 4.1. If a single intake shaft is used, what would be its size to keep the velocity of air below 2000 ft/min?

4.3 Calculate the optimum spacing of gob wells on a 1200-ft-wide longwall panel in a moderately gassy mine. Create a graph similar to Fig. 4.10. Specific gob emission is 8 MMCF/acre. Use 8-inch gob wells producing 2 MMCFD/well.

References

[1] Thakur PC. Optimum widths of longwall panels in highly gassy mines – Part I. In: Mutmansky J, editor. 11th U.S. Mine Ventilation Symposium. University Park, PA, USA: Penn State University; 2006. p. 433–7.
[2] Thakur PC. Methane drainage from gassy mines — a global review. In: Proceedings of the 6th International Mine Ventilation Congress; 1997. p. 415–22.
[3] Schatzel S, et al. Methane sources and emissions on the longwall panels in VA. In: Annual Pittsburgh Coal Conference; 1992. p. 991–8.
[4] Thakur PC. Advanced reservoir and production engineering for coal bed methane. Elsevier Publishing; 2016. p. 216.
[5] Kissell FN, Beilica RJ. An in-situ diffusion parameter for the Pittsburgh and Pocahontas #3 coalbed. USBM Research Report 7668. 1972.
[6] Leeming JR, Yates CP. Current British practice for methane ignition prevention in coal mine headings. In: DeSouza E, editor. Proceedings of the U.S. Mine Ventilation Symposium. Balkema Publishers; 2002. p. 487–96.
[7] Thakur PC. Mathematical modelling of tunnel air pollution. In: Pattison HC, editor. Proceedings of the Rapid Excavation and Tunneling Conference; 1974. p. 283–94.
[8] Thakur PC. Optimum width of longwall panels in northern Appalachian basin, USA. In: U.S. Longwall Symposium, Pittsburgh, PA; June 2015.
[9] Thakur PC. Optimum width of longwall faces in highly gassy coal mines – Part II. In: The 12th U.S. Mine Ventilation Symposium. Reno, Nevada, USA: University of Nevada; 2008. p. 81–6.
[10] Parsall BR. In: Roberts A, editor. Ventilation planning, mine ventilation. Cleaver Hume Press Ltd; 1980. p. 244–62.

Ventilation Network Analysis

5

Chapter Outline

Air flows in a single airway as well as in several airways in parallel or in series were discussed earlier. A typical mine is, however, a network of hundreds of airways with all kinds of combinations. Analytical solutions of simple ventilation networks exist in literature [1,2]; but for larger networks, computers must be used. Many computer programs were developed to just calculate the airflow and ventilation pressure losses in coal and metal mines [3,4] in 1970s. Thakur [5] developed an advanced program that not only calculated the airflow and pressure losses in each branch of a ventilation network but also calculated the steady-state distribution of diesel exhaust pollutants, such as, carbon monoxide, carbon dioxide, and nitrogen compounds in the network. It also calculated the air index utilizing the concentrations of gases and their respective threshold limit values (TLVs).

To illustrate a model problem, a hypothetical mine is illustrated in Fig. 5.1A.

To be able to solve any mine ventilation network mathematically, it must be converted into a linear network and mathematical model in that order. Fig. 5.1B shows a linear network analogue of the hypothetical mine under discussion.

The main assumptions, in this case, are as follows:

1. Turbulent flow of air in the roadways with a uniform velocity distribution. Consequently, the gaseous pollutants are assumed to be uniformly mixed across the whole cross section of roadways and junctions in the network. Components of diesel exhaust are used in this example.
2. The flow of air and diesel exhaust is steady.

Advanced Mine Ventilation. https://doi.org/10.1016/B978-0-08-100457-9.00005-5

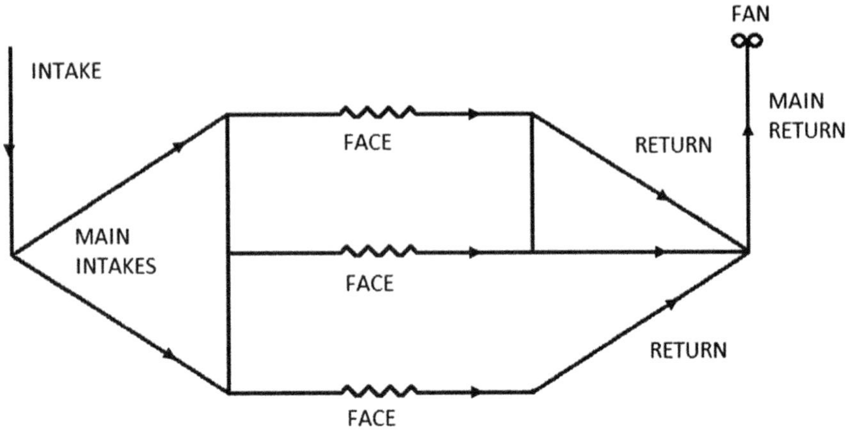

Figure 5.1A A hypothetical mine layout.

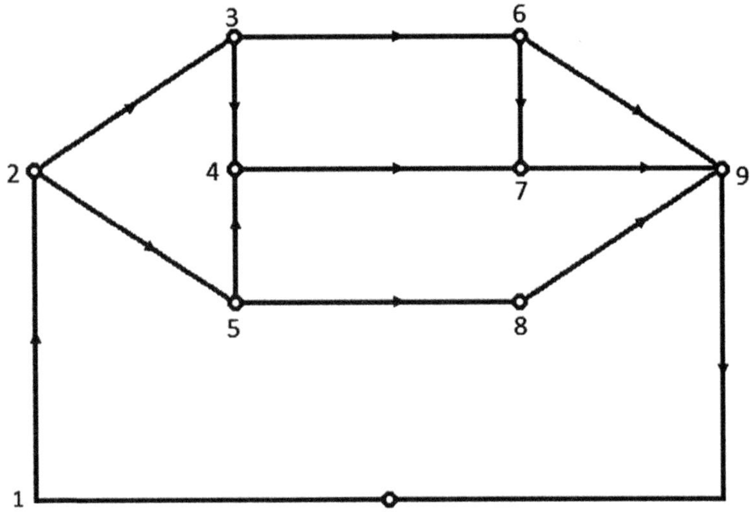

Figure 5.1B Linear network analogue of the hypothetical mine.

3. The volume of diesel exhaust is so small in comparison with the volume of circulating air that no change in the latter is anticipated.
4. Variations in temperature and pressure are insignificant to cause any density change.

The problem of turbulent dispersion of exhaust from multiple diesel engines in a network of roadways can be divided into three parts: (1) solution of the network for quantity of air in each branch, (2) emission rates, and (3) calculation of exhaust concentration in each branch and at each junction.

5.1 Network Analysis for Air Quantities and Pressure

All computer programs that analyze the air quantity and corresponding pressure loss basically use Kirchoff's first and second laws for transmission of electric power. This is possible because pressure and quantity relation in mine ventilation ($P = RQ^2$) is quite similar to electric current flow where:

$$V = RI^2 \tag{5.1}$$

where: V is the voltage, R is the resistance, and I is the current.

5.1.1 *Kirchoff's First Law*

Translated into mine ventilation terms, it states that the quantity of air reaching a junction in a network must equal the quantity of air leaving the junction. It is the principle of mass conservation. It is illustrated in Fig. 5.2A.

$$Q_1 + Q_2 = Q_3 + Q_4 \tag{5.2}$$

If we designate the airflow as vectors and assign positive sign to incoming flows and a negative sign to outgoing flows, $Q_1 + Q_2 - Q_3 - Q_4 = 0$
or

$$\sum_{i=1}^{4} Q_i = 0 \tag{5.3}$$

5.1.2 *Kirchoff's Second Law*

Again using the terms of mine ventilation network, the second law says that the sum of the pressure drops around any closed path must be equal to zero. Referring to

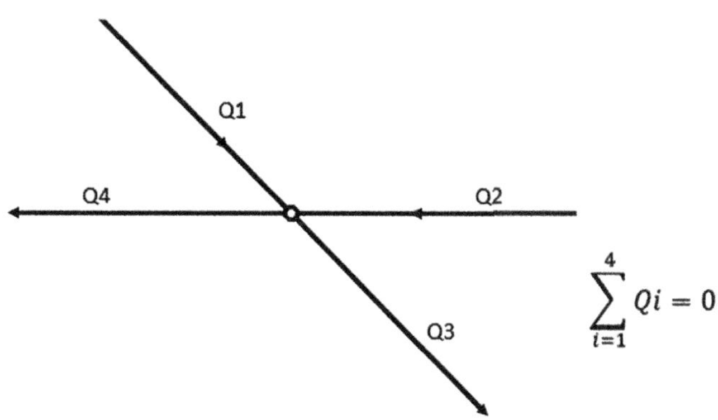

Figure 5.2 A Illustration of Kirchoff's first law.

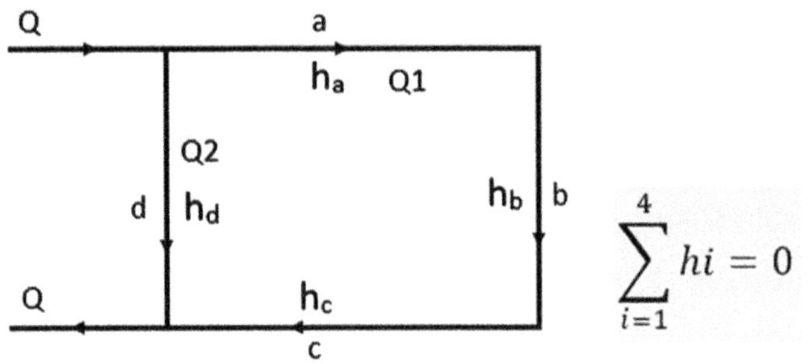

Figure 5.2 B Illustration of Kirchoff's second law.

Fig. 5.2B, the vector sum of pressure losses in branches a, b, c, and d is zero. Mathematically:

$$\sum_{i=1}^{4} h_i = 0 \tag{5.4}$$

The head losses where air flows in a clockwise direction (branches a, b, and c) are positive but it is negative in branch d where air flows in an anticlockwise direction.

Details of ventilation network analysis are available in literature [2,3]. Reference can be made to them for additional discussions.

5.1.3 Emission Rates

In a mine, the branches and junction where sources of diesel exhaust are located are known a priori. These are the haulage roadways and face areas. For example, in Fig. 5.1B, let the branches (2, 3) and (3, 4) and junction (4) have sources of diesel exhaust.

The emissions of all gases (the pollutants) are now generally available in cubic feet per minute (cfm) for each branch and junction. Methane emissions and diesel exhaust emissions will be quantified in next two sections of the book. These are used as direct input into the program.

5.1.4 Dilution and Distribution Model

5.1.4.1 Branch Calculations

For diesel exhaust emitted in a roadway, the dilution and distribution mechanisms are shown in Fig. 5.3A.

The exhaust is assumed to be discharged axisymmetrically and is designated by the symbol Q_{dwall}. Making mass balance over the length of the branch we have:

$$\text{mass in} = \text{mass out}$$

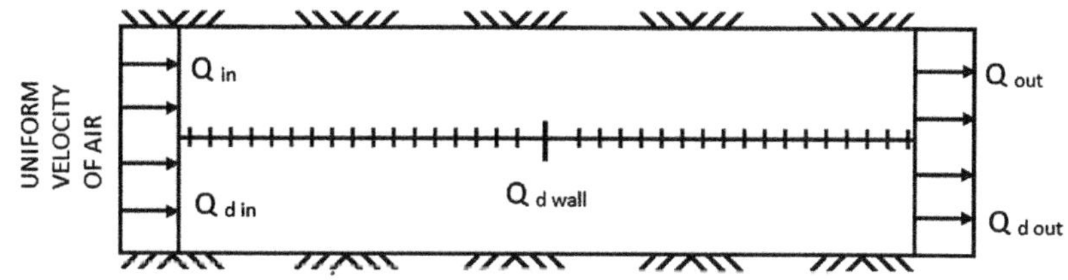

Figure 5.3A Diesel exhaust dilution and distribution model for a branch.

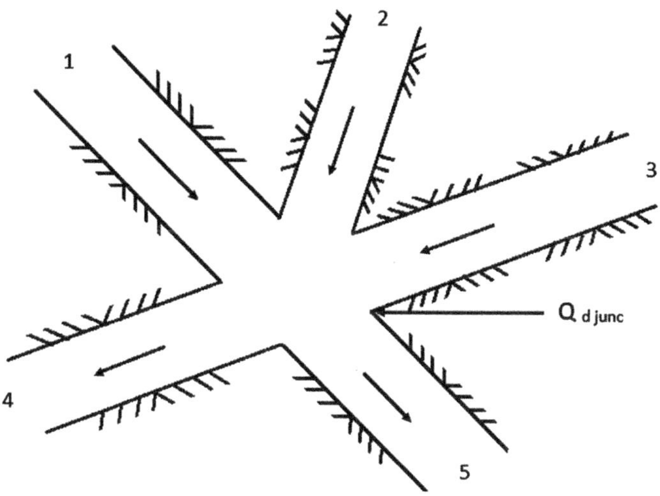

Figure 5.3B A typical junction in a network.

or

$$g\,\rho_{in}\,Q_{in} = g\,\rho_{out}\cdot Q_{out} \text{ for air} \tag{5.5}$$

and

$$g\,\rho_{din}\,Q_{din} + g\,\rho_D Q_{dwall} = g\,\rho_{dout} Q_{dout} \text{ for diesel exhaust} \tag{5.6}$$

where: Q, volume flow of air (ft^3/sec); Q_d, volume flow of diesel exhaust (ft^3/s); $g\cdot\rho$, density (weight) of the air (lb/ft^3); and Q_{dwall}, emission of diesel exhaust (ft^3/s) in the roadway.

Subscripts "in" and "out" refer to the start and the end points, respectively, of a branch. Assuming that the temperature and pressure of both gases are equal throughout the roadway, the density terms can be dropped from the mass balance equations. Hence, we have:

$$\begin{aligned} &Q_{in} = Q_{out} \text{ for air and} \\ &Q_{din} + Q_{dwall} = Q_{dout} \text{ for diesel exhaust.} \end{aligned} \tag{5.7}$$

The concentration of diesel exhaust at the end of the roadway (branch), C_B is given by:

$$C_B = \frac{Q_{dout}}{Q_{out}} \tag{5.8}$$

Eqs. (5.7) and (5.8) are referred to as roadway calculations in the computer program.

5.1.4.2 Junction Calculation

The concentration of diesel exhaust at a junction can be solved in a similar fashion. Fig. 5.3B shows a typical junction in a mine network. Here roadways (1), (2), and (3) have flows into the junction; and in roadways (4) and (5), the flow is away from the junction. Continuous mining machines can only operate at locations in a network designed as junctions. At such places the diesel exhaust generation is quite substantial. Let the quantity of diesel exhaust generated at a junction be Q_{djunc} ft^3/s. The total airflow at the junction is shown in Fig. 5.3B.

$$SQ = Q_1 + Q_2 \cdots\cdots\cdots Q_n = \sum_{i=1}^{n} Q_i \tag{5.9}$$

where SQ is the summation of the air quantities flowing through a junction and n is the total number of branches having flows into the junction.

Taking a mass balance for diesel exhaust,

$$\begin{aligned} SQ_d &= Q_{djunc} + Q_{dl} + Q_{d2} + \cdots\cdots\cdots + Q_{dn} \\ &= Q_{djunc} + \sum_{i-1}^{n} Q_{di} \end{aligned} \tag{5.10}$$

where SQ_d is the summation of all the quantity of diesel exhaust flowing into the junction and Q_{di} is the quantity of exhaust in branch i.

The concentration of diesel exhaust at the junction, C_j, is given by:

$$C_j = \frac{SQ_d}{SQ} \tag{5.11}$$

In the computer program, Eqs. (5.9)–(5.11) are referred to as junction calculations.

The concentration of diesel exhaust in the branches (4) and (5) as shown in Fig. 5.3B is the same as C_j; and hence the quantity of diesel exhaust flowing in these branches can be calculated from the following relation:

$$Q_{din} = C_j \cdot Q_{in} \tag{5.12}$$

This process can now be repeated again till the entire network is covered. Solutions of concentrations of diesel exhaust in a network of roadways with multiple diesel engines can, now, be obtained by repeated use of Eqs. (5.7)–(5.11). It is obvious that manual calculations are very tedious except for very small networks; and therefore, a computer program was developed to carry out these computations.

5.2 Introduction to the Ventilation Network Analyzer

Variable definitions and program listings are given in Appendices B and C along with a brief description of the program statements. Appendix B program uses Fortran IV compiler, which is rather old yet workable. Appendix C gives listings for C++, which is currently popular.

Input parameters needed for the execution of ventilation network analyzer (VNA) can be broadly classified as below:

1. System Size Parameters: These are the total number of branches, junctions, maximum number of iterations, etc. Detailed description of each parameter is given in the next section where a hypothetical mine layout is solved using VNA.
2. Roadway Parameters: These are junction numbers, frictional factor, height, width, and length of a roadway. One card is provided for each roadway. Roadways with special attributes, e.g., a fixed quantity of air, a fan, etc., are arranged in a specific manner as described in the next section.
3. Fan Parameters: These are the values of quantities of air and corresponding pressures, which describe a fan characteristic.
4. Emission Parameters: These are the values of emissions of various species of diesel exhaust in cfm along with locations of such emissions. A typical illustration is given in the next section.

5.2.1 Output Results

The program prints out the following:

1. All input parameters as detailed above.
2. The quantities of air and head loss in each branch of the network.
3. The size of the regulator, or the capacity of the booster fan as the case may be, for branches with fixed quantities.
4. The operating point on the fan characteristics.
5. Diesel exhaust concentrations for each species in each branch and at each junction of the network.
6. Mixture TLV[1] for each branch and junction in the network.

[1] The mixture TLV is defined as $\sum_{j=1}^{N} \frac{C_j}{TLV_j}$. Where C_j and TLV_j are the concentration and TLV, respectively, for the species j, and N is the number of species in the exhaust. For a nonreacting mixture of gases, its value should not exceed 1.0. This is recommended by the American Conference of Governmental Hygienists [6].

Diagnostic messages, whenever the TLV is exceeded, are also printed out. Further details of output results are given in the next section where a hypothetical mine layout is solved using VNA.

5.2.2 Program Testing on a Hypothetical Case

To illustrate the use of this program, the hypothetical mine shown in Fig. 5.1A is analyzed. A small network such as this provides an opportunity for exact solution providing a check on machine computations. It is assumed here that branches (4) and (6) and junction (4) have diesel exhaust emissions.

The Input: The description of the input deck and numerical values of parameters are given in Appendix D.

The Output: The program prints out all the input data for all the branches in the network, the mesh tables, data on fan characteristics, coefficient of the polynomials approximating the fan curves, and final solutions of quantity of air and head loss in each branch in order as given here. It also prints out the operating point on the fan characteristic. Output of concentrations of carbon monoxide, nitric oxide, etc., follow in sequence. TLV of all gases except hydrocarbons is the same as given in American Conference of Governmental Industrial Hygienists [6]. For hydrocarbons, a TLV of 15 ppm is assumed to force the computer program to print diagnostic messages. As can be seen from the Table, the TLV is exceeded by carbon monoxide concentration in branch (6). A message to this effect is printed by the computer. Similar diagnostic messages are printed out for NO, NO_2, and hydrocarbon (HC). Finally, it prints out the mixture TLV. Whenever it exceeds the value 1.0, a message is printed to that effect.

To keep the concentration of diesel exhaust components, and mixtures thereof below their respective TLV, increased quantities of air should be assigned to branches wherein these are exceeded. In general, for a given equipment deployment, several runs of the simulator will provide enough information to decide on the ventilation requirements.

Table 5.1 shows a typical output from the VNA. Emission sources and concentrations of CO, NO, NO_2, and HC are listed. A complete list of all inputs and output printouts are listed in Appendix D. The results were verified by hand calculations.

5.3 Verification of the Ventilation Network Analyzer in a Working Mine

The VNA was used to simulate a working mine with several diesel equipment in operation. Only CO was tracked by actual measurement in designated branches. Drager-handheld instrument (with test tubes) and an ecolyzer instrument were used for actual CO measurements in the mine. The mine line diagram is shown in Fig. 5.4.

The input data for this problem are listed in Table 5.2.

Table 5.1 Summary of Diesel Exhaust Concentrations in a Hypothetical Mine

	Diesel Exhaust Components								
	CO		NO		NO_2		HC		
	Emission (cfm)	Conc[a] (ppm)	Emission (cfm)	Conc[a] (ppm)	Emission (cfm)	Conc[a] (ppm)	Emission (cfm)	Conc[a] (ppm)	Mixture (TLV)
Branch Number									
1	0.0	27.7	0.0	21.6	0.0	16.9	0.0	77.8	9.99
2	0.0	14.0	0.0	10.7	0.0	4.5	0.0	24.0	3.22
3	0.0	0.0	0.0	0.0	0.0	0.0	0.0	0.0	0.0
4	1.0	18.6	0.75	13.9	0.05	0.9	0.65	12.1	1.92
5	0.0	0.0	0.0	0.0	0.0	0.0	0.0	0.0	0.0
6	0.4	75.3	0.30	56.4	0.50	71.8	2.0	295.6	37.83
7	0.0	0.0	0.0	0.0	0.0	0.0	0.0	0.0	0.0
8	0.0	18.6	0.0	13.9	0.0	0.9	0.0	12.1	1.92
9	0.0	0.0	0.0	0.0	0.0	0.0	0.0	0.0	0.0
10	0.0	18.6	0.0	13.9	0.0	0.9	0.0	12.1	1.92
11	0.0	18.6	0.0	13.9	0.0	0.9	0.0	12.1	1.92
12	0.0	25.8	0.0	20.0	0.0	13.5	0.0	63.9	8.28
13	0.0	0.0	0.0	0.0	0.0	0.0	0.0	0.0	0.0

Continued

Table 5.1 Summary of Diesel Exhaust Concentrations in a Hypothetical Mine—cont'd

	Diesel Exhaust Components								
	CO		NO		NO_2		HC		
	Emission (cfm)	Conc[a] (ppm)	Emission (cfm)	Conc[a] (ppm)	Emission (cfm)	Conc[a] (ppm)	Emission (cfm)	Conc[a] (ppm)	Mixture (TLV)
Junction Number									
1	0.0	0.0	0.0	0.0	0.0	0.0	0.0	0.0	0.0
2	0.0	0.0	0.0	0.0	0.0	0.0	0.0	0.0	0.0
3	0.0	18.6	0.0	13.9	0.0	0.9	0.0	12.1	1.92
4	0.3	27.7	0.25	21.6	0.0001	16.9	0.25	77.8	9.99
5	0.0	0.0	0.0	0.0	0.0	0.0	0.0	0.0	0.0
6	0.0	18.6	0.0	13.9	0.0	0.9	0.0	12.1	1.92
7	0.0	25.8	0.0	20.0	0.0	13.5	0.0	63.9	8.28
8	0.0	0.0	0.0	0.0	0.0	0.0	0.0	0.0	0.0
9	0.0	14.0	0.0	10.7	0.0	4.5	0.0	24.0	3.22

cfm, cubic feet per minute; *HC*, hydrocarbon; *ppm*, parts per million; *TLV*, threshold limit value.
[a]Concentration.

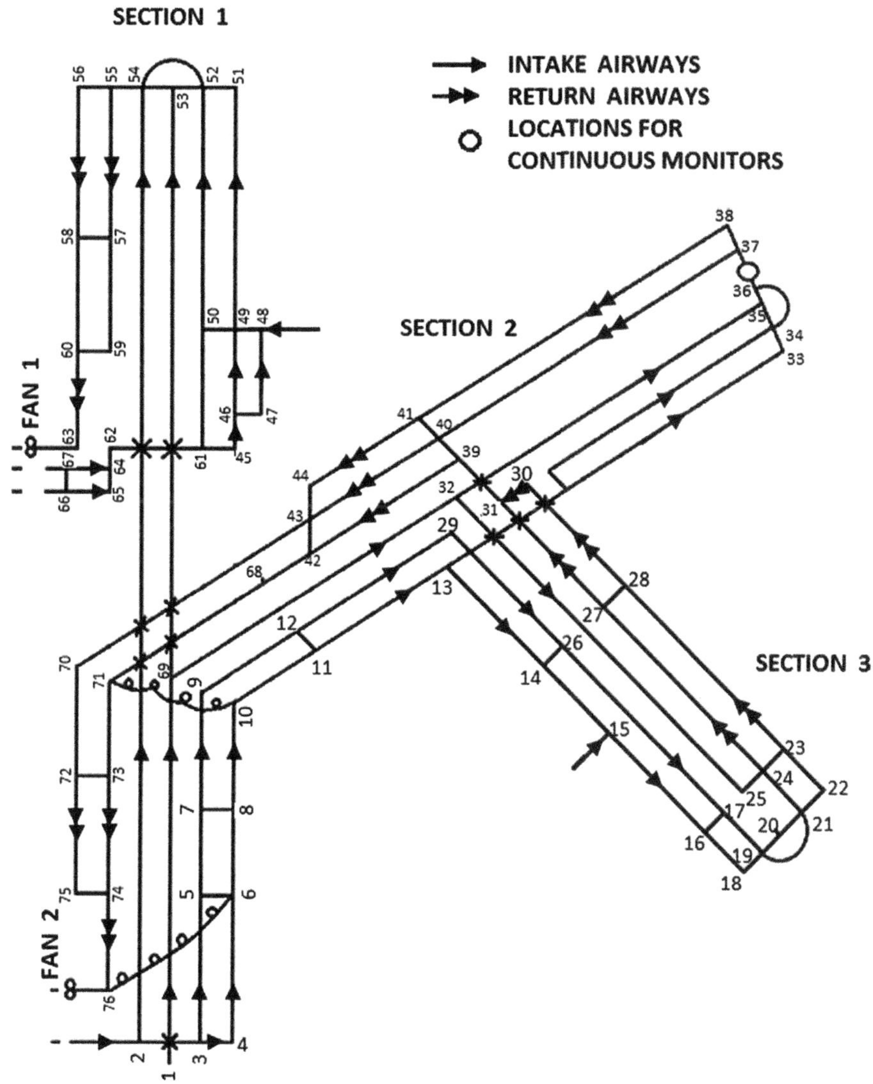

Figure 5.4 Schematic layout of the working mine.

Table 5.2 Input Data Listing for a Working Mine

Simulation of Diesel Exhaust Contamination of the Working Mine							Card 1
116	79	79	2	300 0		10 0.001	Card 2
1	15	80.0	8.0	80.0	600.0		Card 3
1	48	80.0	8.0	40.0	200.0		Card 4
6	76	100.0	8.0	20.0	400.0		Card 5
10	71	100.0	8.00	20.0	400.0		Card 6
19	21	50.0	8.0	40.0	0.1		Card 7
25	24	90.0	8.0	20.0	80.0		Card 8
32	35	90.0	8.0	20.0	1300.0		Card 9
34	36	50.0	8.0	40.0	0.1		Card 10
52	54	100.0	8.0	20.0	200.0		Card 11
69	53	90.0	8.0	20.0	2800.0		Card 12
63	1	100.0	8.00	20.0	200.		Card 13
76	1	100.0	8.0	20.0	180.0		Card 14
1	2	80.0	8.0	20.0	170.0		Card 15
1	66	80.0	8.00	20.0	280.0		Card 16
1	67	90.0	8.00	20.0	280.0		Card 17
1	69	80.0	8.0	20.0	1800.0		Card 18
2	3	80.0	8.00	20.	150.0		Card 19
2	54	80.0	8.00	20.	6400.0		Card 20
3	4	80.0	8.00	20.	75.0		Card 21
3	5	80.0	8.00	20.	800.0		Card 22
4	6	80.0	8.00	20.	800.0		Card 23
5	6	80.0	8.00	20.	75.0		Card 24
5	7	80.0	8.00	20.	400.0		Card 25
6	8	80.0	8.00	20.	400.0		Card 26
7	8	80.0	8.00	20.	75.0		Card 27
7	9	80.0	8.00	20.	250.0		Card 28
8	10	80.0	8.00	20.	130.0		Card 29
9	12	80.0	8.00	20.	740.0		Card 30
10	11	80.0	8.00	20.	750.0		Card 31
11	12	80.0	8.0	20.	75.0		Card 32

Table 5.2 Input Data Listing for a Working Mine—cont'd

11	13	80.0	8.0	20.	440.0		Card 33
12	29	80.0	8.0	20.	500.0		Card 34
13	14	90.0	8.0	20.0	1000.0		Card 35
13	77	90.0	8.0	20.0	75.0		Card 36
14	15	90.0	8.0	20.0	600.0		Card 37
14	26	90.0	8.0	20.0	75.0		Card 38
15	16	90.0	8.0	20.0	600.0		Card 39
16	17	90.0	8.0	20.0	80.0		Card 40
16	18	90.0	8.0	20.0	800.0		Card 41
17	19	90.0	8.0	20.0	800.0		Card 42
18	19	90.0	8.0	20.0	80.0		Card 43
19	20	150.0	8.0	20.0	80.0		Card 44
20	21	150.0	8.0	20.0	80.0		Card 45
21	22	100.0	8.0	20.0	80.0		Card 46
21	24	100.0	8.0	20.0	600.0		Card 47
22	23	100.0	8.0	20.0	600.0		Card 48
23	28	100.0	8.0	20.0	1500.0		Card 49
24	23	100.0	8.0	20.0	80.0		Card 50
24	27	100.0	8.0	20.0	1500.0		Card 51
26	17	100.0	8.0	20.0	1200.0		Card 52
27	28	100.0	8.00	20.0	80.0		Card 53
27	31	100.0	8.00	40.0	550.0		Card 54
28	30	100.0	8.00	20.0	550.0		Card 55
29	77	100.0	8.00	20.0	80.0		Card 56
30	31	100.0	8.00	20.0	80.0		Card 57
31	39	100.0	8.00	20.0	250.0		Card 58
32	25	90.0	8.0	20.0	2300.0		Card 59
33	34	150.0	8.0	20.0	80.0		Card 60
34	35	150.0	8.0	20.0	80.0		Card 61
35	36	150.0	8.0	20.0	80.0		Card 62
36	37	150.0	8.0	20.0	80.0		Card 63
37	38	100.0	8.0	20.0	80.0		Card 64

Continued

Table 5.2 Input Data Listing for a Working Mine—cont'd

37	40	100.0	8.0	20.0	1500.0		Card 65
38	41	100.0	8.0	20.0	1500.0		Card 66
39	40	100.0	8.0	20.0	80.0		Card 67
39	42	100.0	8.0	20.0	1200.0		Card 68
40	41	100.0	8.0	20.0	80.0		Card 69
40	43	100.0	8.0	20.0	1200.0		Card 70
41	44	100.0	8.0	20.0	1200.0		Card 71
42	71	100.0	8.0	20.0	500.0		Card 72
43	42	100.0	8.0	20.0	80.0		Card 73
43	70	100.0	8.0	5.0	500.0		Card 74
44	43	100.0	8.0	20.0	80.0		Card 75
45	46	90.0	8.0	20.0	200.0		Card 76
46	47	90.0	8.0	20.0	80.0		Card 77
46	49	90.0	8.0	20.0	700.0		Card 78
47	48	90.0	8.0	20.0	700.0		Card 79
48	49	90.0	8.0	20.0	80.0		Card 80
49	50	90.0	8.0	20.0	80.0		Card 81
49	51	90.0	8.0	15.0	1500.0		Card 82
50	52	90.0	8.0	20.0	1500.0		Card 83
51	52	150.0	8.0	20.0	80.0		Card 84
52	53	150.0	8.0	10.0	80.0		Card 85
53	54	150.0	8.0	10.0	80.0		Card 86
54	55	150.0	8.0	20.0	80.0		Card 87
55	56	150.0	8.0	20.0	80.0		Card 88
55	57	100.0	8.0	20.0	1200.0		Card 89
56	58	100.0	8.0	20.0	1200.0		Card 90
57	58	100.0	8.0	20.0	80.0		Card 91
57	59	100.0	8.0	20.0	1200.0		Card 92
58	60	100.0	8.0	20.0	1200.0		Card 93
59	60	100.0	8.0	20.0	80.0		Card 94
60	63	100.0	8.0	10.0	300.0		Card 95
61	45	100.0	8.0	20.0	80.0		Card 96

Table 5.2 Input Data Listing for a Working Mine—cont'd

61	50		100.0	8.0	20.0	900.0		Card 97
62	61		100.0	8.0	20.0	300.0		Card 98
64	62		100.0	8.0	20.0	200.0		Card 99
65	64		100.0	8.0	20.0	80.0		Card 100
66	65		100.0	8.0	20.0	100.0		Card 101
66	67		100.0	8.0	20.0	80.0		Card 102
67	64		100.0	8.0	20.0	100.0		Card 103
68	32		90.0	8.0	20.0	750.0		Card 104
69	68		90.0	8.0	20.0	700.0		Card 105
70	72		100.0	8.0	20.0	500.0		Card 106
71	70		100.0	8.0	20.0	80.0		Card 107
71	73		100.0	8.0	20.0	500.0		Card 108
72	75		100.0	8.0	20.0	600.0		Card 109
72	73		100.0	8.0	20.0	80.0		Card 110
73	74		100.0	8.0	20.0	600.0		Card 111
74	76		100.0	8.0	20.0	400.0		Card 112
75	74		100.0	8.0	20.0	80.0		Card 113
77	26		100.0	8.0	20.0	1000.0		Card 114
77	78		100.0	8.0	20.0	250.0		Card 115
78	33		100.0	8.0	20.0	1400.0		Card 116
78	79		100.0	8.0	20.0	80.0		Card 117
79	34		100.0	8.0	20.0	1400.0		Card 118
6	0							Card 119
67.0	1.2	78.0	1.0	89.0	0.8	98.0	0.6	Card 120
107.0	0.4	113.0	0.2					Card 121
7	0							Card 122
106.0	10.0	115.0	9.0	123.0	8.0	129.0	7.0	Card 123
136.0	6.0	140.0	5.0	144.0	4.0			Card 124
20.0	15.0	20.0	20.0	40.0	10.0	10.0	40.0	Card 125
40.0	10.0							Card 126
1								Card 127

Continued

Table 5.2 Input Data Listing for a Working Mine—cont'd

The Following Results Are for Carbon Monoxide								Card 128
1								Card 129
		50.0		6				Card 130
18	0.04							Card 131
31	0.08							Card 132
32	0.08							Card 133
40	0.16							Card 134
81	0.16							Card 135
116	0.16							Card 136
								Card 137
								Card 138
								Card 139
								Card 140
/*								Card 141

The results of carbon monoxide concentrations and mixture TLV as calculated by VNA are well within the allowed concentrations. Computed values with observed concentrations are presented in Table 5.3. The agreement between the computed and observed values of carbon monoxide concentration is satisfactory and within the limits of experimental errors.

Table 5.3 Carbon Monoxide Concentrations in the Experimental Mine

	Computed			**Average Observed**	
	J1	**J2**	**Concentration (ppm)**	**Concentration (ppm)**	
				Drager	**Ecolyzer**
Branch					
46	22	23	4.7	4.0	4.1
61	36	37	4.5	5.0	4.8
Junction					
23			4.7	4.0	4.1
37			4.5	5.0	4.8

ppm, parts per million

Problems

5.1. Using the VNA program in Appendix B and C, analyze the network in Fig. 5.4 but change the input data.

References

[1] Thakur PC, Chopra AS. Analysis of mine ventilation problems. Industry Press; 1962. 120 pp.
[2] Hartman HL, et al. Mine ventilation and air conditioning. 3rd ed. John Wiley and Sons, Inc.; 1997. p. 730.
[3] Wang YJ, Saperstein LW. Computerized solution of complex ventilation networks. Transactions SME/AIME 1971;250.
[4] McPherson MJ. Ventilation network analysis by digital computers. The Mining Engineer 1966;126:12–29.
[5] Thakur PC. Computer-aided analysis of diesel exhaust cntamination of mine ventilation systems [Ph.D. Thesis]. The Pennsylvania State University; 1974. 233 pp.
[6] Threshold limit values for chemical substances and physical agents in the working environment with intended changes. In: American Conference of Governmental Industrial Hygienists, Cincinnati, Ohio, USA; 1973.

Mechanical and Natural Ventilation

6

Chapter Outline

In the old days, mines were mostly naturally ventilated. Temperature differences induced air flow, but it was not reliable. People used furnaces in the upcast shafts to induce air flow or forced air in the downcast shafts with bellows or paddle wheels operated by wind, hydropower, or animals. Mechanical ventilation using fans developed during the second half of the 19th century to replace all old unreliable devices. The early fans were of simple radial flow design, which were gradually improved as experience was gained. Backward and forward bladed radial flow fans were widely used.

Advanced Mine Ventilation. https://doi.org/10.1016/B978-0-08-100457-9.00006-7

Around 1920, axial flow fans became popular. Its design improved with better knowledge of aerodynamics. Even the radial flow fans improved their efficiency by designing the blades aerodynamically. Both types of fans are in use today, and hence both types will be discussed in detail.

6.1 Radial Flow Fans

Radial flow fans, also known as centrifugal fans, are basically designed to contain radial blades with arrangements to let air enter at the base, at the center, and exit at the periphery of the rotors. Two independent actions produce pressure in a radial flow fan: the centrifugal force due to the rotation of air and the kinetic energy imparted to the air as it leaves the tip of the impeller blades. The magnitude of kinetic energy depends mainly on the tangential velocity (tip speed) of the blades, V_t. The centrifugal energy imparted to air is a function of the change in the radial velocity, V_r, of air entering and leaving the impeller.

The ideal performance characteristic of a centrifugal fan can be derived from the energy transfer and slip relationship. The theoretical head, H, is given by the following equation:

$$H = \frac{V_{t_2}}{g}\left[V_{t_2} - \frac{Q}{A_2}\cot\theta_2\right] - \frac{V_{t_1}}{g}\left[V_{t_1} - \frac{Q}{A_1}\cot\theta_1\right] \quad (6.1)$$

where V_t is the tangential velocity; Q is the air flow rate; A_1 and A_2 are areas perpendicular to radial velocity at the inlet and outlet; θ is the blade angle; subscripts 1,2 refer to the inlet and outlet of the impeller; Q/A is the radial air velocity.

The second term in Eq. (6.1) becomes zero if we assume that there is no inlet whirl. Hence,

$$H = \frac{V_{t_2}}{g}\left[V_{t_2} - \frac{Q}{A}\cot\theta\right] \quad (6.2)$$

This equation will be used to calculate the theoretical head developed by different types of centrifugal fans.

Fig. 6.1 shows the velocity diagram for the three types of centrifugal fans: (a) radial bladed, (b) backward bladed, and (c) forward bladed fans.

6.1.1 Head Developed by Radial Bladed Fans

Because the blade angle, θ, is 90 degrees in this case, and cot 90 degrees is equal to zero, Eq. (6.2) reduces to:

$$H = \frac{V_t^2}{g}$$

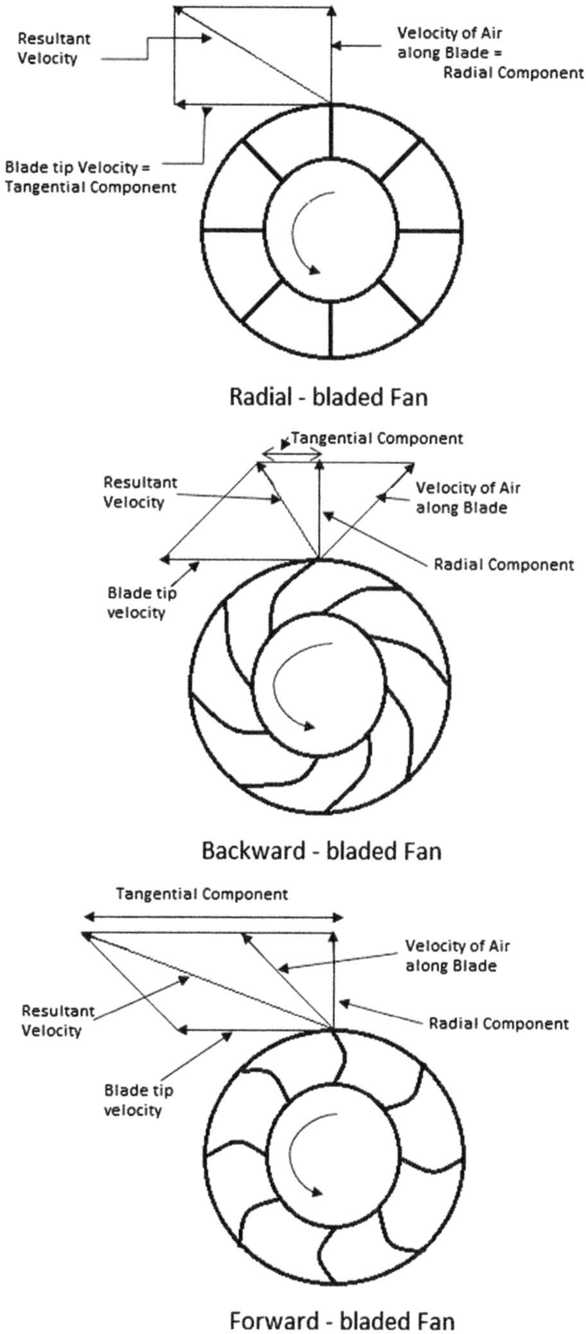

Figure 6.1 Velocity diagram for centrifugal fans.

Expressing the head into inches of water gauge,

$$H = \frac{WV_t^2}{5.2\ g} \tag{6.3}$$

where W is the density of air and g (acceleration due to gravity) = 32.2 ft/s^2

6.1.2 Backward Bladed Fans

Blade angles for backward bladed fans seldom exceed 45 degrees. In the velocity diagram, $\theta = 90 + \theta$.

Hence,

$$H = \frac{WV_t}{5.2\ g}\left[V_t - \frac{Q}{A}\cot\theta\right] \tag{6.4}$$

if $\theta = 45$ degrees; $\cot\theta = 1.00$

$$H = \frac{WV_t}{5.2\ g}\left[V_t - \frac{Q}{A}\right]$$

6.1.3 Forward Bladed Fans

Here, θ is an acute angle. It seldom exceeds 60 degrees.

Hence,

$$H = \frac{WV_t}{5.2\ g}\left[V_t + \frac{Q}{A}\cot\theta\right] \tag{6.5}$$

It is to be noted that for the same diameter, speed and air flow; forward bladed fans develop the most pressure and the backward bladed fans the least.

An example:

Given the following conditions, calculate head developed by each of the above type blades in a centrifugal fan:

Q = 50,000 CFM; n (rpm) = 500; D (diameter) = 8 ft, Width = 2 ft at outlet; $\theta = 45$ degrees; W (air density) = 0.075 lb/ft^3

Calculate:

$$V_t = \frac{\Pi\ D\ n}{60} = 209.4\ \text{ft/s}$$

$$A = \Pi\ D\ b = 50.26\ \text{ft}^2$$

Hence,

$$\frac{Q}{A} = V_r = \frac{50,000}{60 \times 50.26} = 16.58\ \text{ft/s}.$$

Using Eq. (6.3) for radial bladed fan,

$$H = \frac{0.075 \times (209.4)^2}{5.2 \times 32.2} = 19.6\ \text{in.}$$

Using Eq. (6.4) for backward bladed fan,

$$H = \frac{0.075(209.4)(209.4 - 50.26 \times 1)}{5.2 \times 32.2} = 14.9\ \text{in.}$$

Using Eq. (6.5) for forward bladed fans,

$$H = \frac{0.075(209.4)(209.4 + 50.26 \times 1)}{5.2 \times 32.2} = 24.3\ \text{in.}$$

The theoretical head, H, is never realized in practice because some energy is uselessly spent in overcoming the following conditions.

6.1.3.1 *Frictional and Shock Losses*

The fan provides a narrow path for air flow, and hence the velocities are higher. Consequently, some of the pressure generated is consumed in overcoming these frictional and shock losses.

6.1.3.2 *Conversion Losses*

A considerable part of the theoretical head developed by the fan is in the form of velocity pressure in the air leaving the blade tips. This must be converted to static pressure, otherwise it will be lost. This is done by letting the air go through a "volute casing" of uniform cross-sectional area. The volute ends up in an "evasée," which has a gradually increasing cross-sectional area. As the air velocity decreases, the velocity pressure is converted to static pressure. The conversion is, however, incomplete because the size of the evasée has to be practically limited and air will leave it with some velocity pressure unconverted to static head.

6.1.3.3 *Recirculation*

There is a tendency for air at the blade tip to flow back to the center and get recirculated. It mostly happens when the air quantity is much lower than the optimum quantity for the fan. The energy required to maintain these eddy currents of air reduces the theoretical head and is mostly dissipated as heat.

6.1.3.4 Bearing Friction

The bearing friction causes loss of energy provided to the fan by a prime mover. It has no impact on the static head developed by the fan. When the fan is working at its optimum (most efficient) point, the power loss is less than 1%.

6.1.3.5 Fan Efficiency

The work done by a fan is measured by the air horsepower, which is the product of air flow rate (ft^3/min) and total pressure (lb/ft^2) divided by 33,000. Air horsepower divided by horsepower to the fan shaft and multiplied by 100 is the percent efficiency of the fan.

6.2 Fan Characteristics

Because of many measurable and some immeasurable variables, it is not possible to mathematically relate the pressure generated by a fan to the different quantities, but it can be done graphically by running tests on the fan. The pressure–volume curve is called the "fan characteristic" curve. For a given speed, pressure, efficiency, and horsepower consumed can be plotted against air quantities. Fig. 6.2 shows a typical characteristic curve for all three types of centrifugal fans.

In all cases, pressure declines as the air quantity goes up. The efficiency has a peak in each case, which is the ideal operating point for the fan. The power consumption continues to increase with increasing volume for radial bladed and forward bladed fans, but it begins to decline for backward bladed fans after reaching a high point. This is called a "nonoverloading" characteristic of the fan, and it is very desirable.

6.3 Axial Flow Fans

The idea of a fan where the air flows axially and does not suffer change of direction is obviously attractive. The early axial flow fans were manufactured in the 1910s using aircraft propeller blades. Modern aerodynamically designed axial flow fans were developed soon afterward.

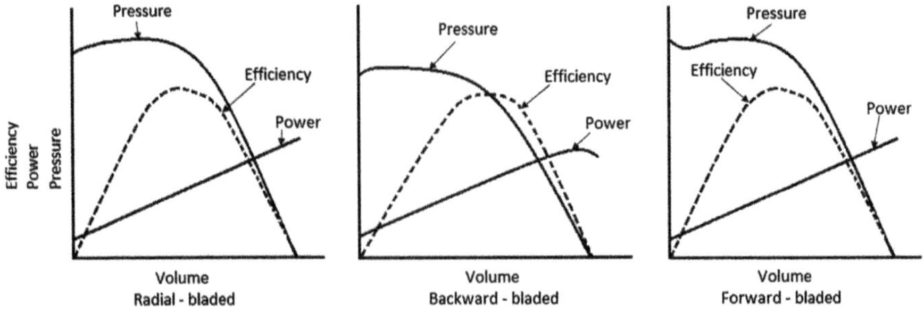

Figure 6.2 Pressure, power, and efficiency versus air quantity.

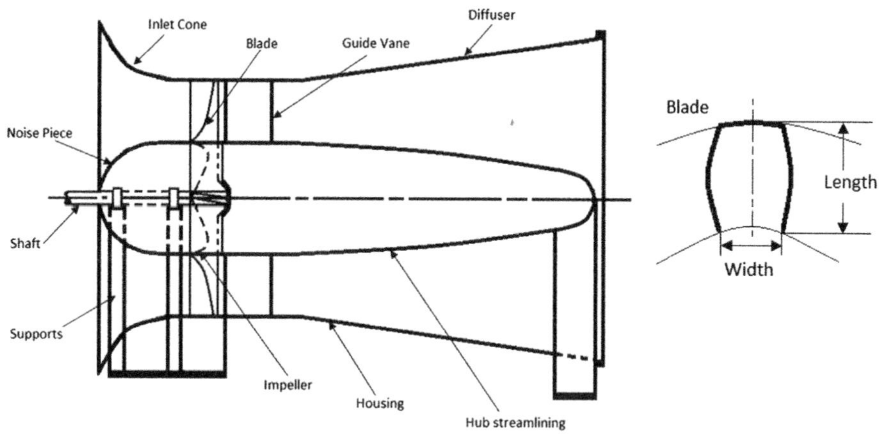

Figure 6.3 Axial flow fan parts.

The fan consists of a number of aerofoil-sectioned blades mounted on a streamlined hub that revolves within a cylindrical casing. The air reaches the blades at high velocity, having been accelerated in the streamlined approaches to the casing. It leaves the blades at roughly the same velocity, and the gain in pressure is essentially one of static pressure. The blades tend to rotate the air in a helical motion. Most of the energy of the swirl is recovered by fixed guide vanes placed downstream of the rotor. The air then enters the diffuser, or an evasée, where the velocity pressure is converted to static pressure. Fig. 6.3 shows various parts of an axial flow fan.

6.3.1 Performance Characteristics of an Axial Flow Fan

Eq. (6.1) can again be used to calculate the theoretical head developed by an axial flow fan, but the blade angle, θ, must be replaced by actual fluid angle, α. Further assuming that there is no inlet whirl,

$$\text{Theoretical head } H = \frac{V^2}{g} - \frac{VQ}{gA}\cot\alpha \tag{6.6}$$

Here, A is the area through which the fluid flows, that is, the area of the annulus between the hub and tip minus the area taken up by the blades. Because α is always less than 90 degrees, the actual head developed will always be less than the ideal head of $\frac{V^2}{g}$. The actual head is further reduced by some of the same losses as for centrifugal fans discussed earlier except for shock losses.

Fig. 6.4 shows the characteristic of a typical axial flow fan. The nonoverloading characteristic of the axial flow fan is noticeable and very desirable [1].

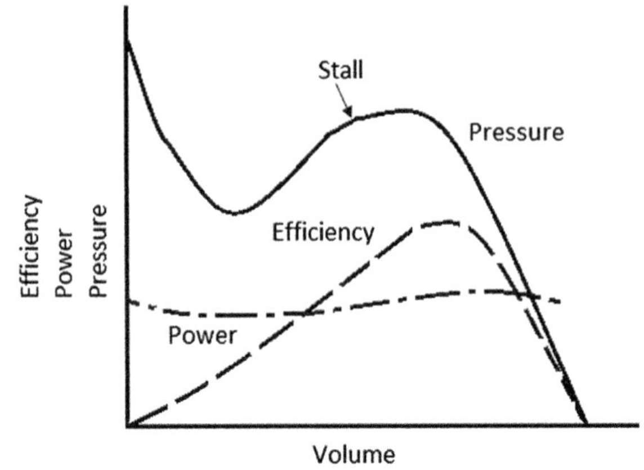

Figure 6.4 Characteristics of an axial flow fan.

6.4 Fan Laws

A detailed discussion of fan laws is available in literature [1], hence only a summary discussion will be provided here.

Assuming the density of air is constant, fan performance is controlled by only two variables: speed (n) and fan diameter, D. The efficiency is usually a constant.

Table 6.1 summarizes the impact of variation in (n) and D; on quantity, Q; head, H; and power, P.

The changes in pressure and power are directly proportional to air density, but the quantity is not impacted.

An example:

A fan running at 500 rpm produces a head of 1″ W.G. (water gauge), with 40,000 CFM and a horsepower of 7.00 and an efficiency of 0.9.

Calculate the new Q, H and horsepower if (a) the speed is changed to 1000 rpm or (b) the diameter is doubled, keeping tip speed constant.

1. From Table 6.1; speed is doubled to 1000 rpm.
   a. $\frac{Q_2}{Q_1} = \frac{n_2}{n_1}$; hence, $Q_2 = 80{,}000$ CFM
   b. $\frac{H_2}{H_1} = \frac{n_2^2}{n_1^2}$; hence, $H_2 = 4$ in.
   c. $\frac{H.P._2}{H.P._1} = \left(\frac{n_2}{n_1}\right)^3$; hence, new $H.P_2. = 56$ H. P.

   Note that efficiency, η, does not change.

2. If tip speed, ΠDn, stays constant and $D_2 = 2\ D_1$
   a. $\frac{Q_2}{Q_1} = \left(\frac{D_2}{D_1}\right)^2$; $Q_2 = 160{,}000$ CFM
   b. $H_2 = H_1 = 1$ in.
   c. $\frac{H.P._2}{H.P._1} = \left(\frac{D_2}{D_1}\right)^2$; hence $H.P._2 = 28$ H.P.

Table 6.1 Fan Law Summary (Air Density Remains Constant)

Performance Criteria	Speed, n D, Constant	Diameter, D Dn, Speed Constant
Quantity, Q	Q is proportional to n	Q is proportional to D^2
Pressure, H	H is proportional to n^2	H does not vary
Power, P	P is proportional to n^3	P is proportional to D^2

6.5 Fan Testing

It is necessary to test all main fans in a mine after installation and at intervals thereafter to make sure they are working efficiently. Changing mine resistance also makes it often necessary to readjust the fan blade angles. The quantities that need to be measured and the procedures are described in the following section.

6.5.1 *Air Quantity*

The area behind the fan is divided into 2 × 2 ft by wires, and an anemometer is kept in each tiny square for 10 s to get an average air velocity. If the air velocity is very high, a pitot tube can be used. Air quantity is the product of average velocity and the area.

6.5.2 *Total and Static Pressures*

Each fan house has a static pressure reading gauge. Several readings are taken and averaged over the test period. The velocity head is calculated from the average velocity obtained with anemometers. The total pressure is the sum of static pressure and velocity pressure.

6.5.3 *Air Density*

If the pitot tube is used for velocity measurement, air density must be calculated from barometric pressure and wet- and dry-bulb temperatures as discussed in Chapter 1.

6.5.4 *Fan Speed*

Typically, the fan rpm is measured with a revolution counter and stop watch, but a tachometer can also be used. Several readings should be taken to get a good average. If the fan is run by a synchronous electric motor, the fan speed is the same as the motor speed.

6.5.5 *Fan Shaft Horsepower*

The power of the main drive corrected by the efficiency of the motor and the efficiency of the drive is known as the fan-shaft horsepower.

6.5.6 *Air Horsepower*

Finally, the air horsepower (static or total) is calculated (refer to article 2.8) and divided by fan-shaft horsepower to obtain the static or total efficiency of the fan. It should be reasonably high as indicated by the operating range prescribed by the manufacturer.

6.6 Matching a Fan to Mine Characteristics

As discussed earlier, the air volume, Q, and consequent pressure required, P, are established for a mine at various stages of its development. To select an efficient fan, the mine characteristic ($P = R\ Q^2$) is plotted on the same axes as the fan characteristic as shown in Fig. 6.5. The intersection point of the two characteristics indicates the pressure and volume at which the fan will work in ventilating the mine. The fan must have a high efficiency at the operating point. Usually, a very high or a very low mine resistance is unsuitable. It is a lot easier to design a fan that will suit a mine characteristic than the other way around.

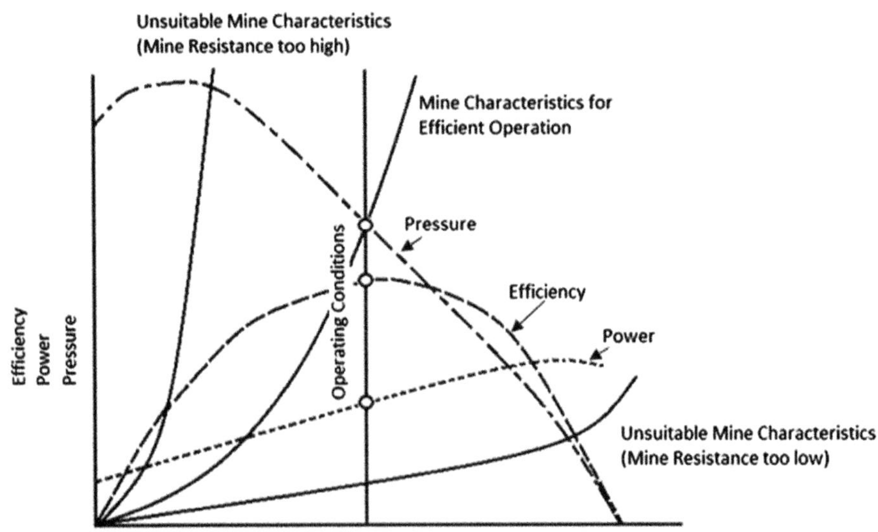

Figure 6.5 Matching mine characteristics with fan characteristics.

6.7 Natural Mine Ventilation

An underground coal mine can have a natural circulation of air in it without any fans. This happens because the air in the mine extracts heat from the mine workings and gets lower in density, creating a natural flow of air. This is called "natural ventilation." Usually, it is not strong enough to allay the use of mechanical mine fans, but it does assist and hence it is worth our study.

Consider a mine ventilated naturally without fans, and assume that the downcast and upcast shafts are of equal depth and the shaft tops are at the same level on surface as shown in Fig. 6.6.

The points 1, 2, 3, and 4 are the top and bottom of downcast shaft (intake air) and the bottom and top of the upcast shaft (return air), respectively. The temperature, pressure, and density of air are denoted by t in degree Fahrenheit, p, in lb/ft^2, and w in lb/ft^3.

From Eq. (1.5),

$$W = \frac{1.3258\ (B - 0.378\,'e')}{460 + t}$$

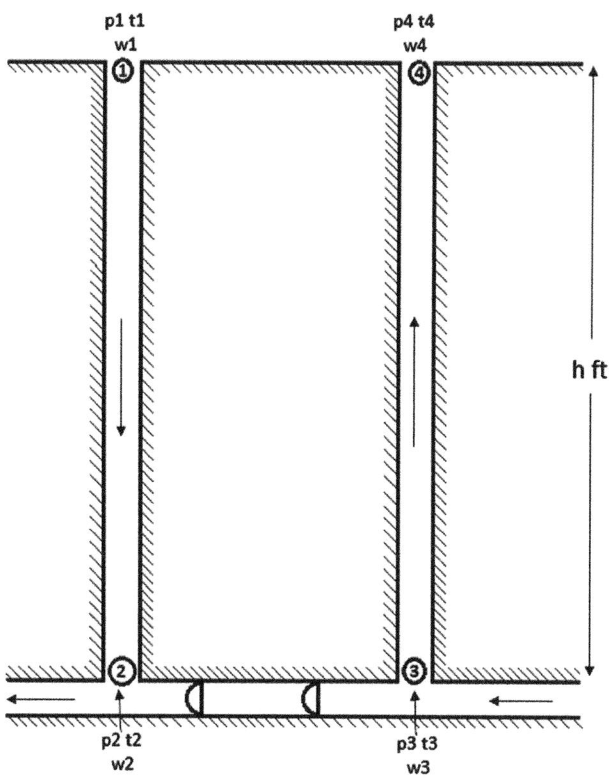

Figure 6.6 Natural ventilation.

where B is the barometric pressure in inches of mercury and 'e' is the water vapor pressure.

Let the depth of the shafts be D, in feet.

Pressure at the downcast shaft bottom caused by the air column = $D(W_1 + W_2)/2$.

Pressure at the upcast shaft bottom, likewise, is = $D(W_3 + W_4)/2$.

Then NVP (natural ventilation pressure) =

$$D\left(\frac{W_1 + W_2}{2}\right) - D\left(\frac{W_3 + W_4}{2}\right)$$

$$\text{or } \frac{D}{2}((W_1 + W_2) - (W_3 + W_4)) \quad \text{lb/ft}^2$$

$$= \frac{D}{2}(W_1 + W_2 - W_3 - W_4)/5.2 \text{ in. of water.}$$

An example:

Calculate the NVP for the following conditions:

$D = 3000$ ft; $W_1 = W_2 = 0.075$ lb/ft^3 t
$t_3 = t_4 = 90°$F; $e = 0.5$ in. of Hg.
$B = 30$ in. of Hg.
Here, $W_1 = W_2$ and $W_3 = W_4$, hence
$NVP = D\ (W_1 - W_3)/5.2$

$$W = \frac{1.3253(30 - 0.378(0.5))}{460 + 90}$$

$$= \frac{39.51}{550} = 0.0718 \text{ lb/ft}^3$$

hence NVP = 3000 (0.075–0.0718)/5.2 = 1.85 in. of water.

The net NVP will always be less than the calculated NVP because of pressure losses in the downcast and upcast shafts. If the mine workings are inclined and they trend downward, additional NVP is created that adds to the shaft NVP. On the other hand, if mine workings trend upward, the NVP becomes negative and works in opposition to the shaft NVP.

6.7.1 Historical Review of Natural Ventilation Pressure Calculations

6.7.1.1 Weeks Method

The earliest derivation of NVP was by Weeks [2]. The pressures P_2 and P_3 (Fig. 6.6) were calculated using the basic gas laws, Eq. (1.3).

In a shaft of depth, D, the relation between P1 and P2 can be expressed as:

$$\ln\left(\frac{P_2}{P_1}\right) = \frac{D}{R\overline{T}_d} \tag{6.7}$$

where $\overline{T}_d$ is the average temperature $= \frac{T_1+T_2}{2}$ in the downcast shaft.

Likewise, $\ln(\frac{P_3}{P_4}) = \frac{D}{R\overline{T}_u}$ where $\overline{T}_u = \frac{T_3+T_4}{2}$, average temperature on upcast shaft. Because P_1 and P_4 are known and often equal, P_2 and P_3 can be determined.

$$NVP = (P_2 - P_1)$$

6.7.1.2 McElroy Method [3]

It is an empirical formula but surprisingly accurate.

$$NVP = 0.03/10^\circ F/100 \text{ ft in. of water} \tag{6.8}$$

6.7.1.3 Rees Method [4]

It uses only the dry-bulb temperature in the downcast and upcast shafts.

$$NVP \text{ (in. of water)} = \left(\frac{\overline{T}_u - \overline{T}_D}{5.2\,\overline{T}}\right) W.D. \tag{6.9}$$

where $\overline{T}_D$, $\overline{T}_u$ are the average temperatures in upcast and downcast shafts, respectively T (with bar on top) is the average surface temperature.

6.7.1.4 Hinsley Method [5]

This is the most recent and refined measure of NVP:

$$NVP \text{ (in. of water)} = \frac{P_1}{5.2}\left[\left(\frac{T_2}{T_1}\right)^{\frac{D}{R\Delta T_D}} - \left(\frac{T_3}{T_4}\right)^{\frac{D}{R\Delta T_U}}\right] \tag{6.10}$$

where ΔT_D is the $(T_2 - T_1)$ and $\Delta T_u = (T_3 - T_4)$. (Refer to Fig. 6.1)

Problems

1. Calculate the head developed by the (a) radial bladed, (b) backward bladed, and (c) forward bladed fans, given the following data:
   $Q = 100{,}000$ CFM
   $n = 800$ rpm
   $D = 8$ ft
   Width at outlet $= 3$ ft
   $\theta = 35^\circ$ for backward and forward bladed fans
   $\theta = 90^\circ$ for radial bladed fans.

2. Calculate the pressure developed by an axial flow fan, given the same conditions as above but blade angle, $\alpha = 35°$.
3. Calculate the NVP for a deep gold mine, given the following conditions:
   $D = 10{,}000$ ft; $W_2 = 0.075$ lb/ft^3
   $t_3 = t_4 = 100°F$; $e = 0.8$ in. of Hg.
   $B = 30$ in. of Hg.

References

[1] Jorgensen R. Fan Engineering. Howden Buffalo Inc.; 1999.
[2] Weeks WS. Ventilation of mines. New York, NY: McGraw Hill; 1926.
[3] McElroy GE. Engineering factors in the ventilation of metal mines. USBM Bulletin No. 385; 1935.
[4] Rees JP. Ventilator calculations, transvaal chamber of mines. Johannesburg: South Africa; 1950.
[5] Hinsley FB. New method of calculating the natural ventilation in mines. Colliery Guardian 1965;211(5454).

Section Two

Respirable Coal Dust Control

Health Hazards of Respirable Dusts

7

Chapter Outline

Coal is a brittle material, and therefore dust is produced when it is mined, crushed, or reduced in size by any method. Dust has been defined as solid particles smaller than 100 μm (1 μm = 10^{-4} cm) in size, which can be disseminated and carried by air. Coal dust when raised into a cloud creates two well-known hazards in coal mines, namely, coal dust explosion and coal workers' pneumoconiosis (CWP). For an explosion, a coal dust concentration of at least 86.5 gm/m^3 of air is required [1]. Maintaining a dust concentration in mine air below this level is relatively easy because the coarse dust particles normally settle in the roadways almost immediately after dissemination. Explosion hazard of dust and its mitigation will be discussed later in the book.

The fine dust particles, particularly the ones below 10 μm, based on unit density in size, do not settle easily. In the minus 10 μm range, coal particles are called respirable dust because these can be inhaled and can contribute to the CWP. It is mostly composed of coal and silica, but there are 50 other elements and their oxides in it in very small proportions.

7.1 Growth of Coal Workers' Pneumoconiosis

The health effects of respirable particles on the lungs are better understood in the light of the knowledge of the components and functions of the human respiratory system. The respiratory system consists of a series of branching passages decreasing in size but increasing in number as shown in Fig. 7.1.

Thus starting with inhalation either through the nose or mouth, air passes in succession through the trachea, bronchi, bronchioles, alveolar ducts, and finally into the alveolar air sacs where gaseous exchange of oxygen takes place [2]. The hairs and small bones of the nose act as a filter and almost completely filter out particles larger than 10 μm. Particles between 2 and 10 μm in size usually settle on the walls of the trachea, bronchi, and bronchioles. The particles finer than 2 μm reach the inner parts of the lungs. The respiratory system above the bronchioles are lined with a hairlike structure,

Advanced Mine Ventilation. https://doi.org/10.1016/B978-0-08-100457-9.00007-9

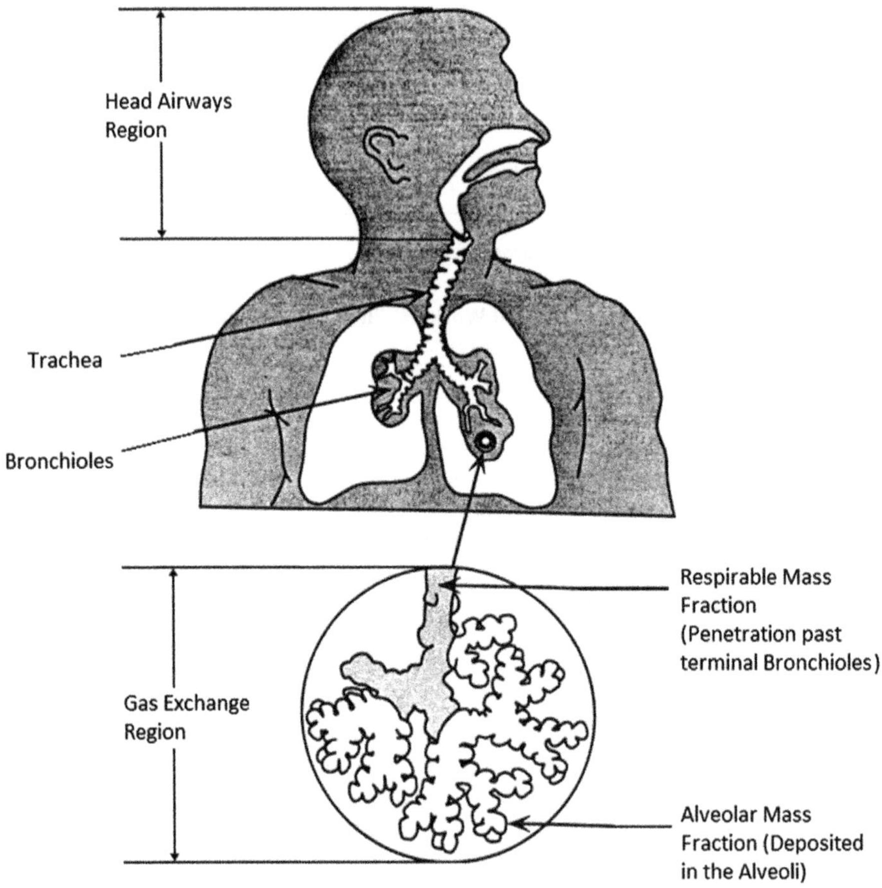

Figure 7.1 The human respiratory system.

called cilia, which trap and transport to the mouth any insoluble particle deposited in that region. These particles are spat, sneezed out, or swallowed. Because the digestive system is much sturdier than the respiratory system, the swallowed particles seldom do any harm. Particles deposited in the alveoli and alveolar sacs are ingested by migratory cells, the phagocytes, and transported to lymph nodes to be eventually rejected through the body waste disposal system.

Notwithstanding the protective mechanism outlined earlier, deposition of dust in lungs builds up with continued breathing in dusty air, which leads to the breakdown of the elimination system and pathological harm ensues. In the earlier stages of the disease, the dust particles accumulated in the lymph nodes produce local inflammation. This is known as simple pneumoconiosis, and given better environment at this stage, a victim can recuperate in time. Continued breathing in the dusty air, however, can eventually lead to the fibrosis of the lung tissues. A tissue affected by fibrosis subsequently enlarges and coalesces to form a dense mass of fibrous tissues. The alveolar walls swell up, become less distensible, and break down, leaving cavities 1–4 mm in diameter. This is called pulmonary massive fibrosis and implies grave prognosis, with death frequently resulting from pulmonary failure or infection with tuberculosis.

In view of the human suffering and financial costs involved, the control of pneumoconiosis has become essential. On the basis of the studies made so far, the following factors seem to influence the disease:

1. Concentration of dust in the air.
2. Size distribution of the dust particles.
3. Chemical and physical composition of the dust.
4. Life style of the subject, such as, smoking habits.
5. Individual's resistance to pneumoconiosis/lung diseases.
6. Presence of electrical charge on dust particles. More particles are deposited in the respiratory system if they are charged than if they are uncharged [3]. Freshly mined dust has more free radicals in it than old dust.
7. Thermal repulsive and evaporative effects in the lungs. The presence of these forces in the lungs are known but their effects on dust deposition are not fully established [4]. Control of the disease, therefore, should basically center around the following factors:
   a. The maintenance of dust concentration in the mine air below a given level that minimizes the risk of CWP.
   b. Medical supervision of miners on a regular basis.
   c. Lifestyle of mine workers.

Engineering control of the disease is thus concerned mainly with the control of dust concentration in mine air. Most of the major coal-producing countries have established respirable dust standards for coal mines. The concentration of dust was measured by the number of particles per unit volume of air in the past. This standard has now been replaced by the mass per unit volume of mine air on the basis of studies which indicate that the mass of coal dust deposited in the lungs bears a better correlation with the total damage caused by the disease [5,6]. Postmortem studies conducted on a miner's lungs indicate a strong correlation between the average weight of dust in the lungs of a miner and the radiological category of pneumoconiosis contracted by him [7]. To establish a relationship between the dust dose and the growth of pneumoconiosis in coal miners, the Pneumoconiosis Field Research group of the National Coal Board (NCB), United Kingdom, surveyed a population of 30,000 coal miners over a period of 18 years, beginning in 1952. The dust dose was expressed as mean coalface dust concentration during the period of observation. The response of a miner to dust inhalation was measured as the amount of radiological change determined from chest radiographs taken at the beginning and at the end of the observation period. To measure the radiological changes quantitatively, CWP has been divided into four categories, namely, 0, 1, 2, and 3, by the International Labour Office in increasing order of severity [8]. Each of these categories has been further subdivided into three stages by the NCB to account for even minor changes in lung conditions [6]. The progress of disease from one stage to the next has been considered as one step of progression. The response of a miner to dust inhalation was measured as the number of steps of progression per million working shifts.

The number of steps of progression per million shifts bore the following coefficients of correlation with the dust concentration [6].

Concentration Parameter	Correlation Coefficient[1]
Mass concentration of dust	0.80
Mass concentration of coal in dust	0.82
Number concentration of dust particles	0.44

This proved that the mass concentration is by far the most reliable dust standard. Dust standards for some of the major coal-producing countries based on mass concentration are shown in Table 7.1.

The mass of dust for the purposes of these standards refer to the mean mass of respirable dust. Respirable dust literally means any dust that can enter the respiratory system. The size of respirable dust has been defined by an analogy to the mechanism of

Table 7.1 Major Coal-Producing Countries' Dust Standards

Country	Parameter	Dust Standard
United Kingdom	Mass of respirable dust	3.8 mg/m^3 (personal exposure). 7 mg/m^3 (average) as measured in the return airway.
United States	Mass of respirable dust	1.5 mg/m^3 (average on a shift basis). $\frac{10}{\%\ SiO_2}$ mg/m^3 if silica concentration is higher than 5%.
Russia	Mass of respirable dust	1 to 4 mg/m^3 depending upon silica content of coal.
Australia	Mass of respirable dust	3 mg/m^3; silica $\leq$5%.
Germany	Mass of respirable dust	4 mg/m^3; silica 0.15 mg/m^3.

dust deposition in the lungs, which is mainly by sedimentation. The size of a respirable dust particle is represented by the diameter of a sphere having the same density and settling velocity as that of the particle. This is called the Stoke's diameter. The British Mining Research Council (BMRC), United Kingdom, regards all particles below 7.1

[1] The correlation coefficient is defined as follows. If x and y are two random variables jointly distributed over a population, then correlation coefficient, ρ_{XY}, is defined as:

$$\rho_{XY} = \frac{E(XY) - \mu_X \bullet \mu_Y}{\sigma_X \sigma_Y}$$

where E(XY) = expected value of X•Y; μ_x = expected value of X; μ_Y = expected value of Y; σ_X = standard deviation of X; σ_Y = standard deviation of Y.

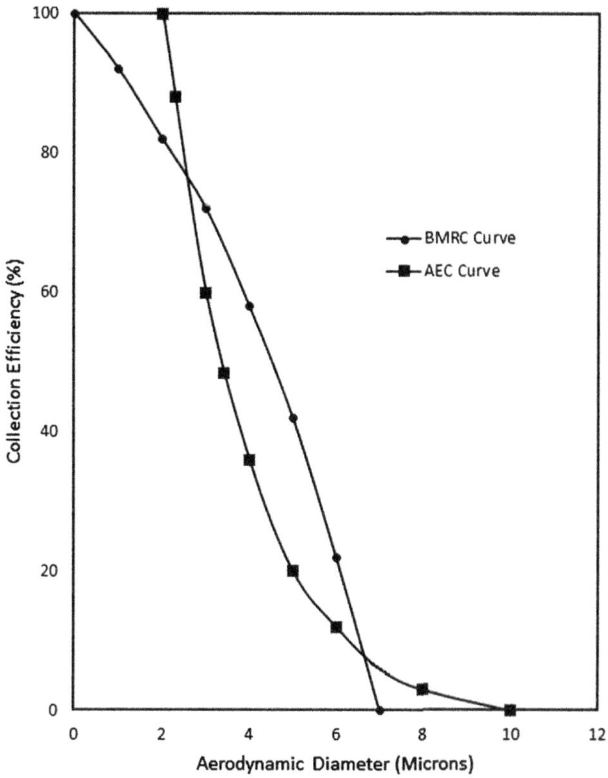

Figure 7.2 Definition of respirable dust as collected by sampling instruments.

equivalent diameter[2] as respirable. This has been agreed upon and adopted by the Pneumoconiosis Conference in Johannesburg [9]. The Atomic Energy Commission, United States, regards particles below 10 equivalent diameter as respirable. Fig. 7.2 shows these hypothetical depositional curves [10].

From a health point of view, the particle size distribution determines (discussed in Chapter 8) the following:

1. The overall retention of the particles in the respiratory system.
2. Locations within the respiratory system where the particle would be deposited.
3. The rate of growth of disease, especially in the case of physiologically and chemically active dust.

Fig. 7.3 shows a real lung deposition efficiency curve [11]. The full line shows the efficiency obtained for the particle size range 1–6 μm equivalent diameter experimentally. Because it is generally agreed that 10 μm (unit density) is the upper limit of respirable size, the curve is extrapolated to zero efficiency at 10 μm equivalent diameter. As

[2] Equivalent diameter = Stoke's diameter x (density)$^{1/2}$. It is equivalent to particle with unit (1.0) density that has the same settling velocity.

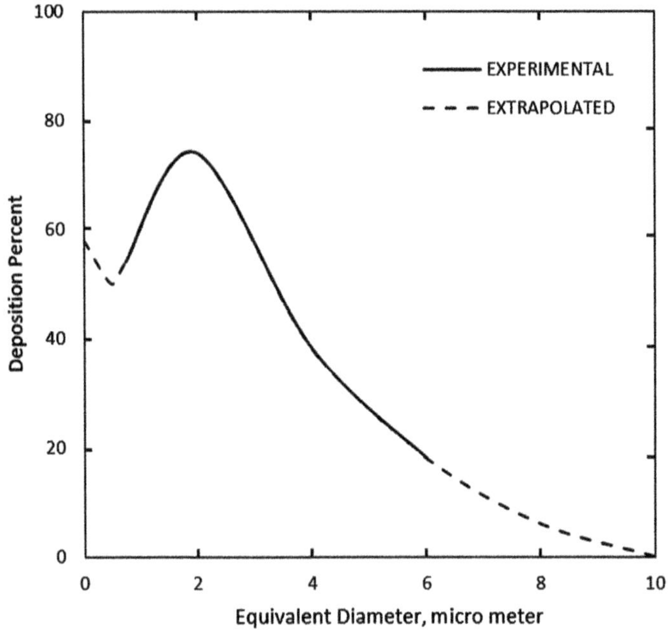

Figure 7.3 Lung deposition efficiency curve.

the particle size decreases below one micron, the efficiency of deposition decreases until it reaches a minimum at 0.3 μm [3]. With further decrease in size, the efficiency of deposition increases as shown by the dotted line in Fig. 7.3. This extrapolation is also justified by the fact that submicron coal particles are known to be electrically charged, which increases their deposition efficiency considerably. The mean size of the dust recovered from the lungs of dead miners is of the order of 1 μm [12].

Control of dust concentration is very much dependent on the size of dust particles. From this viewpoint, respirable dust particles can be divided into two classes, namely, settling dust and nonsettling dust. Settling dust consists of the relatively larger sizes of respirable dust. Technology for effective suppression of such particles is available, and by and large such particles can be precipitated in the vicinity of the coal face. On the other hand, nonsettling dust particles being very fine in size remain relatively unaffected even by the best dust suppression techniques available to date. These particles seem to show little tendency to settle by gravity and therefore are finally discharged into the outside atmosphere, leading to surface air pollution. A number of pneumoconiosis cases among the residents in mining areas but not working in mines have been reported [13]. These particles also seem to be contributing appreciably toward the fouling of mine intake air. A survey conducted in West Virginia coal mines showed a respirable dust concentration of 0.4–1.4 mg/m^3 in intake airways [14]. Surveys conducted by the US Bureau of Mines have also indicated presence of respirable dust in considerable quantities in the intake airways [15]. The average diameter of nonsettling coal particles is 0.42–0.6 μm [16].

7.2 A Basis for Respirable Dust Standard

To set a scientific basis for respirable dust standard, one must determine the dust exposure versus health response curve, preferably for each coal basin. This provides a basis for setting a personal exposure limit (PEL), such that when exposed to this level of respirable dust for a working life of 30–40 years, the probability of a worker developing a CWP of category 2 or above is zero. The only scientific study in this area was carried out in the United Kingdom over a large cohort [17], and the results are shown in Fig. 7.4.

It clearly shows that the probability of contracting category 2 CWP is zero at 2 mg/m^3 of respirable dust concentration. The United States' PEL for respirable coal dust was thus set at 2.0 mg/m^3 in 1972. It was further reduced to 1.5 mg/m^3 in 2015 to lower the health risk. However, a monatomic decrease in PEL is not going to guarantee the elimination of CWP. This will be discussed further in the next article.

7.3 Prevalence and Cessation of Coal Workers' Pneumoconiosis

Before the enforcement of the new US Coal Mine Health and Safety regulation of 1970, the incidence of CWP was very high, but it has declined drastically since then. Figs. 7.5 and 7.6 show the progressive decline in respirable dust concentrations and the prevalence of CWP in US coal mines, respectively [10].

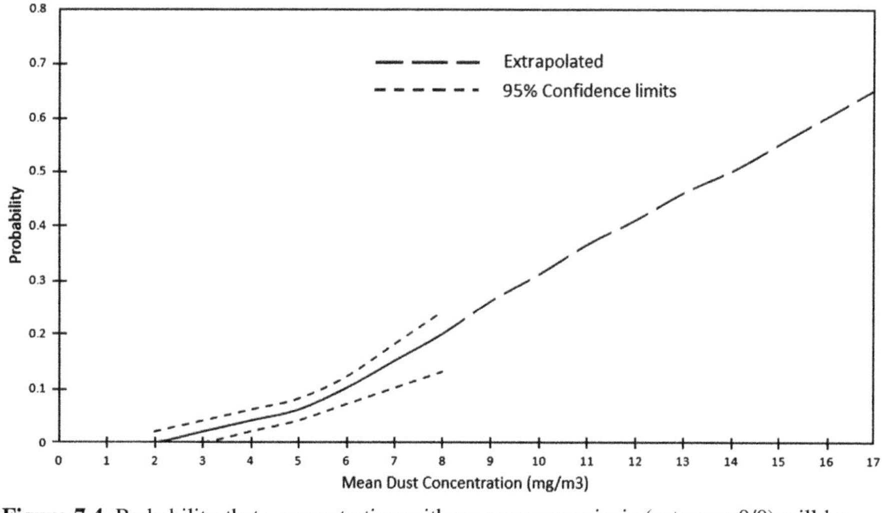

Figure 7.4 Probability that a man starting with no pneumoconiosis (category 0/0) will be classified as category 2 or higher after 35 years of exposure to various concentrations of coal mine dust.

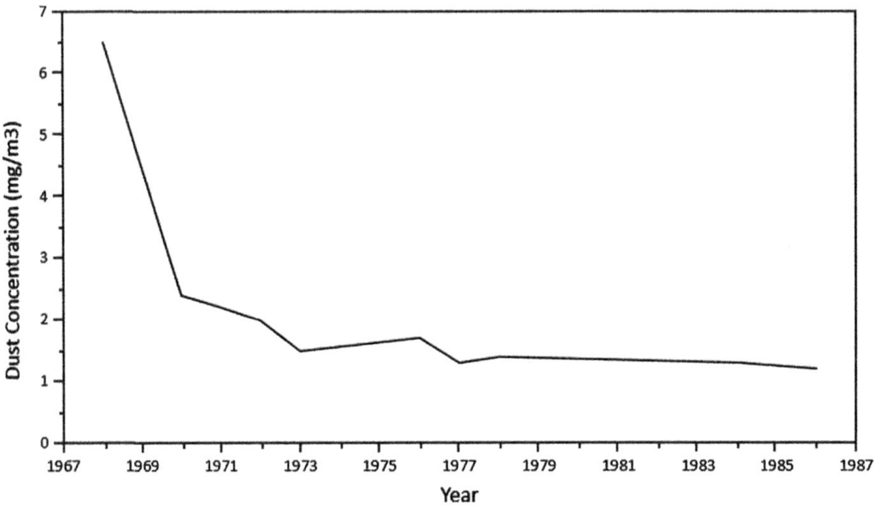

Figure 7.5 Reported trends in dust concentrations for continuous miner operators.

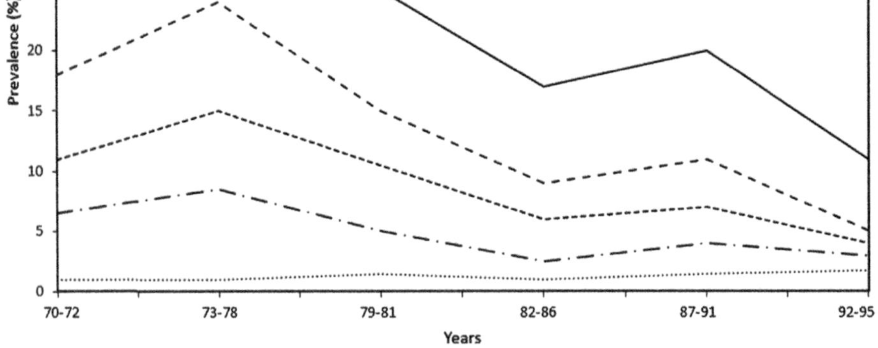

Figure 7.6 Prevalence of coal workers' pneumoconiosis category 1 or higher identified in the coal workers' X-ray surveillance program by tenure in coal mining.

In Australia and Germany, the disease is practically eliminated even with a PEL of 3–4 mg/m^3. The elimination of CWP is possible mainly based on the following:

1. Setting a scientific respirable dust standard.
2. Measuring the respirable dust concentration accurately.
3. Engineering control of dust to achieve this PEL.

4. Recognizing characteristics of various coal types, e.g., the rank effect, silica contents, and bioavailability of silica.
5. Lifestyle intervention programs including cessation of smoking and routine medical surveillance.

Items 2, 3, and 4 will be discussed in the following chapters, but item 5 is discussed here.

7.4 Lifestyle Intervention Program

It is now well established that the lifestyle of the miner and his or her personal susceptibility to coal mine dust injury play important and determining roles in the occurrence of CWP [18,19]. In a typical year, a miner may spend about a quarter of his or her time in the work environment. The remainder of the hours in the year is spent away from the mine. Lifestyle practiced outside the mine is likely to have a greater impact on the miner's health now because respirable dust levels are quite low in the work environment.

By emphasizing healthy lifestyles, offering wellness programs, and providing for certain preventative examinations, the employers can hope to reduce the cost of health care by intervening with the employee before the disease becomes manifest. Essential parts of a lifestyle intervention program should include cessation of smoking, good nutrition, and medical surveillance for early detection of personal susceptibility. Cessation of smoking is perhaps the most important step in elimination of CWP. Coal miners must not smoke either on the job or at home because there is overwhelming evidence to prove that smoking and exposure to dust particles (coal and other particles) have a synergistic effect on the development of occupational respiratory diseases [10]. Finally, all coal miners should participate in routine medical surveillance that not only includes radiographic chest examinations and spirometry but also looks for early symptoms of high rate of dust deposition and retention in the lungs. For example, measuring the presence of iron in the lungs using sophisticated equipment, such as superconductive quantum interference device or acoustic reflection techniques [20], could provide a measure of individual susceptibility. Individuals with diminished ability to clear the inhaled dust must use protective devices, e.g., air helmets, or be located in outbye areas where ambient dust concentration is substantially lower. A holistic approach is needed to finally eradicate this dreaded disease. It consists of the following four steps:

1. Better personal exposure monitoring.
2. Better engineering control of respirable dust.
3. Recognizing unique characteristics of coal (such as, rank, silica content, and iron content).
4. Enforcement of lifestyle intervention programs.

The latter involves cessation of smoking, good nutrition, and medical surveillance to detect individual susceptibility to CWP and provide adequate safeguards.

A concerted effort as proposed has eliminated CWP in some countries and may reduce or even eliminate it in the rest of the coal-mining countries of the world.

References

[1] Whitaker JW, Willet HL. Colliery explosions and recovery work. London: Sir Isaac Pitman and Sons Ltd.; 1946. p. 72.
[2] Davies CN. A formalized anatomy of the human respiratory tract. Inhaled Particles and Vapours 1961;1:82–7.
[3] Fuchs NA. The mechanics of aersols. Pergamon Press; 1964. p. 236.
[4] Green HL, Lane WR. Particulate clouds; dusts, smokes, and mists. London: E. and F. N. Spon Ltd.; 1957. p. 313.
[5] Walton WH. The measurement of respirable dust—the basis for gravimetric standard. U.S. Bureau of Mines Information Circular, 8458; 1970. p. 5–19.
[6] Jacobsen M, et al. New dust standards in British coal mines. Nature 1970;227:445–7.
[7] Rossiter CE, et al. Dust content, radiology and pathology in simple pneumoconiosis of coal workers. Inhaled particles and vapours, vol. 11. New York: Pergamon Press; 1967pp.419–434.
[8] Record of Proceedings, Third International Conference of Experts on Pneumoconiosis, Sydney, 1950. Geneva: I.L.O.; 1953.
[9] Orenstein AJ. In: Proceedings of the Pneumoconiosis Conference, Johannesburg. London: J. and A. Churchill Ltd.; 1960. p. 619.
[10] U.S. Department of Health and Human Services, CDC, NIOSH. Occupational exposure to coal mine dust. 1995. p. 336.
[11] Jacobson M, Lamonica JA. Personal respirable dust sampler. U.S. Bureau of Mines, Information Circular 8458; 1970. p. 39.
[12] Cartwright J, Nagelschmidt G. The size and shape of dust from human lungs and its relation to selective sampling. Inhaled particles and vapours, vol. 1. New York: Pergamon Press; 1961. p. 445–52.
[13] Enterline PE. Mortality rates among coal miners. American Journal of Public Health 1964; 54(5):758–67.
[14] Private Communications.
[15] Doyle HN. Dust concentration in the mines. U.S. Bureau of Mines, Information Circular 8458; 1970. p. 27–32.
[16] Thakur PC. Mass distribution, percent yield, non-settling size and aerodynamic shape factor of respirable coal dust particles [M.S. Thesis]. The Penn State University; 1971. p. 133.
[17] Jacobson M, Walton WH, Rozan JM. The relationship between pneumoconiosis and dust exposures in British coal mines. In: Walton WH, editor. Inhaled particles, vol. III. England: Old Woking, Unwin Brothers; 1971. p. 901–3.
[18] Morgan WKC, Lapp NL. State of the art: respiratory diseases in coal mines. AM Reviews of Respiratory Diseases 1976;113:531–59.
[19] Roy TM, Collins LC, Snider HL, Anderson WH. Cigarette smoking and federal black lung benefits in bituminous coal miners. Journal of Occupational Medicine 1989;31:98–101.
[20] Legg JK. Clinical assessment and random error reduction related to acoustic reflection technique [Ph. D. Thesis]. West Virginia University; 1996. p. 191.

Characteristics of Respirable Coal Dust Particles

8

Chapter Outline

Although fine dust particles behave like gases, there are significant differences that warrant a separate treatment. Respirable dust particles range from 10^{-3} to 10 μm. (1 μm = 10^{-4} cm). We need to define a few terms here before we proceed further.

1. Fine particles: These particles have an aerodynamic diameter of 10 μm or less. Micron is the unit most commonly used in particle mechanics.
2. Stoke's diameter: It is the diameter of a spherical particle of the same density and the same settling velocity as the nonspherical particle.
3. Aerodynamic diameter: It is equal to the diameter of a sphere of unit density that has the same settling velocity as the "Stoke's diameter" of the particle with a specific gravity of ρ.

$$\textit{Aerodynamic diameter} = \textit{Stoke's diameter} \times (\rho)^{\frac{1}{2}}.$$

4. Settling velocity: It is the displacement of a particle under gravity per second. It will be derived and discussed later.
5. Diffusivity: Small particles have a diffusivity coefficient like gases. It is given by D in the equation

$$\frac{dQ}{dt} = -D\left(\frac{dc}{dx}\right)dy \cdot dz \tag{8.1}$$

Advanced Mine Ventilation. https://doi.org/10.1016/B978-0-08-100457-9.00008-0

where dQ is the amount passing through an area dy·dz in the direction of x in time, dt, creating an increase of volume concentration by $\frac{dc}{dx}$.

6. Displacement by Brownian motion: Fine dust particles have a random Brownian motion too. The finer the size of a particle, the greater the Brownian displacement. Thus, below a certain size (typically around 0.5 μm in diameter) the dust particle may not settle and that fraction is called "nonsettling fraction" of respirable dust [1].
7. Relaxation time, (T): Each particle has a characteristic "relaxation time", T, that has a unit of time (s). The settling velocity, v_S, $= T \cdot g$, where "g" is the acceleration due to gravity. It will be mathematically derived later in the text.
8. Mean free path: It is the average distance travelled by a moving particle between successive impacts (collisions) that modify its direction or energy or other particle properties.
9. Particle Reynolds number: It is defined as R, where $R = \frac{d \cdot v}{\sigma}$, where d is the particle diameter, v is the settling velocity, and σ is the kinematic viscosity of air.

These characteristics of particles ranging from 10^{-3} to 10 μm are shown in Table 8.1 [2].

It is to be noted that mean free path, ℓ does not change significantly with the size of the particle in this range.

8.1 Settling Velocity of Small Particles Due to Gravity (Stoke's Formula)

When a small dust particle is floated in a medium such as air, it is pulled down by gravity, Fg, and the motion is resisted by the viscous force of the medium, Fm. The settling velocity, v_S, is the final constant rate at which the particle is pulled down by gravity.

Table 8.1 Characteristics of Fine Particles

Radius, r (μm)	Diffusivity, D ($cm^2 s^{-1}$)	Relaxation Time, T (s)	Mean Free Path, ℓ (cm)	Settling Velocity, $\overline{X}_S$ (cm/s)	Mean Brownian Movement, $\overline{X}_B$ (cm/s)
10	1.38×10^{-8}	1.23×10^{-3}	6.08×10^{-6}	1.21	1.23×10^{-4}
5	2.38×10^{-8}	3.08×10^{-4}	4.32×10^{-6}	3.02×10^{-1}	1.74×10^{-4}
1	1.27×10^{-7}	1.31×10^{-5}	2.06×10^{-6}	1.28×10^{-2}	4.02×10^{-4}
0.5	2.74×10^{-7}	3.54×10^{-6}	1.53×10^{-6}	3.47×10^{-3}	5.90×10^{-4}
0.2	8.32×10^{-7}	6.87×10^{-7}	1.21×10^{-6}	6.73×10^{-4}	1.03×10^{-3}
0.1	2.21×10^{-6}	2.28×10^{-7}	1.13×10^{-6}	2.24×10^{-4}	1.68×10^{-3}
0.01	1.35×10^{-4}	1.40×10^{-8}	2.20×10^{-6}	1.37×10^{-5}	1.31×10^{-2}
0.001	1.28×10^{-2}	1.33×10^{-9}	6.59×10^{-6}	1.31×10^{-6}	1.28×10^{-1}

Adapted from Fuchs NA. The mechanics of aerosols. Pergamon Press; 1964. p. 408.

The mathematical equations to predict v_S varies depending on the size of the particle, r, in relation to its mean force path, ℓ. The following three cases will be considered:

Case 1. $r \gg \ell$
Case 2. $r \ll \ell$
Case 3. $r \approx \ell$

In all cases, the derivation of mathematical equation is based on the following assumptions:

1. Incompressibility of the medium (air).
2. Infinite extent of the medium (no boundary influence).
3. Very small rate of movement.
4. Constant rate of movement.
5. Rigidity of particle (always true for solid particles).
6. Absence of slipping at its surface.

For a detailed discussion of the above factors, a reference can be made to Fuchs [2].

8.1.1 Derivation of Settling Velocity When $r \gg \ell$

For a force balance, the rate of change of momentum, $M'v$ equals force due to gravity, Fg, less resistance due to medium, Fm. Let the particle radius be r, particle density, ρ, and air density equals ρ_{air}. M' is the apparent weight of particle and equals $\frac{4}{3}\pi r^3(\rho - \rho_{air})$

$$Fm = -6\pi\, rV\eta \tag{8.2}$$

Hence,

$$\frac{M'dv}{dt} = Fg - Fm$$
$$\text{or} \quad \frac{M'dv}{dt} = M'g - 6\pi\, rV\,\eta \tag{8.3}$$

Substituting for M' and transposing, $\frac{dv}{dt} = g - \frac{v}{\left(\frac{2}{9}r^2(\rho-\rho_{air})\right)/\eta}$.

Let us call $T = \frac{2}{9}\frac{r^2(\rho-\rho_{air})}{\eta}$; it is known as the relaxation time of the particle. Hence,

$$\frac{dv}{dt} = g - \frac{v}{T} \tag{8.4}$$

The constant settling velocity of the particle is called $V = V_S$ at $t = T$. Thus in Eq. (8.4), $\frac{dv}{dt} = 0$ at $t = T$.

Hence,

$$v_S = g \cdot T = g \cdot \frac{2r^2(\rho - \rho_{air})}{9\eta} \tag{8.5}$$

Table 8.1 shows the values of T for particle radii in the range 0.001–10 μm.

8.1.2 Settling Velocity When r ≪ ℓ

In this case, the resistance of the medium, Fm, is derived as [3]:

$$Fm = -\frac{6\pi\eta\, r^2\, v}{(A + Q)\ell} \tag{8.6}$$

The value of ℓ is given in Table 8.1. (A + Q) has a value of 1.175, 1.091, or 1.131 depending on the type of reflection of particle (s^2) [2].

Settling velocities can be derived in the same manner as shown in Article 8.1.1.

8.1.3 Settling Velocity When r ≈ ℓ

For a more rigorous analysis of Fm in Article 8.1.1, Cunningham [4] provided a better estimate of resistance due to the medium as:

$$Fm = -\frac{6\pi\eta r v}{\left(1 + A\frac{\ell}{r}\right)} \tag{8.7}$$

and the settling velocity is calculated as:

$$v_S = \frac{mg}{6\pi\eta\, r}\left(1 + A\frac{\ell}{r}\right) \tag{8.8}$$

A has a value of 0.864 [5].

Eq. (8.7) was further modified to allow for cases where r ≈ ℓ. There is no theoretical solution, but the experimental data fit Eq. (8.9) well [5].

$$Fm = -\frac{6\pi\eta r v}{\left(1 + A\frac{1}{r} + Q\frac{1}{r}e^{-b\frac{r}{\ell}}\right)} \tag{8.9}$$

This is the most generalized estimate for Fm, and both estimates of Fm (r ≫ ℓ and r ≪ ℓ) can be derived from it. For r ≫ ℓ; $\frac{\ell}{r} \rightarrow 0$, and for r ≪ ℓ; $e^{-b\frac{r}{l}} \rightarrow \ell$.

Most accepted values of A, Q, and b are 0.879, 0.23, and 2.61, respectively [2].

The term $\frac{\left(1+A\frac{\ell}{r}+Q\frac{\ell}{r}e^{-b\cdot\frac{r}{\ell}}\right)}{6\pi\eta r}$ is called mobility of the particle, B such that:

$$Fm\cdot B = -v \tag{8.10}$$

Fuchs [2] makes a recommendation for the use of various equations depending on the particle size to obtain better accuracy (Table 8.2).

Table 8.2 Regions of Applicability of Various Equations for Fm

	Permissible Error	
Equation	**1%**	**10%**
Eq. (8.1) (Stoke's)	$8 < r < 15$ μm	$0.8 < r < 35$ μm
Eq. (8.8) (Cunningham)	$0.18 < r < 8$ μm	$0.05 < r < 0.8$ μm
Molecular kinetic (8.6)	$0.001 < r < 0.002$ μm	$0.0005 < r < 0.02$ μm

Adapted from Fuchs NA. The mechanics of aerosols. Pergamon Press; 1964. p. 408.

8.2 Aerodynamic Shape Factor for Dust Particles

The size of a respirable dust particle in the range 0.01−10 μm is mainly measured with a light or electron microscope. These particles are usually a platelike structure [1] of very irregular shape. The measured surface area of the particles is equated to a circle of the same area, and the diameter of the circle is called a "projected diameter" of the particle.

It is related to the Stoke's diameter of the particle by a term, "aerodynamic shape factor". Thakur [1] defines it as:

$$\text{Aerodynamic Shape Factor} = \frac{\text{Projected Diameter}}{\text{Stoke's Diameter}} \tag{8.11}$$

In an experiment, he measured the projected diameter of coal particles for given Stoke's diameters of 4.87, 3.40, 2.40, and 1.81 μm for 10 different coal dusts. The aerodynamic shape factors ranged from 1.31 to 1.5 with a mean of 1.42 and a standard deviation of 0.16. These values agree well with values obtained for British coal using different experimental setups [6,7]. Table 8.3 shows some of the data for various coal and glass particles.

Statistical analysis of all data further showed that the variations in ASF were significant at 0.05 level of significance for both the Stoke's diameter and material type, including different ranks of coal [1]. This indicates that the shape of particles created is material dependent.

Table 8.3 Aerodynamic Shape Factors (ASFs)

Author	Material	Instrument	ASF
Timbrell [6]	Ground glass	Elutriator	1.49
Timbrell [6]	Coal	Elutriator	1.35
Sawyer and Walton [8]	South Wales coal	Conifuge	1.35
Sawyer and Walton [8]	Lancashire coal	Conifuge	1.31
Sawyer and Walton [8]	Cumberland coal	Conifuge	1.35
Sawyer andWalton [8]	China clay	Conifuge	1.61
Sawyer and Walton [8]	Quartz	Conifuge	1.67
Thakur [1]	Bituminous coal	Elutriator	1.42 (1.34−1.48)[a]

[a]ASF = 1.34 for Stoke's diameter of 3.98 μm.
ASF = 1.37 for Stoke's diameter of 2.77 μm; ASF = 1.38 for Stoke's diameter of 1.97 μm; ASF = 1.48 for Stoke's diameter of 1.48 μm.

8.3 NonSettling Fraction of Respirable Dust

Thakur [1] postulated that a small portion of respirable dust (under 1 μm of unit density) may not settle by gravity or be affected by water sprays. As the particle size goes down, its settling velocity reduces but its random movement owing to Brownian motion increases. The nonsettling dust size upper limit is defined as the size for which settling velocity equals Brownian displacement [1].

Settling velocity for a particle was already derived in article 8.1. Brownian motion of particle suspended in air is given by Einstein [9] in Eq. (8.12).

$$\overline{X_B^2} = 2\,Dt \tag{8.12}$$

where $\overline{X_B^2}$ is the mean square displacement with respect to any coordinate axis in time, t. D is the diffusion coefficient of the particle.

Einstein defined,

$$D = KTB \tag{8.13}$$

where K is the Boltzmann constant; T, absolute temp in Kelvin; B, particle mobility defined in Eq. (8.10).

Fuchs [2] expanded Eq. (8.13) further and concluded:

$$\overline{X} = \text{Average displacement/s} = \left(\frac{4D}{\pi}\right)^{\frac{1}{2}} \tag{8.14}$$

Hence, for upper limit for nonsettling particle size,

$$\text{Fm B} = v_S = \text{mg B} = \left(\frac{4D}{\pi}\right)^{\frac{1}{2}}$$

$$\text{or} \qquad \frac{4}{3}\pi\, r^3 \rho\, gB = \left(\frac{4\,KTB}{\pi}\right)^{\frac{1}{2}} \tag{8.15}$$

Rearranging,

$$r^3 = \frac{3}{2}\frac{(KT)^{\frac{1}{2}}}{\pi^{\frac{3}{2}}\rho\, gB^{\frac{1}{2}}} \tag{8.16}$$

Assuming $r \gg \ell$:
Eq. (8.10) for B, reduces to $B = \frac{1}{6\pi\eta r}$
Substituting the value of B and squaring both sides in Eq. (8.16):

$$r^6 = \frac{9}{4}\cdot\frac{KT\,6\pi\eta\, r}{g^2\,\rho^2\pi^3}$$

or

$$r^5 = \frac{13.5\,KT\,\eta}{\pi^2\,g^2\,\rho^2} \tag{8.17}$$

An example:
Using Eq. (8.17), calculate the nonsettling size of a coal particle with a density of 1.3: assume T = 300° Kelvin

η for air $= 181 \times 10^{-6}$ P
$K = 1.3708 \times 10^{-16}$ erg. Deg^{-1}
$g = 981$ cm/s^2

Hence,

$$r^5 = \frac{13.5 \times 1.3708 \times 10^{-16} \times 181 \times 10^{-6} \times 300}{\pi^2 \times (981)^2 (1.3)^2}$$

$$= 66.3 \times 10^{-25}$$

$$\therefore r = 2.31 \times 10^{-5}\ \text{cm} = 0.23\ \mu\text{m}$$

$$\text{or non} - \text{settling particle diameter} = 0.46\ \mu\text{m}$$

$$\text{Aerodynamic equivalent diameter} = 0.52\ \mu\text{m}$$

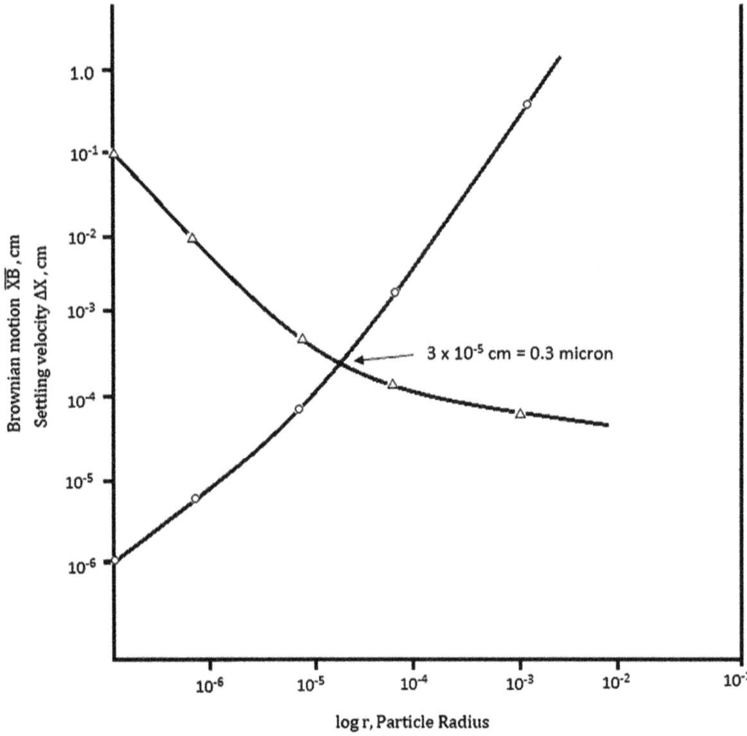

Figure 8.1 A plot of settling velocity and Brownian motion for respirable dust particles.

To confirm this value from other published data shown in Table 8.1, the average movement per second by Brownian motion and particle settling velocity were plotted against log r and shown in Fig. 8.1.

The diameter of particle that has the same velocity as settling velocity as the displacement by Brownian motion is 0.6 μm (unit density). This is in fair agreement with the calculated value of 0.52 μm from Eq. (8.17).

8.4 Size Distribution of Respirable Dust Particles

The fine particle universe consists of particles of different sizes. The number of particles, surface area of particles, or the mass (weight) of the particle associated with different sizes is called the number distribution, surface distribution, or mass distribution of the particles, respectively. It can be expressed as a frequency distribution or a cumulative distribution. In frequency distribution, the percent number or mass (frequency) associated with particle size is plotted on the y-axis against the size on the x-axis, respectively. In cumulative distribution, all frequencies below the particle x or above it are plotted against the particles size. Different particle universes follow different distributions. We will discuss only three distribution laws here.

8.4.1 Normal Distribution

This is found to be most applicable to only particles created by chemical processes, such as condensation and precipitation. It usually does not represent well the size distribution of solid particles.

Mathematically,

$$y = \frac{1}{\sigma_n \sqrt{2\pi}} \cdot \exp\left(-\frac{(x - x)^2}{2\,\sigma_{n^2}} \right) \tag{8.18}$$

where y is the probability density; x is the particle size; $\bar{x}$ is the arithmetic mean; particle size σ_n is the number standard deviation.

The number of particles with a diameter between x_1 and x_2 is given by:

$$n_i = \frac{\Sigma n_i}{\sigma_n \sqrt{2\pi}} \int_{x_1}^{x_2} \exp\left(-\frac{(x - \bar{x})^2}{2\,\sigma_n^2} \right) dx \tag{8.19}$$

The normal distribution has very limited application for particles created by broken solid materials.

The number distribution of Eq. (8.19) can be easily converted to a mass (weight) distribution as shown in Eq. (8.20).

$$w_i = \frac{\rho\, S_V\, \Sigma\, n_i\, x_i^3}{\sigma w \sqrt{2\pi}} \int_{X_1}^{X_2} \exp\left(-\frac{(x - \bar{x})^2}{2\sigma_w^2} \right) dx \tag{8.20}$$

where ρ, S_V and σ_w are density, volume shape factor, and weight standard deviation, respectively, and w_i is the weight fraction between size x_1 and x_2.

8.4.2 The Log Normal Distribution

This distribution appears to fit many fine solid particles created by comminution (milling, grinding, crushing). Pulverized silica, clay, granite, limestone, and quartz yield size distributions that satisfactorily fit the log normal distribution [10].

Mathematically, it is obtained by replacing x in Eq. (8.18) by log x as shown in Eq. (8.21).

$$y = \frac{1}{\log\,\sigma_g \sqrt{2\pi}} \exp\left[-\frac{(\log x - \log x_g)^2}{2\,\log^2\,\sigma_g} \right] \tag{8.21}$$

where x_g is the geometric mean size; σ_g is the geometric standard deviation.

The number of particles between x_1 and x_2 is given here by:

$$n_i = \frac{\Sigma n_i}{\log \sigma g \cdot \sqrt{2\pi}} \int_{\frac{x_1}{\log \sigma_g}}^{\frac{x_2}{\log \sigma_g}} \exp\left[-\frac{(\log x - \log x_g)^2}{2\log^2 \sigma_g}\right] \cdot d \log x \tag{8.22}$$

For the mass (weight) distribution, weight fraction between x_1 and x_2 is

$$w_i = \frac{\rho S_v \Sigma xi^3 \cdot n_i}{\log \sigma_g^1 \sqrt{2\pi}} \int_{\frac{x_1}{\log \sigma_g^1}}^{\frac{x_2}{\log \sigma_g^1}} \exp\left[-\frac{(\log x - \log x_g^1)^2}{2\log^2 \sigma_g^1}\right] \cdot d \log x \tag{8.23}$$

where x_g^1 and σ_g^1 are the weight mean and standard deviation, respectively.

8.4.3 The Rosin–Rammler Distribution

Broken coal behaves a bit differently than silica or limestone. The cumulative mass frequency of large or fine coal particles is best described as the Rosin–Rammler (RR) distribution [11]. It also fits very well for fine particles obtained from cement, gypsum, magnetite, quartz, and glass. Herdan [10] recommends its use to (a) distributions that deviate significantly from log normal distributions and (b) where particle sizing is done by a series of sieves.

Let us consider the distribution of broken coal obtained by sieves. Calling the quantity (in percentage) which passes the sieve, that is, the weight percentages of particles smaller than the sieve opening, Y, and the quantity retained on the sieve, that is percentage by weight of particles bigger than the sieve opening, R, we obtain by plotting either Y or R against the particle size a straight line called the "fineness characteristic curve" of the material. Y + R is always 100%. The weight distribution curve is mathematically expressed as:

$$dy = \frac{d\,G(x)}{dx} = n\left(\frac{x}{k}\right)^{n-1} e^{\left(\frac{-x}{k}\right)^n} \tag{8.24}$$

where n and k are constants. k decreases with fineness but "n" remains independent of fineness and is a characteristic of the material. "n" is also independent of the device used for comminution [10].

Integrating Eq. (8.24) we get:

$$y = G(x) = \left(1 - e^{-\left(\frac{x}{k}\right)^n}\right) \tag{8.25}$$

where y is the cumulative mass fraction below size, x.

Taking the log to the base, we have:

$$\ln \ln(1/1 - y) = n\ln x - n\ln k \tag{8.26}$$

Hence a plot of $\ln \ln\left(\frac{1}{1-y}\right)$ against ln x yields a straight line. The gradient of the straight line is the constant, n, that is a characteristic of the broken material. The RR graph paper enables direct plotting of the cumulative mass fraction undersize, y against particle size, x. On this graph, k is the characteristic size parameter corresponding to 63.21% undersize. This is easily derived by putting x = k, in Eq. (8.25). Typical values of n and k for some coal types are shown in Table 8.4 for −37 μm particles [1].

A smaller value of n means a flatter gradient of mass distribution and hence a higher proportion of mass in finer sizes. Fig. 8.2 shows a plot of mass distribution of respirable dust for four types of coals. It can be clearly seen that the fraction of submicron

Table 8.4 Typical Values of n and k for Different Coal Types

Coal Type	n, (Nondimensional)	k, (μm)
Anthracite	0.76	19
Low-volatile coal	0.84−1.05	15.5
High-volatile coal	1.17	15−19

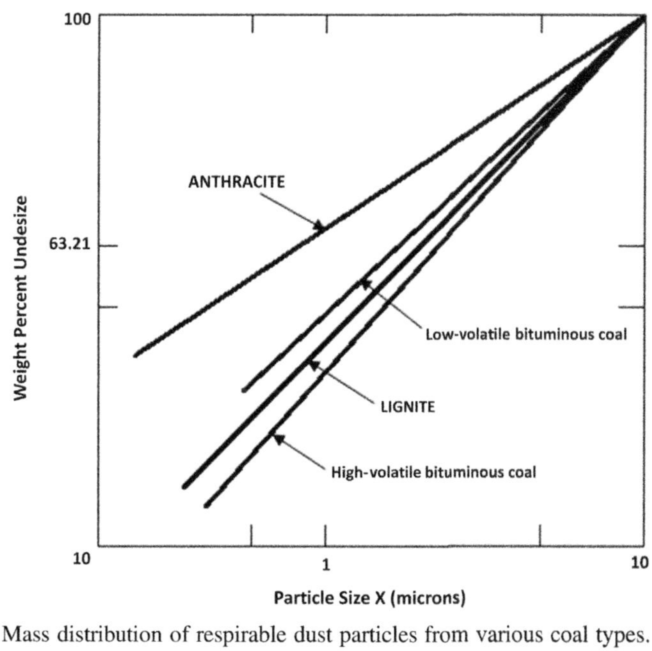

Figure 8.2 Mass distribution of respirable dust particles from various coal types.

Table 8.5 Dust Deposition in Human Lungs

Coal Type	Percentage of All Respirable Dust Deposited in Lungs (%)
Anthracite	76
Low-volatile bituminous coal	58
High-volatile bituminous coal	53
Lignite	56

particles (below 1 μm) constitutes 72% of respirable dust for anthracite compared with only 39% for high-volatile bituminous coal.

This has a serious health implication. Using the lung deposition curve of Chapter 7, it can be concluded that "lung dose" of respirable dust is significantly higher for anthracite mines compared with high-volatile bituminous coal mines even if the total respirable dust is maintained at 1.5 or 2.0 mg/m^3 in both mines.

Although a rigorous quantitative estimate of percentage deposition of dust in lungs is not feasible for lack of lung deposition efficiency values for particles below 0.5 μm, a close approximation can be made by assuming that all dust particles below 0.5 μm have a 100% deposition efficiency. Thakur [1,12] has calculated the percentage of total respirable dust deposited in human lungs as shown in Table 8.5.

Thus for the same level of respirable dust in the mine air, a mine worker in an anthracite mine will receive a dust dose that is 1.43 times higher than that of a mine worker in a high-volatile bituminous coal mine. This indicates that anthracite miner's lungs will be damaged faster and partially explains why the rate of CWP is globally higher in anthracite mines. There are other reasons (chemical composition) for this anomaly that will be discussed later in the chapter.

8.5 Determination of Mass Distribution for Fine Coal Dust Particles

In recent years, many sophisticated instruments have appeared in the market that measure the number or the surface area of particles (such as light scattering instruments), but they fail to produce true mass distribution of the fine respirable dusts. Instruments that measure the electrical conductivity claim to measure the volume of the particles but do not yield a true mass distribution. The only reliable device that gives good mass distribution is the Andreasen pipette [13]. Fig. 8.3 shows the Andreasen pipette.

The apparatus consists of a graduated cylinder, w; with a stopper, L, that has a hole in it; a pipette, P, that can hold 10 cc of fluid; and a three-way cock, T, that enables the sample to be withdrawn and discharged into a container, A. Thakur [12] used this apparatus extensively to measure the mass distribution of 18 coal samples with excellent results. If the concentration of solids is kept less than 1% by volume, 2%–5% accuracy can be achieved [10].

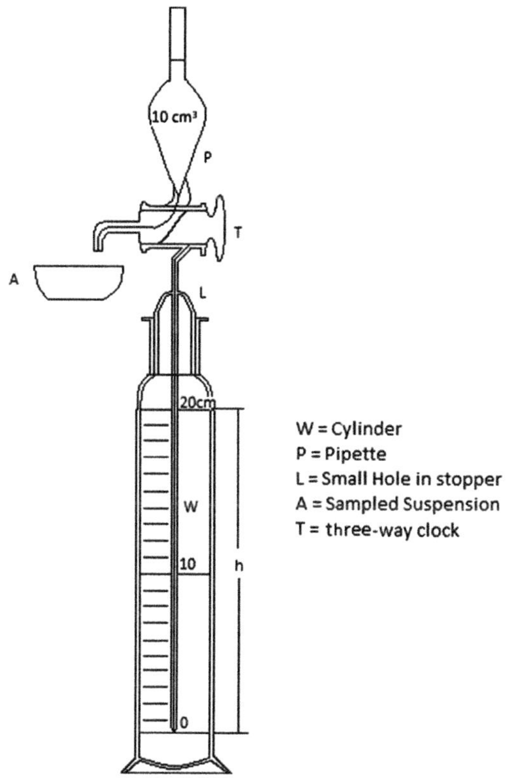

Figure 8.3 Andreasen sedimentation pipette.

8.5.1 Experimental Procedure for Mass Distribution Measurement

The following procedure was followed and is recommended for such work [12]:

1. The density of the coal dust (−37 μm) was determined with a pycnometer and distilled water.
2. Sufficient dust was weighed out so that on dilution with 550 cc of water in the cylinder (to reach the 20 cm mark on the cylinder), the concentration of solids would be 0.5% by volume.
3. A suspension of the weighed coal sample was made in 100 cc of distilled water using 5− 6 drops of Aerosol O-T, a wetting agent.
4. The suspension was next transferred to the cylinder and fluid volume was made up to the 20 cm mark.
5. The stopper, L, was inserted, and the apparatus was insulated with glass wool. About 20 min were allowed for the fluid temperature to stabilize.
6. The small vent in the stopper was closed with a finger, and the apparatus was shaken for 2 min.
7. A clock was started as soon as the shaking was stopped, and the apparatus was put on a firm stand.

8. The first sample was taken immediately by drawing 10 cc into the pipette slowly. It took 20 s to withdraw the sample.
9. The sample was drained into a preweighed aluminum container and dried in an oven at 105 ± 2°C for 24 h. It was then weighed with an accuracy of ±0.25 mg on a Mettler analytical balance. The weight of the first sample gives the initial dust concentration, and it agreed well with the calculated concentration.
10. Subsequent samples were withdrawn at precalculated intervals to have experimental data uniformly spread over the range 0−37 μm.

8.5.2 Calculation of Stoke's Diameter for a Given Time, t

Settling velocity was earlier calculated in Eq. (8.5) as:

$$v_S = \frac{2\,r^2(\rho - \rho_f)g}{9\eta}$$

where ρ_f is the fluid density.

v_S is also equal to h/t, where h is distance shown in Fig. 8.2 and "t" is the time elapsed since the clock was started.

Hence,

$$h/t = \frac{2\,r^2(\rho - \rho_f)\,g}{9\eta}$$

or

$$r = \left[\frac{9h\,\eta}{2(\rho - \rho_f)gt}\right]^{\frac{1}{2}} \tag{8.27}$$

The r values for each time interval, t, was calculated for plotting on an RR graph.

The weight of each successive sample was expressed as a percentage of the first sample that gave the cumulative mass (weight) frequencies. The results were plotted on an RR graph as shown in Fig. 8.4.

Table 8.6 lists the n and k values of −37 μm dust from various coal ranks [12].

It is clear from Table 8.6 that the "characteristic n" is very much rank dependent. Rank is decided by the volatile content of coal on a dry ash-free (DAF) basis.

Fig. 8.5 shows a plot of 16 data points. This reconfirms that in general, high rank coals, but particularly anthracite, yields more submicron dust leading to higher "respirable dust dose" and consequently a higher rate of CWP in anthracite mines.

Fig. 8.5 shows a variation of distribution parameter, n, with volatile matter (DAF).

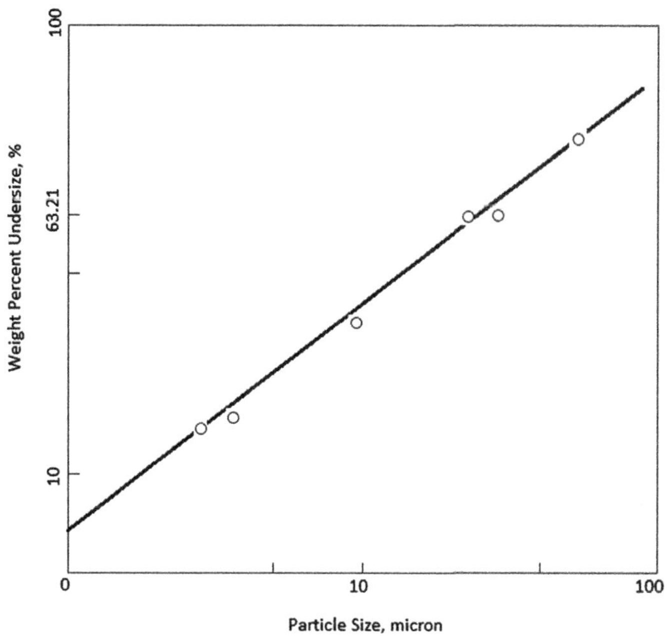

Figure 8.4 Graph of mass distribution for a low-volatile coal; n = 1.00, k = 15.5 μm.

The submicron (most dust particles found in human lungs) fraction of dust from various coal types are as follows:

Anthracite	72%
Low-volatile coal	48%
High-volatile bituminous coal	39%
Lignite	44%

Lignite is mostly mined by surface mining and is not of concern for CWP.

8.6 Chemical Composition of Respirable Coal Dust

Coal seams are not all coal; it also contains rock bands in it. Besides, the roof and floor are also mined to a little extent with coal. Calcium carbonate dust is profusely used in mines to neutralize coal dust. Diesel engines, if used, add diesel particulate matter to respirable coal dust. The resulting coal dust, therefore, contains many minerals and elements in small proportions besides carbon.

Table 8.7 shows a typical analysis of mine dust.

Table 8.6 The "n" and "k" Values for Various Ranks of Coal Particles (Below 37 μm)

Coal Type	Density (gm/cc)	n	K (μm)	Average (n)
Anthracite				
Sample 1	1.6	0.88	20	(0.76)
Sample 2	1.6	0.89	18.5	
Sample 3	1.5	0.52	11.5	
Low-Volatile Bituminous				
Sample 1	1.32	1.00	15.5	(1.05)
Sample 2	1.34	1.09	6.7	
High-Volatile Bituminous A				
Sample 1	1.28	1.20	16	(1.17)
Sample 2	1.31	1.21	16.5	
Sample 3	1.35	1.12	15.5	
Sample 4	1.32	1.20	13	
Sample 5	1.34	1.16	11.8	
Sample 6	1.38	1.15	14	
High-Volatile Bituminous B				
Sample 1	1.38	1.23	19	(1.23)
Lignite				
Sample 1	1.40	1.15	14.5	(1.11)
Sample 2	1.43	1.06	13.3	

As far as the growth of CWP is concerned, only a few elements, besides carbon, play an important role; namely silicon and iron. Silica may play the most important role. It occurs both as an oxide and in the crystalline form. Trydimite and cristobalite forms of crystalline silica are more fibrogenic than alpha quartz. Most coal seams contain less than 2% silica, but the coal mine respirable dust in US and British mines contain 4%–5% silica [15,16]. Silica concentrations appear to increase as the coal dust gets finer in size. The PEL for silica varies from 0.1 to 0.2 mg/m^3 and generally appears adequate. Apart from the total concentration of silica, another important issue is the bioavailability of silica. Balsaitis and Wallace [17] and Conner et al. [18] have used electron microscopy coupled with energy-dispersive X-ray analysis to explore the composition of silica particles and found that they were heavily occluded with clay in lower rank coals, making lower rank coal dust less fibrogenic even if they may have a higher total concentration of silica.

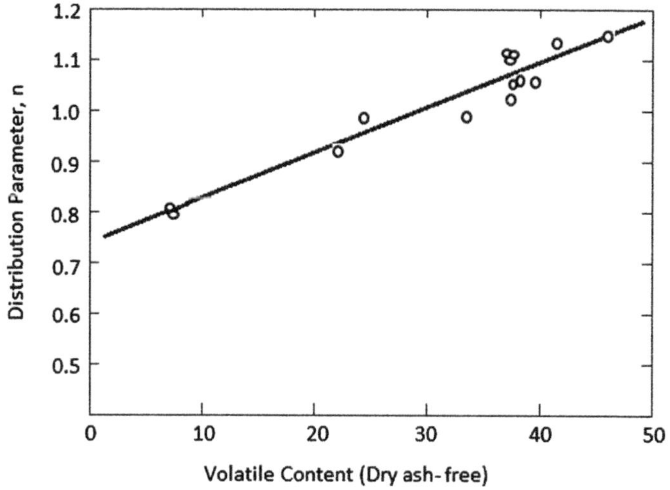

Figure 8.5 Variation of distribution parameter, n, with volatile matter (DAF).

Table 8.7 Typical Composition of Mine Dust [14]

Element	Weight
Silica	1%–2%
Iron	0.4%
Aluminum	0.98%
Chromium	15.8 ppm
Copper	73.1 ppm
Zinc	312.9 ppm
Cobalt	Trace
Carbon	The rest

The role of iron sulfate in coal dust–induced lung injury and injury owing to iron to proteins in human and rat alveolar microphages have been investigated [17,18]. It appears that iron may play a bigger role in causing CWP than was realized heretofore.

Finally, one needs to consider the role of free radicals on the respirable dust particles in the growth of CWP [19]. Freshly cleaved silica and coal particles carry larger amounts of free radicals as compared with stale or aged particles and are, therefore, more harmful to lung tissues [20]. Certain water-soluble chemicals, e.g., organosilanes, appear to suppress the free radicals on freshly mined coal, but no experimental or field data are yet available [21]. It may be easier to remove mine workers away from the active mine faces by using remote operation to better safeguard their health. This requires remote operation of all coal cutting and loading equipment.

References

[1] Thakur PC. Mass distribution, percent yield, non-settling size and aerodynamic shape factor of respirable coal dust particles [M.S. thesis]. The Penn State University; 1971. p. 133.
[2] Fuchs NA. The mechanics of aerosols. Pergamon Press; 1964. p. 408.
[3] Fuchs NA. The mechanics of aerosols. Pergamon Press; 1964. p. 28–9.
[4] Cunningham E. Proceedings of Royal Society 1910;83 A:357.
[5] Millikam R. Physics Review 1923;21:217.
[6] Timbrell V. The terminal velocity and size of airborne dust particles. British Journal of Applied Physics 1954;(Suppl. 3):86–9.
[7] Watson J. British Journal of Industrial Medicine 1953;10:93.
[8] Sawyer KF, Walton WH. The conifuge – a size separating sampling device for airborne particles. Journal of Science Instruments 1950;27:272.
[9] Einstein A. Annalen der Physik 1905;17:549.
[10] Herdan G. Small particle statistics. Butterworths Scientific Publication; 1960. p. 80–2.
[11] Rosin P, Rammler E. Journal of the Institute of Fuel 1933;7:29.
[12] Thakur PC. Mass distribution of respirable dust particles from U.S. coals. Colliery Guardian 1974;227(F):236–9.
[13] Andreasen AHM, Berg S. Angew Z Chemistry 1935;48:203.
[14] Kaya E, Hogg R, Mutmansky JM. Evaluation of procedures for production of dust samples for biomedical research. Applied Occupational and Environmental Hygiene July 1996; 11(7).
[15] Tomb TF, et al. Analysis of quartz exposure data from underground and surface coal mine operations. Applied Occupational and Environmental Hygiene 1995;10:1019–26.
[16] Walton, et al. The effect of quartz and non-coal dusts in coal workers' pneumoconiosis. In: Walton WH, editor. Inhaled particles, vol. IV. England: Old Woking, Unwin Brothers; 1977. p. 669–89.
[17] Balsaitis PB, Wallace WE. The structure of silica surfaces in relation to cytotoxicity. In: Castranova, editor. Silica and silica induced lung disease. CRC Press; 1995. p. 79–89.
[18] Conner JR, et al. Mineral dust promotes alterations in ferritin isoforms and oxidative damage to protein in rats and human alveolar micro phases. Applied Occupational and Environmental Hygiene 1996;II(7):969–72.
[19] Chen LC, et al. The role of ferrous sulfate in coal dust induced lung injury. Applied Occupational and Environmental Hygiene 1996;II(7):973–80.
[20] Vallyathan V, et al. Freshly fractured quartz inhalation leads to enhanced lung injury and inflammation. American Journal of Respiratory and Critical Care Medicine 1995. 152 pp.
[21] Thakur PC. How to eliminate coal workers' pneumoconiosis. In: Proceedings of the 27th international conference on safety in mines. New Delhi, India: Research Institutes; 1997. p. 217–22.

Generation of Respirable Coal Dust

9

Chapter Outline

It has been well established that the amount of respirable dust produced in the mining process is a function of two groups of variables [1,2], namely:

1. Material-dependent variables, such as moisture content, hardness, etc.
2. Breaking mechanism–dependent variables, e.g., type of machine, size of cutting tool, depth of cut, etc.

Studies on size distribution of broken coal have indicated that the breaking mechanism (except for bit sharpness) has a significant impact only in the coarser range (above 100 μm). Their influence on fine dust particle generation is insignificant [3]. It is postulated that respirable dust is not produced by successive breakdown of larger particles but is rather produced by local crushing of the coal by the cutting bits and other objects.

On the other hand, physical properties of coal have considerable influence on the production of respirable dust particles [4,5]. Results of theoretical and experimental studies are presented in the following section.

Advanced Mine Ventilation. https://doi.org/10.1016/B978-0-08-100457-9.00009-2

9.1 A Mathematical Model for Respirable Dust Generation

Panov [5] was the first to derive a mathematical expression for dust generation based on the physical properties of coal. He assumed that the rate of dust generation (minus 100 μm particles) is a function of seven variables and derived an expression for dust generation as shown in Eq. (9.1).

$$N = 35.6\left(a^{6.5} f \rho^{2.5} SA\right)/w^{0.5}\phi \tag{9.1}$$

where N = rate of dust formation in gm/min; a = characteristic dimension of simultaneously broken coal; f = hardness number; ρ = density; gm/cc; S = specific fissuration; A = specific energy of deformation, (erg/gm); w = percent moisture content; ϕ = percent porosity.

Fig. 9.1 shows the effect of some of these parameters on the production rate of fines. The effect of moisture content on the rate of dust production appears to be very significant, particularly at low moisture contents.

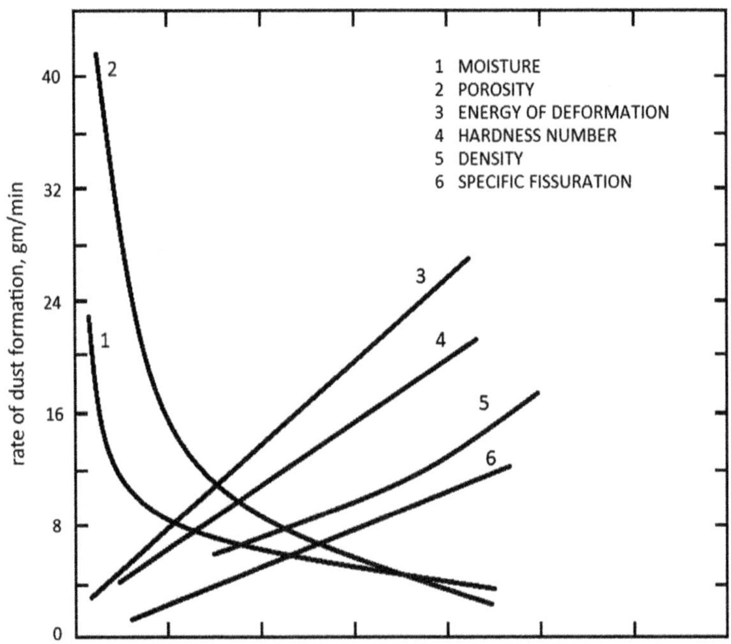

Figure 9.1 Rate of dust formation as a function of the properties of coal.

The physical properties of coal that affect the production of dust, namely, resistance to comminution, microstrength[1], Hardgrove grindability[2], etc., are in turn very much dependent on the chemical and petrological properties of coal [4,5]. Thakur [6] found that the following chemical and petrological properties of coal have significant influence on the physical properties of coal and consequent respirable dust production:

1. Moisture content of coal (as received).
2. Ash content (dry).
3. Volatile matter content (dry ash free [DAF]).
4. Fixed carbon content (DAF).
5. Fixed carbon/volatile matter (DAF).
6. Fusain content.

The various coal types or even coal seams can, thus, be indexed with respect to their proneness to respirable dust production based on their proximate and petrological analyses. Because these coal constituents are easily measurable, such indexing will help the mining engineers in preplanning their dust control programs.

To eliminate the effect of the breaking mechanism, all the samples were subjected to the same comminution process in Thakur's experiment [6]. Equal masses of uniformly sized coal particles were comminuted for a given time, and the yield of respirable dust was measured and expressed as a percentage of the original mass of the sample. This percentage is defined as the respirable dust index (RDI) of the coal.

Thakur [6] derived an expression for RDI as shown in Eq. (9.2).

$$\mathrm{RDI} = \frac{0.95(\mathrm{FC})^{0.714}(\mathrm{F})^{0.191}}{(\mathrm{a})^{0.168}} \tag{9.2}$$

where FC is the fixed carbon (DAF); F is the fusain content; a is the ash content (moisture free).

Moisture was not included in Eq. (9.2), even though it has a great impact on respirable dust production, because it was highly correlated with fixed carbon. Besides, in his experiment, the moisture contents of the samples were not truly representative of the original moisture content of the coal. These samples were air-dried and kept in sealed cans under vacuum to preserve its other chemical compositions. Details of the procedure and derivation of Eq. (9.2) are as follows.

9.2 Sample Preparation and Experimental Details

Eighteen coal samples studied represented the various ranks of coal from anthracite to lignite. Details of locations from which the samples have been drawn, type of coal, and

[1] Microstrength is a measure of the resistance to the degradation of coal.

[2] Hardgrove Grindability Index is a measure of relative ease of grinding coal. For details, reference can be made to the American Society for Testing and Materials, ASTM Standard on Coal and Coke, Philadelphia, PA, 1962, pp. 115–118.

Table 9.1 Rank and Location of Coals Studied

Sample Number	Rank	Seam Name	Locality
1	Anthracite	Buck Mountain	Zerbe, Pennsylvania
2	Anthracite	Buck Mountain	Zerbe, Pennsylvania
3	Anthracite	#8-1/2 Seam	Shamokin, Pennsylvania
4	High Vol. A Bit*.	Elkhorn#3	Deane, Kentucky
5	High Vol. A Bit.	Elkhorn#3	Deane, Kentucky
6	High Vol. A Bit.	Elkhorn#3	Deane, Kentucky
7	High Vol. A Bit.	C. seam	Benham, Kentucky
8	High Vol. A Bit.	Pittsburgh	Marianna, Pennsylvania
9	High Vol. A Bit.	Pittsburgh	Marianna, Pennsylvania
10	High Vol. A Bit.	Tioga	Tioga, West Virginia
11	High Vol. A Bit.	#5 Seam	Bickmore, West Virginia
12	HVB, Bit.	#2 Colchester	Vermont, Illinois
13	HVC, Bit.	Illinois No. 6	Carrier Mills, Illinois
14	Sub. Bit.	Monarch	Sheridan, Wyoming
15	Low Vol. Bit.	Lower Kittanning	Ebensburg, Pennsylvania
16	Med. Vol. Bit.	Pocahontas#3	Gary, West Virginia
17	Lignite	Zap	Zap, North Dakota
18	Lignite	Zap	Zap, North Dakota

*Bit means Bituminous.
*HV means high-volatile.

seam names are given in Table 9.1. The samples were, originally, collected by the organic sediment laboratory, Pennsylvania State University and crushed and sieved to obtain minus 20 mesh particles. This undersize is preserved in vacuum sealed cans.

Table 9.2 shows their chemical and petrological properties. Samples for the purpose of this study were prepared from these by resieving them to obtain a closely sized fraction between 20-mesh (833 μm) and 28-mesh (589 μm) Tyler screens. This closely sized sample was then ground in a Bleuler rotary mill.

9.2.1 Bleuler Rotary Mill

A schematic view of the Bleuler rotary mill is shown in Fig. 9.2. In this mill, an eccentric shaft is driven by a heavy duty enclosed motor. The shaft actuates a flywheel, which makes the upper chamber assume a rotary motion. The sample container, C, is clamped to the chamber wall and contains a heavy cylindrical weight, W, and an

Table 9.2 Chemical and Petrological Properties of Coals

Sample Number	Moisture Percent as Received	Dry Ash Percent	Volatile Percent (DAF)	Fixed Carbon Percent (DAF)	Fixed Carbon/ Volatile	Fusain Percent
1	3.28	7.84	6.13	93.87	14.29	3.70
2	4.07	9.16	6.35	93.65	14.29	27.70
3	1.04	27.74	12.63	87.37	7.15	17.40
4	2.08	3.90	37.62	62.38	1.67	6.00
5	1.74	4.65	37.73	62.27	1.64	11.10
6	1.81	4.50	37.99	62.01	1.64	17.10
7	2.04	1.26	34.10	65.90	1.92	7.50
8	1.88	5.15	37.85	62.15	1.64	5.50
9	1.46	6.23	38.22	61.78	1.61	3.40
10	1.93	6.01	34.01	65.99	1.92	24.90
11	1.73	17.35	39.63	60.37	1.50	14.50
12	4.89	11.83	41.90	58.10	1.39	5.50
13	12.10	10.84	45.66	54.34	1.19	3.20
14	20.47	5.24	46.67	53.33	1.14	10.70
15	1.67	6.34	19.82	80.18	4.04	7.40
16	1.82	6.39	23.01	76.99	3.32	11.20
17	28.07	8.17	45.81	54.19	1.18	12.00
18	28.92	11.92	46.55	53.45	1.15	13.30

annular ring, A. Because of the constant rotary motion of the sample container, the cylindrical weight and annular ring exert a constant pressure on the container wall, and coal particles caught in between are crushed to finer sizes. A fixed amount of energy was expended on a sample of given mass of uniformly sized material. This permitted the evaluation of the effect of the properties of coal alone on the production of respirable coal dust.

A Bleuler mill provides an efficient and rapid means for grinding samples and meets the requirements of reproducibility of particle size distribution for optical and X-ray fluorescence spectrographic studies. The time of grinding can be preset and is controlled electrically.

9.2.2 Procedure for Grinding

Twenty grams of the closely sized material was ground for 1 minute in the Bleuler mill. The crushed material was sieved through a 400-mesh (37 μm) Tyler screen on

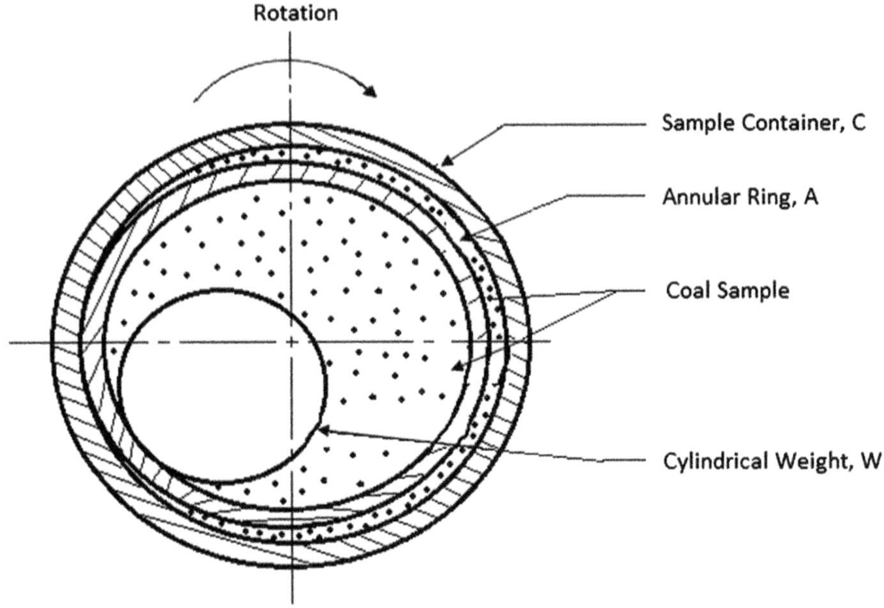

Figure 9.2 Schematic diagram of a Bleuler rotary mill.

a Ro-Tap shaker for 10 minutes to obtain samples for experimentation. The minus 400 mesh (37 μm) material was carefully weighed to 0.1 gm. Three grindings were done for each sample, and an average yield of −37 μm particles was determined. Later, the three samples were mixed and used for experimentation.

9.3 Yield of Respirable Dust

The observations made and the calculated data obtained for all 18 coal samples are shown in Table 9.3. For each sample of coal, the yield of −37 μm particles is shown in the first three columns. The average yield of −37 particles is shown in the fourth column.

The maximum respirable size is taken as 10 μm equivalent diameter in conformity with standards of the Atomic Energy Commission, United States. The corresponding Stoke's diameter for a given coal type is given by $10/(\rho)^{1/2}$, where ρ is the density of coal.

This maximum respirable size was calculated for each coal type. This is shown in Table 9.3. Corresponding to this maximum respirable size, the percentage weight of undersize material is obtained from the respective graphs of mass distribution discussed in Chapter 8. A typical calculation of total yield of respirable dust is given as follows:

From sample number 1:

Average yield of −37 μm particles = 67.3%.

Maximum respirable size for sample number 1 = $10/(\rho)^{1/2} = 10/(1.569)^{1/2} = 8.00$ μm.

Experiment Number	Sample Number	Yield of −37 Particles (Mass Percent) I	II	III	Mean (A)	Maximum[a] Respirable Size 10/Density $^1/_2$ μm	Percent[b] Undersize at 10/Density $^1/_2$ = (B)	Total Yield of Respirable Dust AxB/100% (RDI)
1	1	67.5	67.0	67.5	67.3	8.00	36.0	24.2
2	2	75.5	74.5	75.0	75.0	7.95	38.0	28.5
3	3	85.0	84.0	84.5	84.5	8.10	22.0	18.6
4	4	60.5	61.0	59.5	60.3	8.75	39.0	23.5
5	5	69.5	69.0	71.0	69.5	8.70	39.0	27.1
6	6	58.0	57.5	57.0	57.5	8.64	39.0	22.4
7	7	45.0	45.5	45.0	45.2	8.70	46.0	20.8
8	8	45.5	45.5	45.5	45.5	8.64	51.0	23.2
9	9	57.5	57.0	57.5	57.3	8.52	43.0	24.6
10	10	72.0	73.5	72.0	72.5	8.70	42.0	30.5
11	11	60.5	61.0	60.0	60.5	8.25	44.0	26.6
12	12	70.5	71.0	70.0	70.5	8.52	30.0	21.2
13	14	67.0	67.5	67.5	67.3	8.45	9.5	6.4
14	15	61.5	61.5	61.0	61.3	8.70	29.0	17.8
15	17	64.5	65.0	65.5	65.0	8.70	43.0	28.0
16	18	25.0	25.5	24.0	24.8	8.64	73.0	18.1
17	19	42.5	42.0	42.5	42.3	8.40	42.0	17.8
18	20	35.0	35.0	35.5	35.2	8.25	45.0	15.8

[a]Maximum respirable size is expressed here as Stoke's diameter = 10/(density)1/2μ by definition.
[b]Obtained by referring to the corresponding mass distribution graph in Chapter 8.

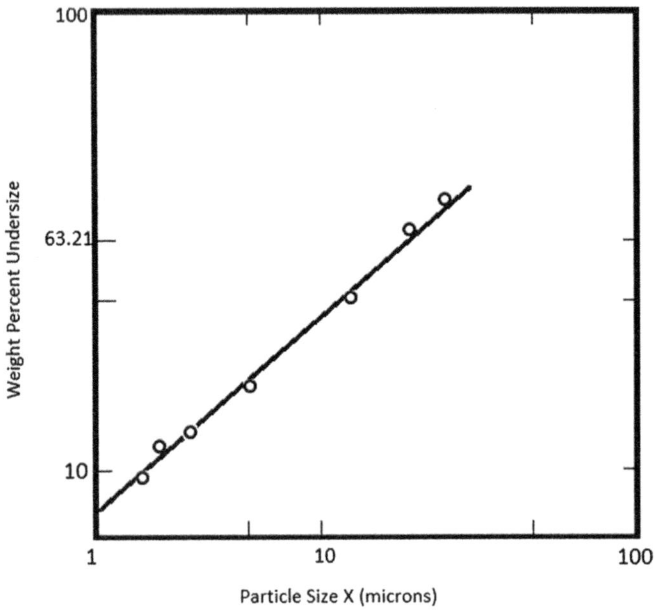

Figure 9.3 Graph of mass distribution.

Referring to the mass distribution curve for sample number 1, shown in Fig. 9.3, the percentage mass undersize corresponding to 8 μm = 36%.

Total percent yield = 67.3 × 36/100 = 24.2% or RDI = 24.2. Similar calculations were done for all the other samples. The total yield of respirable dust is thus obtained as shown in Table 9.3.

9.4 Dependence of Respirable Dust Index on the Properties of Coal

Different ranks of coal appear to yield different amounts of respirable dust (6.4%–28.5%) for the same energy input yielding different RDIs (Table 9.3). It is mainly because of their different physical and chemical properties. Four most influencive properties are discussed here.

9.4.1 Effect of Moisture

Variation of RDI with moisture is shown in Figure 9.4. There is a 40% decrease in dust production as moisture increases from 1% to 12%. Beyond 12%, the fall in dust production does not appear to be so significant. Hence, pretreatment of coal seams with water infusion can drastically reduce respirable dust production.

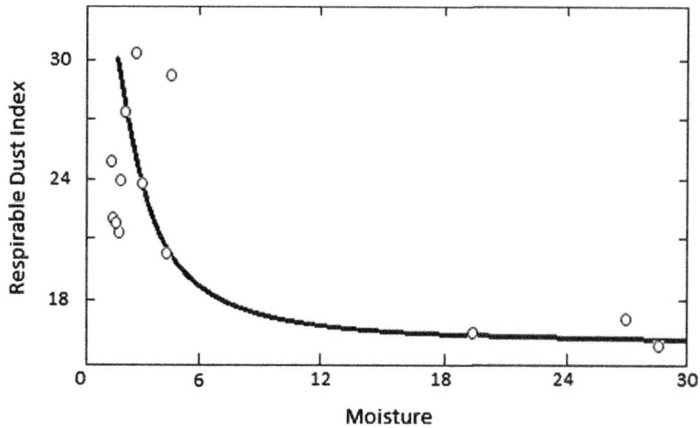

Figure 9.4 Variation of respirable dust index with moisture.

9.4.2 *Effect of Volatile Content or Carbon Content*

The fixed carbon content (DAF) is equal to 100-volatile percentage (DAF). Both of these measure the degree of metamorphosis the coal has undergone. Figure 9.5 shows a graph of RDI against volatile matter content (DAF). For the range of volatile matter content 5% to 35%, there is no great decrease in RDI. The data are somewhat scattered, but a decrease in RDI with increasing volatile content is discernible. With increase in volatile matter content beyond 35%, the RDI appears to fall sharply.

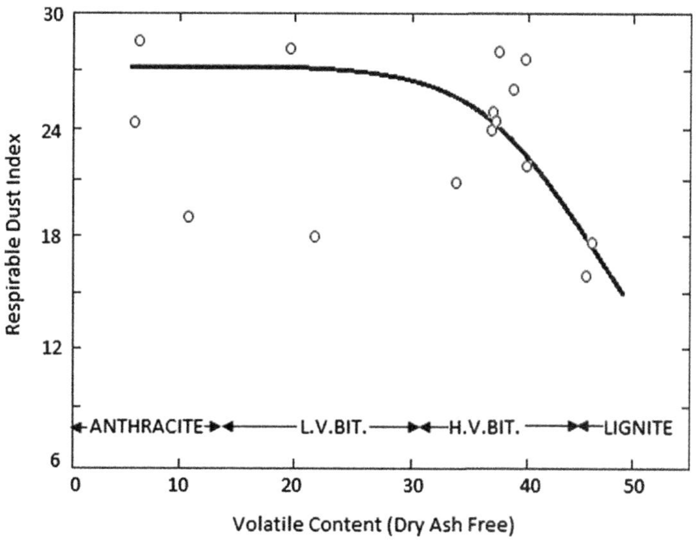

Figure 9.5 Variation of respirable dust index with volatile matter content (DAF).

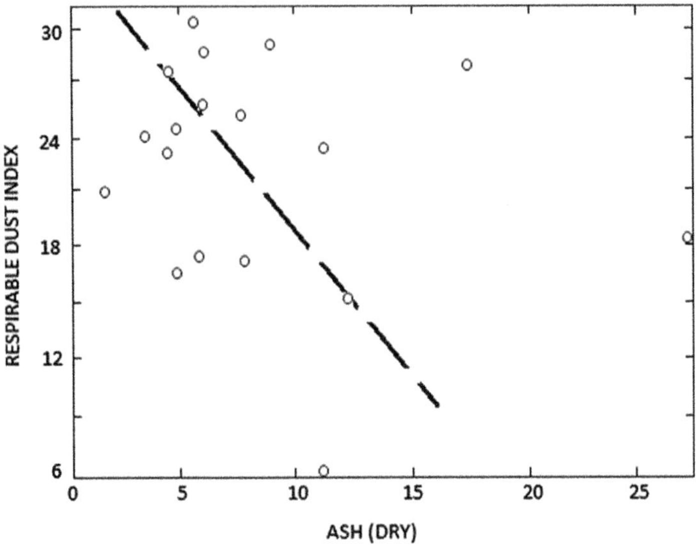

Figure 9.6 Variation of respirable dust index with ash content.

9.4.3 Effect of Ash Content

The variation of RDI with ash content is shown in Figure 9.6. In spite of the scatter in data, a downward trend with increasing ash can be discerned.

9.4.4 Effect of Fusain Content

Figure 9.7 shows the variation of RDI with fusain content. There is a progressive increase in RDI as fusain content increases from 2.5% to 25%.

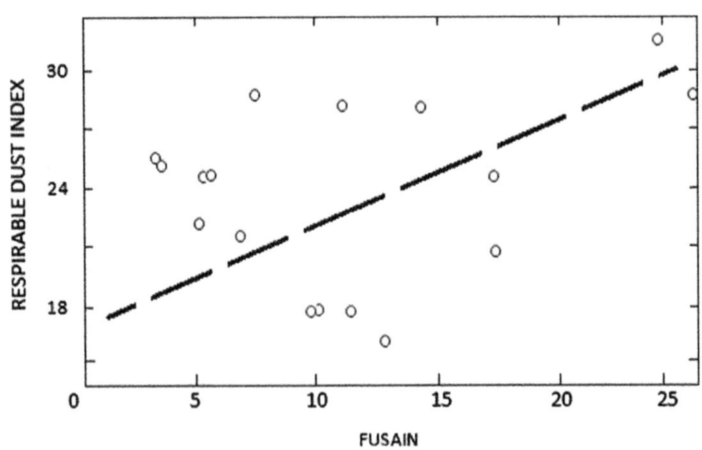

Figure 9.7 Variation of respirable dust index with fusain content.

9.5 Statistical Analysis of Data

Because the data presented and discussed earlier show great scatter, a statistical analysis was done to determine the correlation coefficients and regression coefficients of RDI with respect to the various properties of coal.

Standard library programs were used for this purpose. It was shown earlier [5] that a dimensional analysis of the problem suggests a relationship between the RDI and the various properties of coal, which may be expressed as:

$$\mathrm{RDI} = \mathrm{constant} \cdot \frac{a^x b^y}{c^z}$$

where a, b, and c are properties of coal, and x, y, and z are their exponents.

To obtain such a relationship in this case, the logarithm of the variables were used as input to the statistical analysis program. RDI was the dependent variable, and the moisture content, ash content, volatile content, fixed carbon content, ratio of fixed carbon/volatile and fusain content were used as the independent variables.

9.5.1 *Correlation Between the Logarithm of Variables*

This was obtained using library program "PPMCR" (Pearson product moment correlation coefficient). The results are given in Table 9.4.

Moisture appears to be well correlated with fixed carbon with a correlation coefficient of 0.4224. It was, therefore, not used in the subsequent regression analysis.

9.5.2 *Regression Analysis*

The data from the PPMCR analysis were used as input into another library program called "UPREG," which carried out a regression analysis. The fixed carbon content,

Table 9.4 Correlation of Respirable Dust Index (RDI) With Coal Properties

Variable Number	Independent Variable	Coefficient of Correlation with Dependent Variable, RDI (Variable No. 1)
2	Moisture	−0.1680
3	Ash (dry)	−0.1998
4	Volatile content (DAF)	−0.3043
5	Fixed carbon (DAF)	0.4011
6	Ratio $\frac{\text{Fixed Carbon}}{\text{Volatile Content}}$	0.3226
7	Fusain	0.3664

Table 9.5 Regression Analysis of Data on Respirable Dust Index (RDI)

Independent Variable	Regression Coefficient	Partial Correlation Coefficient
Fixed carbon (DAF)	0.7138	0.4192
Fusain	0.1909	0.3924
Ash	−0.1676	−0.3473
Intercept	−0.0204	

fusain content, and ash percentage were the most significant independent variables in that order. The results of regression analysis were as shown in Table 9.5.

Hence, log RDI = −0.0204−0.1676 Ash +0.7138 Fix C + 0.1909 Fusain

Taking antilog on both sides:

$$\mathrm{RDI} = \frac{0.95(\mathrm{FixC})^{0.714} \cdot (\mathrm{Fusain})^{0.191}}{(\mathrm{Ash})^{0.168}} \tag{9.3}$$

where RDI = Percentage yield of respirable dust; FixC = Percentage fixed carbon content (dry ash-free basis); Fusain = Percentage fusain content; Ash = Percentage of ash content (dry).

The standard error of estimate is 13.6%.

9.6 Results of Similar, Subsequent Studies

Nearly 25 years after the original work done by Thakur [2], the same study was repeated by Srikanth [7] and Jha [8]. Their findings were in agreement with Thakur and reconfirmed all conclusions. A summary of their work is presented here for a good comparison.

9.6.1 Srikanth's Studies

Srikanth [7] used a jaw crusher to create −37 μm coal dust particles. Coal samples were obtained from coal mines in addition to some samples from the same source as Thakur's samples. They used a Microtrac Standard Range Analyzer (SRA) and Small Particle Analyser (SPA), which measured projected area (and hence diameter) using laser scattering and diffraction, respectively. The data were combined and plotted on a Rosin−Rammler graph (discussed in Chapter 8). Their main findings were as follows:

1. Higher rank coals produced more total dust (<15 μm) and respirable dust (<7 μm). Semianthracite coal produced 3.7 times more total dust and 4.2 times more respirable dust compared with high-volatile bituminous coal.

Table 9.6 Matrix of Partial Correlation Between Respirable Dust Production and Coal Seam Properties (ROM Coal)

Coal Properties	Partial Correlation with RDI
Moisture	−0.204
Ln ash	−0.393
Volatile matter	−0.191
Fixed carbon	0.523
Fixed carbon/volatile matter	0.575
Hardgrove Grindability Index	0.216

2. The Rosin−Rammler graph distribution parameter, n, was also rank dependent. The value for n was 0.68, 0.84, 0.90, and 0.95 for semianthracite, low-volatile coal, high-volatile bituminous coal, and subbituminous coals, respectively. This is similar to findings by Thakur (refer to Chapter 8 in the book).
3. High-rank coals produce a higher proportion of submicron dust particles.

9.6.2 *Jha's Studies*

Jha [8] collected coal samples and airborne respirable dust samples from the coal mines for his study. He basically correlated the rate of coal mining with the rate of respirable dust production. The respirable dust concentration was measured with a personal, gravimetric sampler. Statistical analyses of data showed the following correlation matrix.

Data in Table 9.6 compare well with similar data in Table 9.4 and reconfirm Thakur's [6] findings. The only difference is that Jha [8] found a significant correlation with the Hardgrove Grindability Index, which the other two studies [6,7] did not.

9.7 Impact of Cutting Bit Wear on Respirable Dust Production

Although it is generally true that the breaking machinery does not have much influence on respirable dust production, the cutting bit design and consequent specific energy of producing a ton of coal has a direct impact. New, sharp bits have a lower specific energy consumption (Kwh/ton) compared with older, damaged bits. Figure 9.8 shows how a new cutting bit can get damaged in the course of operation.

Khair [9] did a laboratory test to confirm this and found a 26% increase in respirable dust production as the bit got dull and damaged (with a weight loss of 15%). It is recommended that a cutting bit should be replaced when it loses 7% weight.

Figure 9.8 Stages of wear on a cutting bit.

References

[1] Evans I, Pomeroy CD. Strength, fracture and workability of coal. New York: Pergamon Press; 1966. p. 195.

[2] Thakur PC. Mass distribution, percent yield, non-settling size and aerodynamic shape factor of respirable coal dust particles [M.S. Thesis]. The Penn State University; 1971. p. 133.

[3] Leonard JW, Mitchell DR. Coal preparation, the American Institute of mining. New York: Metallurgical and Petroleum Engineers, Inc.; 1968. p. 6–24.

[4] Gomez M, Hazen K. Prediction of coal grindability from exploration data, U.S. Bureau of Mines. Report of Investigation 1970;7421:34.

[5] Panov GE. Dust formation kinetics as a function of the principal mechanical properties of coal. Soviet Mining Science, (English Translation 1966:511–4.

[6] Thakur PC, Sinha AK. Respirable dust index for U.S. coals. Colliery Guardian August 1973:294–8.

[7] Srikanth R, Ramani R. Single-breakage studies to determine the relationship between respirable dust generation and coal seam characteristics. Applied Occupational and Environmental Hygiene 1996;II(7):662–8.

[8] Jha P. et.al. Relationship between respirable dust generation and coal seam characteristics. Applied Occupational and Environmental Hygiene 1996;II(7):655–61.

[9] Khair AW. In: Principles of bit wear and dust generation, new technology in mine health and safety, SME Annual Meeting, Phoenix, USA; 1992. p. 175–83.

Respirable Dust Control

10

Chapter Outline

Engineering control of respirable coal dust follows the classical sequence of control techniques, listed in descending order of importance and effectiveness:

1. Prevention of generation of dust,
2. Suppression of dust on site and prevention of suspension in air,
3. Collection of dust that could not be suppressed and got suspended in air, and
4. Dilution of remaining dust with ventilation air to safe levels.

The use of protective equipment/administrative control in special situations is the last resort, but it is not considered a dust control technique. The following steps illustrate an ideal strategy for respirable dust control in underground coal mines.

Advanced Mine Ventilation. https://doi.org/10.1016/B978-0-08-100457-9.00010-9

1. Coal seams are infused with water and cut with very sharp bits to minimize dust generation (discussed in Chapter 9).
2. Water mixed with surfactants is sprayed at high pressures to suppress the dust on site.
3. All cutting and drilling operations use a scrubber to collect the dust that became airborne.
4. Adequate ventilation air is provided to dilute the remaining airborne dust to keep its concentration below the statutory limits.
5. For mine workers working in return airways, well-designed personal protective equipment, such as air helmets, is provided.

10.1 Theory of Dust Suppression and Collection

To optimize the process of dust suppression by water sprays and dust collection by scrubbers (filters), it is necessary to understand the theory behind them. We will first discuss suppression/collection of a single dust particle by a single waterdrop. Next, the collection of dust particles by a single strand of filter will be mathematically analyzed.

The dust suppression/collection efficiency for water droplets, E, is a function of the following:

The radius of the water droplet, R
The radius of the dust particle, a
Velocity of water droplet, U
Density of dust particle, ρ_1
Density of water particle, ρ_2
Viscosity of water, η

Dimensional analysis of the independent variables yields a relationship given by Walton and Woolcock [1] in Eq. (10.1).

$$E = f\left(\frac{U\rho_2 R}{\eta}, \frac{U\rho_1 a}{\eta}, \frac{a}{R}\right) \tag{10.1}$$

It is easy to recognize the first two terms in Eq. (10.1) as water droplet and dust particle Reynolds numbers. In viscous flow, the term $\frac{U\rho_2 R}{\eta}$ can be discarded, hence:

$$E = f\left(\frac{U\rho_1 a}{\eta}, \frac{a}{R}\right) \tag{10.2}$$

E can thus be expressed as a function of k, where:

$$k = \frac{U\rho_1 a^2}{R\eta}, \frac{T}{R} \tag{10.3}$$

by multiplying the two dependent variables in Eq. (10.2).

Experimental data were plotted against k and are shown in Fig. 10.1.

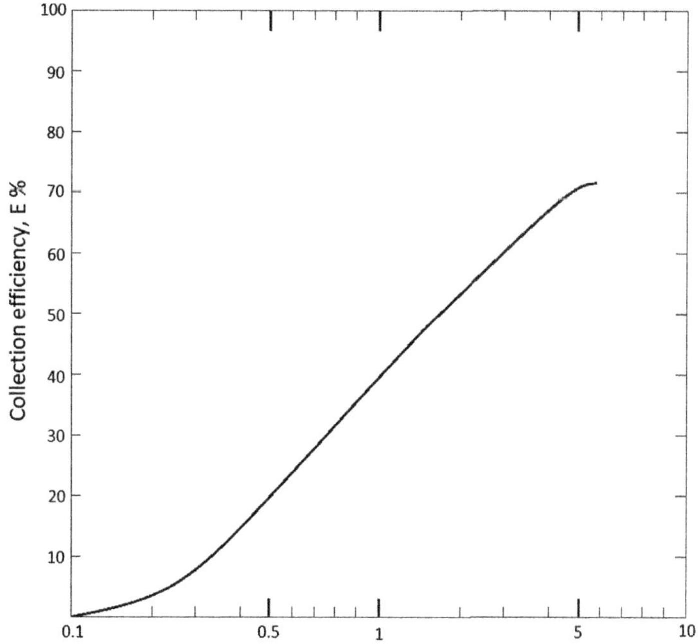

$$k = \frac{2U\rho_1 a^2}{9\eta R}$$

Figure 10.1 Collection efficiency as a function of k.

The relationship between E and k is nearly linear. Some of the theoretical conclusions that can be derived from Eq. (10.3) are as follows:

1. Efficiency of collection goes down rapidly with decrease in particle size.
2. Efficiency goes up with higher waterdrop velocity.
3. Efficiency improves with smaller water particle size.

The combination of the above items (2) and (3) suggests water discharged at higher pressures will have higher collection efficiencies. Table 10.1 shows some of their data [1] that confirm the above conclusions.

Other studies [2,3] have shown that decreasing the water droplet size excessively is not always beneficial. If a water droplet is too small, it may evaporate quickly and capture no dust. Cheng [2] calculated the optimum size of water droplets for various dust particle sizes and concluded that the maximum collection efficiencies for 1 and 2 μm particles is 100 and 200 μm, respectively. Woffinden [3] obtained slightly lower values for optimum droplet sizes. His results are shown in Fig. 10.2. His results indicate that the optimum droplet size for 1 μm dust particle is between 20 and 40 μm, depending on the water droplet velocity.

Table 10.1 provides a basis for designing an optimum spray system for all mining machinery in underground coal mines.

Table 10.1 Collection Efficiency for Different Dust Particle Sizes and Water Particles

Waterdrop Diameter (μm)	Coal Dust Diameter (μm)	E, Efficiency of Collection (%)
500	1	5
	2	24
	3	39
	5	57
	10	78
300	1	10
	2	34
	3	49
	5	66
	10	85
200	1	17
	2	41
	3	57
	5	73
	10	93

Dust Density = 1.37 (coal); Water Velocity = 98 ft/s.

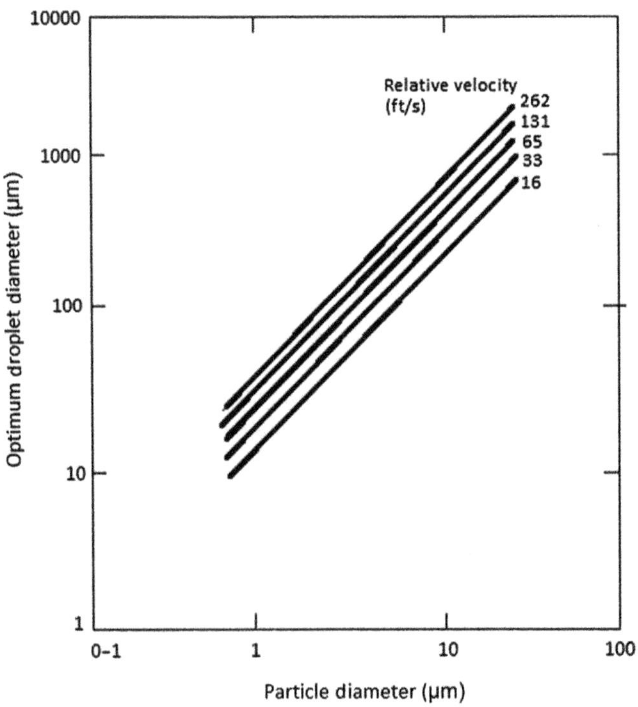

Figure 10.2 The optimum water droplet size for maximum collection efficiency.

10.2 Collection of Dust Particles by Filters

Most scrubbers use some kind of filter for the collection of dust particles after spraying the dust cloud with water. We can, therefore, discard any electrostatic forces acting on dust particles. This leaves the following four mechanisms for dust collection:

1. Inertia,
2. Interception,
3. Sedimentation, and
4. Diffusion.

Fig. 10.3 shows an approximate model of a fibrous filter.

The spacing between fibers is 2h, and the diameter of the fiber strand is 2R. The inlet velocity is U_0. Fuchs [4] gives the efficiency of inertial deposit as $E_i = 1 - \bar{e}^{\lambda_i}$ where:

$$\lambda_i = \frac{\tau\, U_o\, H}{2\sqrt{3}\, h^2} \tag{10.4}$$

where H is the total thickness of the filter bed, and τ is the relaxation time defined earlier (Chapter 8).

Similarly, efficiency of deposit by interception is given by Eq. (10.5).

$$E_h = 1 - e^{-\lambda_h}$$

Where $$\lambda_h = \frac{3\, r^2\, H}{2\sqrt{3}\, h^2\, (R+h)} \tag{10.5}$$

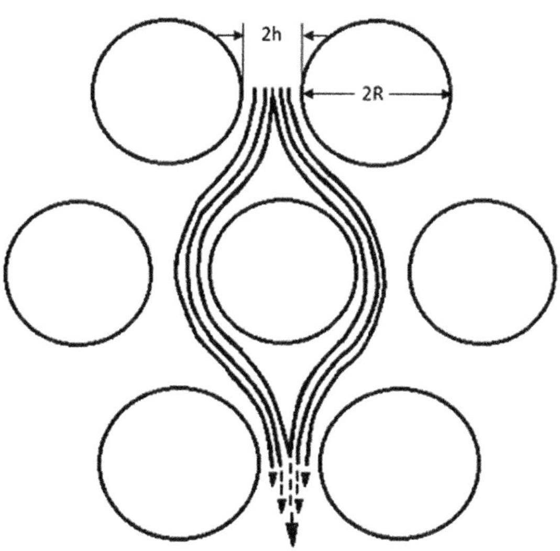

Figure 10.3 Model of a fibrous filter.

Efficiency of collection by sedimentation is given by Eq. (10.6) for one row of filter.

$$E_S = \frac{g\tau}{U_0\left(1+\frac{h}{R}\right)} \tag{10.6}$$

where U_0 is the velocity of air entering the filter.

Finally, the efficiency of collection by diffusion is given by Eq. (10.7).

$$E_D = 2.88\left(\frac{DR}{\overline{U}h}\right)^{\frac{2}{3}} \tag{10.7}$$

The efficiency of a model filter for various dust particle sizes is given in Table 10.2 [4].

Experimental observations confirm that the net efficiency of all filters is usually greater than the efficiency for any one mechanism but is less than their arithmetic sum. It is also clear that the filter efficiency must tend to 1 (100% collection) for very large and very small particles. The curve of efficiency, E, against particle radius, r, therefore must have a minimum for some value of $r = r_{min}$, where r_{min} decreases as the flow velocity increases. A r_{min} of 0.15 μm appears to be the most common experimental value [5,6].

10.3 Dust Control in Continuous Miner Section

A continuous miner is the machine used to drive tunnels/roadways in a mine to create mains, gate roads, and longwall panels. All four techniques to control respirable dust (discussed earlier) have been used, but premining infusion of the coal seam is not commonly done because of high cost. Most common dust control techniques comprise use of water sprays to suppress dust, use of water scrubbers in most cases, and adequate ventilation to dilute the respirable dust concentrations below statutory limits.

Table 10.2 Efficiency of a Model Filter (%)

	U_0 = 1 cm/s				U_0 = 20 cm/s			
Particle Radius (μm)	E_i	E_h	E_S	E_D	E_i	E_h	E_S	E_D
0.01	0	0	0	1	0	0	0	1
0.1	0.03	0.02	0.01	0.95	0.43	0.02	0	0.33
0.3	0.17	0.16	0.06	0.67	0.97	0.16	0	0.14
1.0	0.8	0.86	0.41	0.35	1	0.86	0.03	0.06
3.0	1	1	1	0.17	1	1	0.2	0.03

Assumptions: Fiber radius = 10 μm; Filter thickness = 0.2 cm; h/R = 0.7.

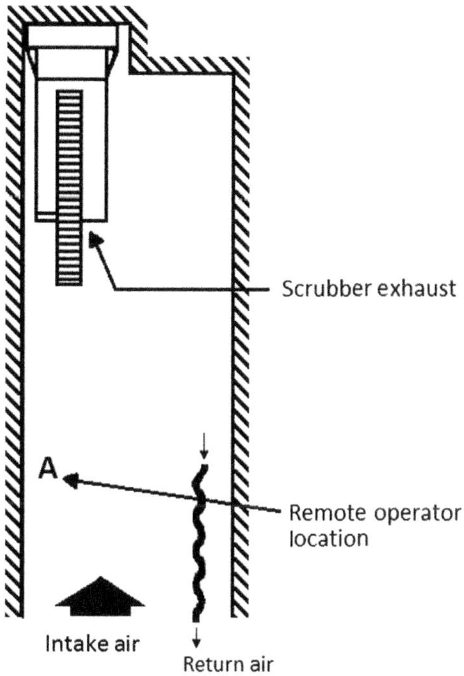

Figure 10.4 Continuous miner section with exhaust ventilation.

Fig. 10.4 shows a typical layout where a continuous miner section is ventilated by the exhaust system and a scrubber is used to collect dust at the source.

10.3.1 Water Spray Systems

A typical continuous miner is equipped with a set of sprays above and below the cutting drum as well as one spray behind some cutting bits. Each spray head delivers 0.5 to 1 gpm of water at 100 to 150 psi. Total water consumption ranges from 30 to 60 gpm. Besides suppressing dust, the sprays also wet the coal, keep the cutting bits cool, and serve as a fire suppression system (in case the cutting bit ignites a methane–air mixture). A well-designed spray system should have the right kind of sprays (a solid cone spray works best) that operates at optimum pressures and consumes the least amount of water. Such a system can reduce the dust concentration by 60%. Typical reductions range from 30% to 50% mainly owing to poor selection of sprays, too low pressures, and bad locations for sprays. Ideally, there should be a spray behind each cutting bit delivering 0.5 to 1 gpm of water at 100 to 150 psi. Such machines are known as wet-head continuous miners, and they achieve the best dust control.

Several attempts were made in the United Kingdom and Germany to equip mining machines with very high pressure water jets that not only suppressed dust but also assisted in cutting hard coals. Such machines are expensive and as such found limited applications. However, they reduced the specific energy of cutting coal (kwh/ton) considerably and also provided effective dust control.

10.3.2 Water Scrubbers

Fig. 10.5 shows the construction details of a typical machine-mounted scrubber. A fan sucks air from inlets under and near the cutting drum. It goes through the filter panel that collects most of the airborne dust. The thickness of the filter bed controls the filter efficiency.

Original scrubbers had 40 layers of stainless steel mesh knit from 85 μm wires, but today thinner filters containing 10 to 30 layers of wire mesh are available. The thinner filter allows more air to be drawn by the fan, improving the efficiency of dust control. Colinet [7] reported an efficiency of 90% for a 30-layer panel, but the efficiency dropped considerably when the 10 and 20 layer filters were tried. The air flow was 8500 CFM. It has to be in proportion to the total intake air to the faces; usually the scrubber capacity is only about 60% of the intake air. This prevents recirculation of scrubber exhaust.

10.3.3 Ventilation Air

The dust that could not be suppressed by water sprays or collected by scrubbers must be diluted by ventilation air. Federal law requires a minimum air velocity of 60 ft/min in the face area, but the average velocity is usually twice as high. The ventilation air is also needed to dilute methane concentrations in the face area to less than 1%. In moderately gassy and highly gassy mines, 15,000 to 20,000 CFM is needed. This is sufficient to dilute the respirable dust concentrations to less than 1.5 mg/m^3, which is the legal US requirement. Best results are obtained when line curtains are kept within 10 ft from the face, and ventilation tubings are kept within 15 ft from the face. When face

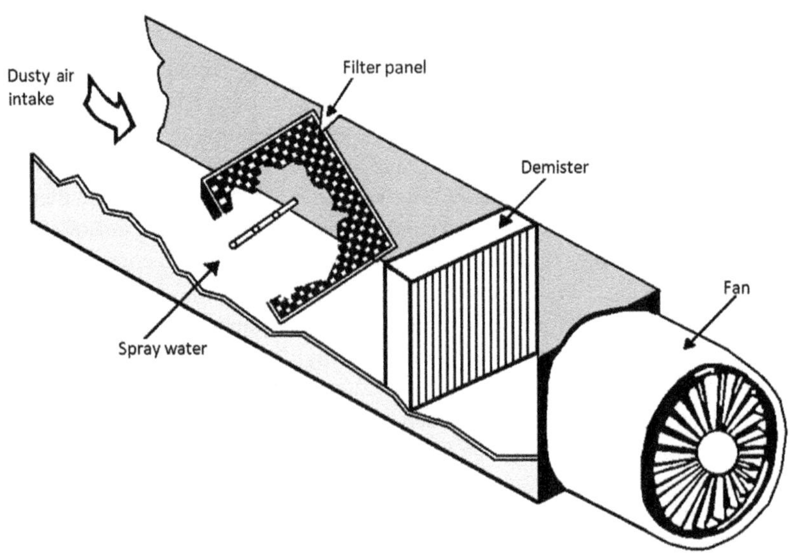

Figure 10.5 A typical machine-mounted scrubber.

ventilation is done with auxiliary fans and tubing, it can be either exhaust or blowing. Sometimes, a combination of both is used for optimum results.

10.3.4 Remote Operation of a Continuous Miner

The last resort for dust control is to operate the continuous miner remotely. With exhaust ventilation as shown in Fig. 10.3, the operator can stay in the intake air and be exposed to less than 1 mg/m^3 of respirable dust. Goodman and Listak [8] measured a reduction of 80% in dust exposure compared with the dust exposure at the mining machine by placing the operator about 50 ft away from the mining machine.

10.4 Dust Control in Longwall Faces

The sources of dust on longwall faces are the intake air, the stage loaders and crushers, the shearers, and the shields. The intake air is a minor source of dust unless the belt air is used on the face. Confinement and water sprays usually at less than 70 psi pressure can generally keep the contribution of stage loader dust to a minimum. The dust created by shield movement is also minimized by mounting water sprays on the shields. The biggest source of dust on longwall faces is the shearer. Table 10.3 gives a summary of data obtained from 13 US longwall faces [9].

Shield movement is always behind the shearer. It comes into consideration when the shearer is cutting from head gate to tail gate and shields are advanced upstream from the shearer operators.

10.4.1 Water Infusion

Excellent dust control can be achieved by drilling the longwall panel horizontally for degasification (to be discussed later in this book) and using the horizontal boreholes for water infusion ahead of mining. Slow infusion of water at 70 to 300 psi has given excellent results, often reducing the dust generation by 80% [10]. Larger and sharper cutting bits also reduce dust generation by reducing the specific energy of cutting coal. Use of very high pressure water (5000 to 10,000 psi) to assist coal cutting has also

Table 10.3 Dust Sources on Longwall Faces

Source	Average (%)	Average Dust Concentration (mg/m^3)
Intake air	9	0.33
Stage loader and crusher	15	0.78
Shields	23	1.80
Shearer	53	3.50

significantly reduced dust generation. It is feasible to modify the shearer and install a high-pressure pump in it to accomplish it. The feed water needs to be very clean.

10.4.2 Water Sprays

Well-designed water sprays are the next best way to control dust on longwall faces. A minimum of 100 gpm at 100 psi is recommended [11]. Ideally, there should be a spray behind each cutting bit, but in general, the larger the number of sprays, the better the dust control. The location of spray is also important. When they are arranged in a manner that directs the dust cloud away from the operator and holds it against the coal face, it can reduce the inhaled respirable dust by 30%–50%. Fig. 10.6 shows a typical layout that is called the shearer-clearer spray system. It is commonly used on all US longwall mining faces.

10.4.3 Scrubbers on Longwall Shearer

Although highly desirable, no successful design has been developed yet owing to height limitations. Efforts to design a longwall scrubber should continue.

10.4.4 Ventilation Air

Longwall ventilation air quantity is more often dictated by gas emissions on the face. It is generally more than adequate to dilute dust concentrations at the operator's station below statutory limits. Federal law requires a velocity of air at 400 ft/min at the tail gate. Assuming a cross section of the longwall face at 65 ft^2 (10′ wide × 6.5′ high), the minimum ventilation air required is 26,000 CFM. Except for a mildly gassy coal seam, it is totally inadequate. Actual ventilation air quantity ranges from 35,000 to 70,000 CFM depending on the gassiness of the coal seam and degree of degasification achieved [12].

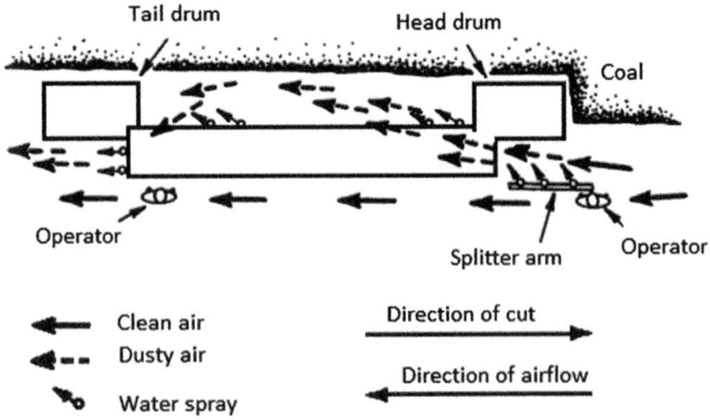

Fig. 10.6 The shearer-clearer system on a longwall face.

10.4.5 Remote Operation

This has become feasible and is an excellent tool in reducing longwall shearer operator's dust exposure. Remote operation allows the shearer operator to stay in the intake air away from much higher dust concentrations downstream from the shearer.

10.5 Dust Control for Roof Bolters

Roof bolters are an essential part of mining operation. It is a drilling rig that can drill 5–10-ft-deep boreholes in the roof for the installation of a steel bolt that strengthens the roof and prevents roof fall. Because it drills into the roof that is mostly shale or sandstone, it creates a dust cloud that is high in silica concentration. Control of the dust at this location is very important.

10.5.1 Wet Drilling

Drilling the roof with a small quantity of water reduces dust generation considerably. As high as 95% reduction has been obtained on certain drill rigs [13]. Typically, less than 1 gpm of water (often mixed with a surfactant) is used for this purpose. The collection efficiency appears to increase as the water flow is increased from 0.25 to 1 gpm [9]. Besides keeping the dust down, wet drilling also prolongs the life of the drill bits by keeping them cool and hence sharp. It is highly recommended even if it may be difficult to do in some situations.

10.5.2 Dust Collection Systems

Almost all roof bolters use a dust collection system. It is mostly a dry collection system. An enclosure is created around the borehole being drilled, which is hooked to a fan and a filtration system. The negative pressure created by the fan helps contain the dust within the enclosure. The filtration system collects most of the dust, and clean air is exhausted to the mine atmosphere.

10.6 Personal Protective Equipment

In many countries, approved respiratory protection equipment is required for those working in an area where the respirable dust concentration exceeds the statutory limit. They are basically a filtration system that cleans the air to be inhaled or they provide filtered air with a positive pressure to the mine workers.

10.6.1 Replaceable Filter Respirator

A respirator consists of a mask that fits a miner's face and forms a seal against the dusty air. Dirty air is drawn through replaceable filters than can collect 80%–98%

of all dust in air. The filter-holding unit is made of plastic but can be made of metal or hard rubber. Although the respirator does an excellent job of dust control, some personal discomfort may arise owing to breathing resistance and facial irritation caused by the face seal. It also interferes with normal voice conversation, as well as eyeglasses and goggles. To eliminate these difficulties, a much lighter single-use respirator is often used. In this case, the whole mask is made of filter material, and it covers the entire nose and mouth area of the wearer. However, it does not form a perfect seal against the dusty air nor is it as efficient as the replaceable, heavier filters.

10.6.2 Air Helmets

The air helmet is a special hard hat (somewhat bigger than the conventional miner's hard hat) that contains a battery-powered fan, a filtration system for dust, and a face visor. It provides protection for head, eyes, and lungs in a single unit. Although the air helmet is much more expensive than mask filters and weighs about three pounds, it is finding greater acceptance in US coal mines.

In the air helmet, dirty air is drawn through a filtration system by a battery-powered fan, and the clean air is directed to the full face visor over the wearer's face. The air exits the wearer's face at the lower end of the visor. The battery for the fan is carried by the wearer on the belt.

If the air velocity is not high, the air helmet performs as well as the replaceable filter mask with a collection efficiency of up to 84% for respirable dust, but at higher air velocity, the collection efficiency appears to go down to about 50% [14]. This contradicts the theory of dust collection, but it may be due to excessive turbulence that can recirculate some collected dust in the visor. Although personal protection devices can be effective in reducing dust exposure, they are vulnerable to human errors in their cleaning, maintenance, fitting, and use. They should not be used for an extended period of time, such as the entire 8-h working shift.

10.7 Optimization of Water Sprays

Water sprays come in many designs to meet specific requirements. It is a good idea to enlist the help of a commercial spray manufacturing company for an optimal design. The main types of sprays are as follows:

1. Hollow cone,
2. Full cone,
3. Flat spray,
4. Solid stream, and
5. Air-atomized.

Each type of spray head had several variations to meet specific needs. The type of spray must be matched with the mining operations for best results.

10.7.1 Drop Size Created by Sprays

Table 10.4 shows how different spray patterns can create different waterdrop sizes for given pressure and flow rates.

In mining application, full cone sprays are preferred, but the flow rate is limited to 1 gpm and pressure maintained above 100 psi. It creates water droplets smaller than 500 μm. Most spray nozzles have a mathematical relationship between flow rate and pressure at the inlet as shown in Eq. (10.8).

$$\frac{Q_1}{Q_2} = \left(\frac{P_1}{P_2}\right)^n \tag{10.8}$$

where Q is the flow rate, in gpm. P is the pressure, in psi. n is a constant with a value of 0.44–0.5 depending on the type of nozzle.

10.8 Use of Surfactant to Improve Dust Control

Thakur [15] carried out extensive research on the ability of surfactants to improve dust control. His main finding was that with the right kind of surfactant (with respect to the type of coal), an improvement of 15% in dust control can be achieved. It is necessary, therefore, to do a laboratory experiment to find the right type of surfactant. Almost all coal dust is hydrophobic—difficult to wet. Surfactants (a short form of surface active agent) are compounds that lower the surface tension (or interfacial tension) between two liquids or a liquid and a solid. They may act as detergents, wetting agents, emulsifiers, foaming agents, and dispersants. Surfactants can be hydrophobic (nonwetting) or hydrophilic (wetting agents). The latter is what is used in coal mines.

Many important surfactants include a polyether chain terminating in a highly polar anionic group. The polyether groups often comprise ethoxylated (polyethylene

Table 10.4 Spray Pattern Type at Various Pressures and Flow Rates

Spray Type	Pressure, 40 psi		Pressure, 100 psi	
	Flow Rate (gpm)	Mean Particle Size (μm)	Flow Rate (gpm)	Particle Size (μm)
Air atomizing	8	200	12	400
Fine spray	0.43	330	0.69	290
Hollow cone	24	1900	38	1260
Flat fan	10	2500	15.8	1400
Full cone	23	2800	35	1720

oxide–like) sequences inserted to increase hydrophilic character of surfactant. All surfactants can be broadly classified as follows:

1. Anionic, such as, sulfates and phosphatesters.
2. Cationic, such as primary and secondary amines.
3. Zwitterionic, such as phospholipids.
4. Nonionic, such as fatty alcohol ethoxylates.

Mostly nonionic surfactants are used in coal mines. The details of Thakur's study [15] are presented in Table 10.5.

Another finding of this study was that surfactants helped create smaller size water droplets in the 200 μm range. This may also have contributed to a better dust collection efficiency.

10.9 Electrostatic Charging of Water Particles for Improved Dust Collection

It was shown in Section 10.1 that dust collection efficiency can be increased by reducing the water droplet size, and increasing the droplet velocity. This is generally achieved by increasing the pressure at the spray inlet. In an experiment, Taylor and Thakur [16] gradually increased the water pressure to 2000 to 4000 psi while adding 0.1% of ALFONIC 10–50 (the best of the bunch in Section 10.8) to water. This created a positive electrostatic charge on the water particles. Most dust particles carry a small negative charge. This improved the collection even further in general but particularly in the under 2 μm range. The dust particles appear to adhere to larger water particles and fall out of air.

10.9.1 Experimental Procedure

Laboratory tests to determine the effects of surfactants and high pressure sprays were conducted on two occasions. During the first series of tests, the reduction in the respirable dust levels and the charge enhancement of the water spray were investigated. The experimental setup used to determine the dust reduction is shown in Fig. 10.7.

An Andersen Cascade Impactor was used to gravimetrically measure the respirable dust concentration in five stages. The sprays were operated at 2900 psi and a flow rate of 0.75 gpm. An intensifying pump was used to supply the high pressure water. The experimental setup shown in Fig. 10.8 was used to determine the charge on the water spray in both series of tests.

In the second series of tests, a small, triplex pump was used in place of the larger, intensifier pump. The original nozzle was also replaced with a higher volume nozzle. The water pressure and flow rates tested varied from 1000 psi and 0.87 gpm to 4000 psi and 1.65 gpm.

The screen shown is a 6.03 ft^2 steel screen, suspended with rubber straps to prevent grounding. A 1000 pF capacitor was used to permit a charge build-up for recording

Table 10.5 Dust Collection Efficiency With Various Surfactants

No.	Surfactant	Initial Dust Concentration (mg/m^3)	Percent Reduction	
			Pure Water	Surfactant (0.1%)
1	ALFONIC 1012-60	13.4	65.0	75.0
2	ALFONIC 1012-60	12.3	63.0	73.0
3	ALFONIC 1012-60	8.0	65.0	62.0
4	ALFONIC 610-50	10.0	67.0	63.0
5	ALFONIC 1214-70	11.6	66.0	65.0
6	ALFONIC 1214-70, at 50° F	16.1	66.0	69.0
7	ALFONIC 1214-70, at 78° F	16.1	66.0	64.0
8	Wen-Don Dust	15.3	65.0	64.0
9	Exxon Correxit 170	15.3	65.0	65.0
10	Sherox[a] Variquat 638	16.2	57.0	61.0
11	Sherox[a] Variquat 66	16.2	57.0	60.0
12	Sherox[a] Variquat CE 100	21.2	59.0	60.0
13	Conoco AXS	41.0	81.5	75.5
14	Conoco C-550-LAS	46.5	77.5	87.0
15	Conoco C-650-LAS	36.5	79.0	83.5
16	ALFONIC 810-40	42.5	83.5	90.0
17	ALFONIC 10-40	48.5	80.0	90.0
18	ALFONIC 6-50	45.5	64.0	62.5
19	ALFONIC 8-50	43.5	72.5	83.5
20	ALFONIC 810-50	47.3	82.5	90.5
21	ALFONIC 10-50	37.0	71.0	86.0
22	ALFONIC 12-50	34.0	81.0	89.5
23	ALFONIC 14-50	48.0	84.0	90.5
24	ALFONIC 10-55	43.5	80.0	90.0
25	ALFONIC 10-60	45.5	48.0	59.5
26	ALFONIC 1012-60	49.0	87.0	91.5
27	ALFONIC 10-70	40.0	72.0	76.5
28	ALFONIC 10-80	48.0	75.5	80.5

ALFONIC brands are alcohol ethoxylates. The two numbers following the name show the number of carbon in the alcohol and the percent ethoxylation respectively. The maximum improvement in dust collection efficiency of 15% is achieved by ALFONIC 10−50. The cationic surfactants show no improvement over the nonionic surfactants.
[a]Cationic.

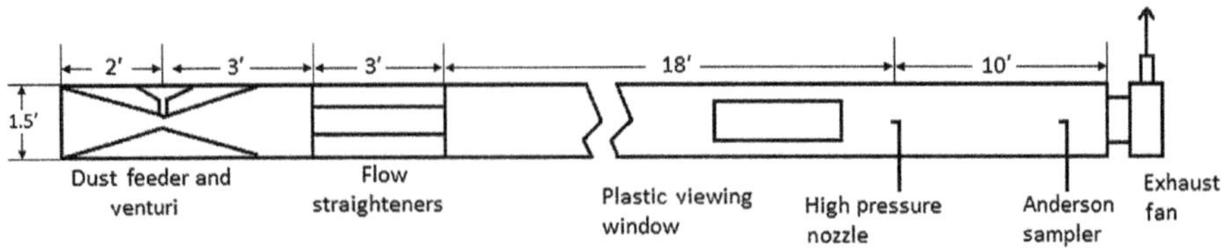

Figure 10.7 The dust suppression test facility.

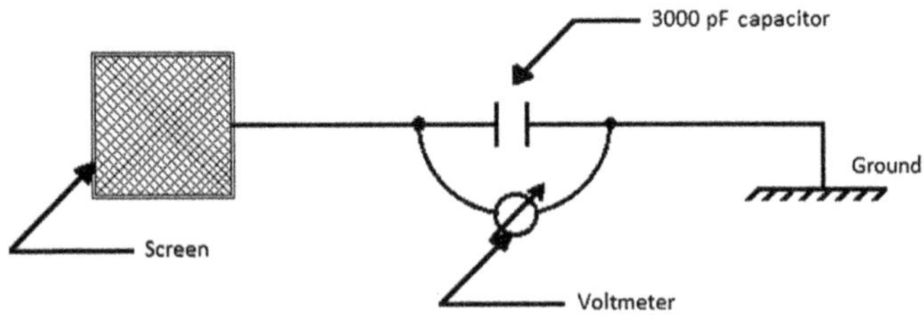

Figure 10.8 Electrostatic charge measurement device.

purposes. A Simpson voltmeter with an input impedance of 10 MΩ was used to measure the DC voltage. ALFONIC 10–50 surfactant which performed best in previous testing was used in all of the surfactant tests at a 0.1% by volume concentration. This surfactant is a nonionic, ethoxylate material produced by reacting ethylene oxide with linear alcohol blends.

10.9.2 *Experimental Results*

A comparison of the dust concentrations measured with and without the surfactant is illustrated in Fig. 10.9. In each size range, the surfactant solution reduced the dust concentration more than the water alone.

The reduction for each size range is shown in Fig. 10.10, which also shows that the reduction was greatest in the size range less than 4.6 μm. In particular, for the 1.8 μm diameter particles, the surfactant increased the reduction from 52% to 80%.

The results of tests to determine the charge on the water sprays are shown in Fig. 10.11.

In Series 1, tests of water only and of the water-surfactant solution were conducted at 2900 psi. The electrostatic foggers were operated with an air pressure of 95 psi and a water supply of 0.05 gpm at 85 psi. These results indicate that the high pressure spray consisting of the water-surfactant solution created a voltage comparable to that produced by electrostatic foggers. The results also show a significant increase in the voltage produced in the second series of tests compared to those in the first series. This increase occurred during tests of water and with the water-surfactant solution. The technique for measuring the voltage was the same for both tests, but a different pump and nozzle were used for the second series of tests. The difference in nozzle design (and a corresponding difference in flow rate) would appear to be the only factor

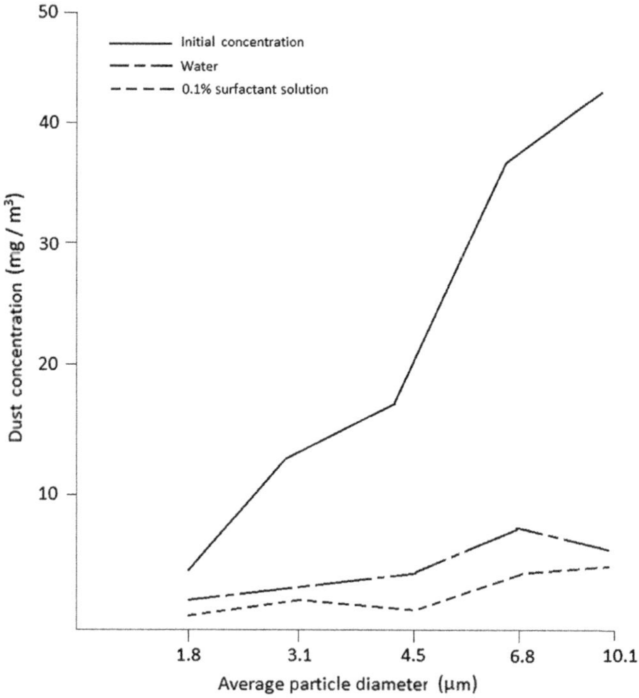

Figure 10.9 Respirable dust control with water and surfactant-water solution.

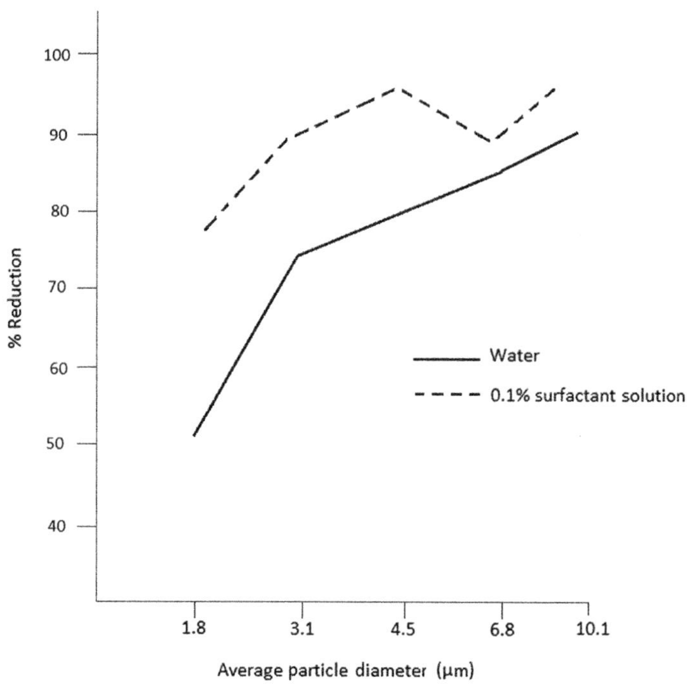

Figure 10.10 Respirable dust reduction with electrostatic charging.

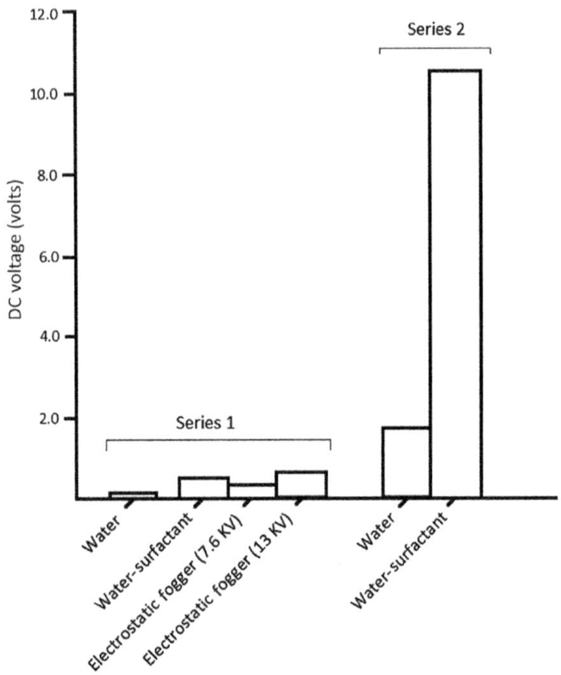

Figure 10.11 Comparison of voltage induced in water sprays.

that would make such a significant difference in the charge. Results of the second series of tests are shown in Fig. 10.12. DC voltage created increased linearly with water pressure from 1000 to 3000 psi.

Increasing the pressure from 3000 psi to 4000 psi resulted in only a slight increase in the voltage. The voltage, measured by placing the screen in the water spray approximately four feet from the nozzle, was higher for the surfactant solution than pure water in every test conducted.

The results of these tests indicate that the use of ALFONIC 10−50 surfactant increases the voltage present on a high pressure spray of the surfactant-water solution. A series of tests conducted to determine the improvement in the dust collection efficiency by adding a surfactant to high pressure sprays also showed a significant improvement. The most pronounced increase in the collection efficiency was found in the smallest size range (−1.8 μm). The collection of particles in this size range is of particular interest since particles in this size range cause the most lung damage.

The experimentation conducted shows a novel technology for suppressing respirable coal dust. The experimental studies did not identify the actual reason for the induced voltage, but the interacting effects of high pressure and the surfactant additive apparently create the electrically charged water particles. They create an electrical attraction for the dust particles, thereby causing agglomeration and consequent increase in collection efficiency. It is assumed that the surfactant, by reducing the surface tension of the water, allows the water to break up into very small droplets at the nozzle

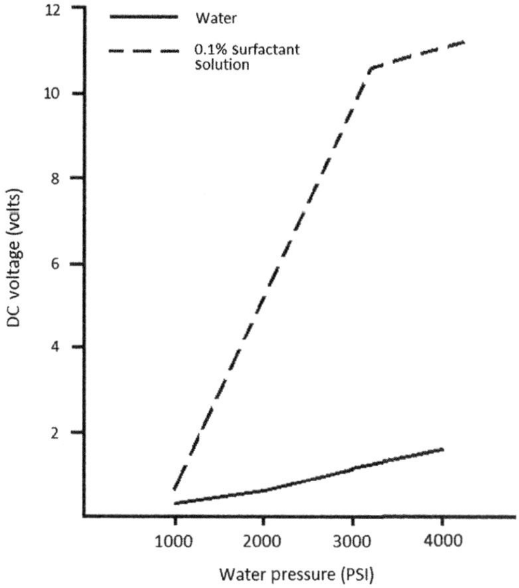

Figure 10.12 Voltage induced with and without surfactant.

walls. A charge is then generated by the friction between the water and the nozzle wall and by viscous interaction between the water and air near the nozzle exit without the use of compressed air or any electrical power supply. This feature may enable the system to be adapted for underground use without additional equipment, such as, a compressor.

References

[1] Walton WH, Woolcock A. The suppression of airborne dust by water.. Aerodynamic Capture of Particles, Pergamon Press; 1960. p. 129–52.
[2] Cheng L. Collection of airborne dust by water sprays. Industrial Engineering Chemistry Process 1973;12(3).
[3] Woffinden G. Effects of surface tension on particle removal. US EPA – 600/7-79-0440. 1977.
[4] Fuchs A. The mechanics of aerosols. Pergamon Press; 1964. p. 213–33.
[5] Washington (DC) Handbook of aerosols. 1950.
[6] LaMer V. Air pollution, New York. 1952. p. 607.
[7] Colinet JF, et al. Silica collection concerns when using flooded-bed scrubbers. Mining Engineering 2000;52(4):49–54.
[8] Goodman GVR, Listak JM. Variations in dust levels with continuous miner position. Mining Engineering 1999;51(2):53–8.
[9] Kissell FN. Handbook of dust control in mining. NIOSH, IC 9465; 2003.
[10] Stricklin JH. Longwall dust control at Jim walter resources. In: The 3rd U.S. Mine Ventilation Symposium, University Park, PA; 1987.

[11] Colinet JF, et al. Status of dust control technology on U.S. longwalls. In: Proceedings of the 6th International Mine Ventilation Congress, Pittsburgh, PA; 1997. p. 345–51.
[12] Thakur PC, Zachwieja J. Methane control and ventilation for 1000 feet-wide longwall faces. In: Proceedings of Longwall USA International Conference and Exhibitions; 2001. p. 167–80.
[13] MSA Research Corporation. Control of dust in non-coal mines and ore processing mills. In: USBM Contract H0220030, NTIS PB 240646, Washington, DC; 1974.
[14] Cecala AB, et al. Protection factors of the airstream helmets. Avondale (MD): USBM Report; 1981.
[15] Thakur PC, Riester JB. The collection of airborne dust with surfactant-water solution, Internal reports of CONOCO R&D. 1983. 16 pp.
[16] Taylor LD, Thakur PC. Electrostatically-charged water sprays improve respirable dust control. In: Presented to SME-AIME Annual meeting, New Orleans, USA, 1986, Published in SME health and safety in coal mining; 1986. p. 227–31.

Diesel Exhaust Control

11

Chapter Outline

During the past 50 years, the US coal industry and particularly underground coal mining have gone through some major changes. Some of these changes are obvious by now, but others are in the making. They are as follows:

1. Restructuring of the mines (fewer but larger coal mines).
2. A productivity of 70–80 tons/manshift versus 20–30 tons/manshift.
3. Longwall mining versus room and pillar mining for improved safety and productivity.
4. Three-entry longwall developments versus four- to six-entry development sections.
5. Diesel personnel and material transport versus trolley wire equipment.

Advanced Mine Ventilation. https://doi.org/10.1016/B978-0-08-100457-9.00011-0

6. Belts for coal transport versus mine cars and locomotives.
7. Mine slopes versus vertical hoists.
8. Trained and skilled personnel versus on-the-job training.
9. Zero accident goals versus fewer accidents than the previous year.

The use of diesel equipment in coal mines in the past was questioned based on the suspicion that exposure to diesel exhaust may affect the health of miners. Diesel engines were introduced in US underground coal mines nearly 50 years ago, and their numbers have steadily increased to approximately 5000 at present. Safety and productivity advantages are the driving forces for the popularity of diesel engines. Diesel equipment improves safety by eliminating shock, fire, and explosion hazards from trolley wires and electrical cable equipment.

An additional safety advantage of diesel equipment use is improved ventilation in the face area. Currently, the belt entry is isolated from intake air in many mines, and the trolley wire entry is regulated for air velocity at 250 ft per minute. The latter restriction is imposed to safeguard against rapid spreading of fire in the trolley wire entry, but all such restrictions can be eliminated with the use of diesel equipment, and air quantities at the face can be maximized to improve safety and health of miners.

11.1 Health Hazards of Diesel Particulate Matter

Diesel exhaust contains some substances that can be potentially harmful to human health at high exposure levels. However, the toxicological effects of any substance are functions of the dose and duration of exposure. For example, carbon monoxide is a deadly substance in very high concentrations. The American Conference of Governmental Industrial Hygienists (ACGIH) has set a threshold limit value for it at 50 ppm. This means it can be inhaled at this concentration for 8 h a day, 5 days a week over the lifetime of a worker without harmful effects. Similarly, if all components of the diesel exhaust are diluted to their respective threshold limit values (TLVs), diesel exhaust does not constitute a hazard to human health.

Table 11.1 shows the major components of diesel exhaust. TLVs for all gaseous components of diesel exhaust have been established by the ACGIH and are incorporated into Federal Mine Safety and Health Administration (MSHA) regulations. Such incorporations are done only after careful examination of their technical and economic feasibility. Many field studies confirm that meeting these TLVs for gaseous components of diesel exhaust has not been a problem in coal or other mines [1,2].

Table 11.1 Major Components of Diesel Engine Exhaust

Diesel Exhaust					
Carbon dioxide	Carbon monoxide	Nitrogen oxides	Unburnt hydrocarbons	Sulfur oxides	Particulate matter (diesel particulate matter)

The National Institute of Occupational Safety and Health (NIOSH) has identified diesel particulate matter (DPM) as a potential human carcinogen, but in-mine experience over the past 50 years does not provide any epidemiological evidence for such risks. A recent study in Australian coal mines, many of which are partially dieselized, determined that the standardized mortality ratio (SMR) for lung cancer in a large cohort of miners was only 78% of that in the general population [3]. A parallel study in highly dieselized German potash mines (with no confounders, such as, silica, radon, arsenic, etc.) found similar results for the incidence of lung cancer [4]. Thus, DPM at prevailing concentration levels does not appear to create any additional health risks. Apart from the lack of substantial epidemiological data, another difficulty in establishing any in-mine personal exposure limit (PEL) in the coal mines is the lack of instruments that can accurately measure the DPM concentrations. Previous experience with the coal mine dust PEL also dictates that any PEL for DPM must be based on gravimetric measurements.

In noncoal mines, the combustible fraction of respirable dust can provide a good measure of DPM, but this obviously will not work in mines where combustible minerals are being mined. The elemental carbon technique that differentiates between the fraction of elemental and organic carbons in coal dust and DPM cannot be utilized universally because neither the composition of the coal mine dust nor that of DPM is always the same. Many other techniques, such as carbon isotope ratio analysis, Raman spectroscopy, and electron spin resonance, also suffer from a similar shortcoming. The size cutoff-based instruments measure everything in mine air below a certain size (0.8 or 1.0 μm). This can be a useful instrument if there was a TLV for all submicron dust particles in the mine atmosphere but is useless if it is used to measure DPM only.

Table 11.2 shows the major components of DPM. It is generally believed that the polynuclear hydrocarbons attached to solid carbon particles are the potential carcinogens in DPM, but their TLVs are yet to be established. Under these circumstances, the most prudent option is to minimize the concentration of DPM in mine air using state-of-the-art technology and introduce diesel engines in all underground coal mines to improve safety by removing ignition, fire, explosion, and tripping hazards related to the use of trolley wire and other electrical equipment.

11.2 Diesel Particulate Matter Standards

In spite of the uncertainties discussed earlier, many countries have set a standard for ambient DPM. Table 11.3 shows the details.

Table 11.2 Major Components of Diesel Particulate Matter (DPM)

DPM			
Solid carbon	Liquid and solid hydrocarbons include polynuclear aromatic hydrocarbons	Sulfates	Moisture

Table 11.3 Current Diesel Particulate Matter (DPM) Standards

Country	DPM Level (mg/m^3)	Analytical Technique
Canada (metal mines)	0.75	Combustible respirable dust.
Germany	0.2	Total carbon by NIOSH 5040 method. Sample collected gravimetrically.
USA (metal mines)	0.16	Total carbon by NIOSH 5040 method. Sample collected gravimetrically.
USA (coal mines)	Limits DPM emission	Gravimetric measurement.
a. Light duty outby engines	<5 g/h	
b. Heavy duty outby engines	<2.5 g/h	
c. Heavy duty permissible engines	<2.5 g/h	
US States		
a. West Virginia	0.12	Integrated system and ventilation.
b. Pennsylvania	0.12	Integrated system and ventilation.

In other countries where they do not have a DPM standard yet, they specify a minimum quantity of ventilation air to dilute the diesel exhaust. Table 11.4 shows some of them.

It is, however, assumed that these ventilation quantities will not only dilute the gases in diesel exhaust to safe levels but will also dilute the DPM to a safe level. Better methods to control DPM will be discussed next.

Table 11.4 Ventilation Air for Diesel Engines

Country	Ventilation Requirement (CFM/BHP)
Australia	80–96
Canada (coal mines)	75–150
South Africa	95
United Kingdom (old)	140

BHP, per brake horsepower; *CFM*, cubic feet per minute.

11.3 Diesel Exhaust Control Strategy

Thakur [5] and Schnackenberg [6] advocated separately a common strategy to minimize diesel exhaust emissions. It consists of the following:

1. Reduce DPM generation.
   a. Clean engines.
   b. Clean fuel.
2. Collect/combust generated DPM.
   a. A catalytic converter.
   b. A soot filter on most engines.
3. Dilute DPM.
   Adequate ventilation for all approved equipment to dilute DPM to safe levels.
4. Monitor DPM emissions.
   Engine performance needs checked predeployment and postdeployment on a periodic basis.
5. Maintenance.
   Proper maintenance of diesel equipment is a must.
6. Training of mechanics and operators.
   The operators and mechanics of diesel equipment must be properly trained, preferably by the equipment manufacturers.

11.3.1 Clean Engines

The amount of DPM emitted in grams per brake horsepower (bhp) is known as "Specific DPM Emission" for the engine and it is a measure of clean engines. Only 20 years back most permissible and nonpermissible diesel engines had very high DPM emission.

Table 11.5 shows the specific emissions of some old and approved diesel engines.

The emissions are, obviously, very high. It required filters with 95% or better efficiency to get an approval for mine use.

Fortunately, in the last 10 years, many diesel engine manufacturers have adopted (1) turbocharging and (2) electronic ignitions that lowered the DPM emissions from 4 to 5 g/h for a typical 100 hp engine (a specific emission of less than 0.04–0.05 g/bhp-h).

Each series of engines is tested by a certified laboratory in an 8-mode test according to ISO 8178, protocol, C1 for off-road vehicles. These 8-modes are as shown in Table 11.6 [7].

All gaseous and DPM emissions for a typical 8-mode test are reported as shown in Table 11.7. All data are nonspecific to protect privacy of the manufacturer.

11.3.2 Clean Fuel

Traditionally all diesel fuels used in coal mines complied with ASTM D975 defined standard for D_2 fuel. It had a sulfur content of 500 ppm (by weight), aromatic contents less than 35%, and a cetane number of 40–48. Because sulfate typically constitutes

Table 11.5 MSHA-Approved Diesel Engines [7]

Engine Type	HP	Specific Emission (g/bhp-h)
Nonpermissible Engines		
CAT 3306	150	0.261
CAT 3304	100	0.255
Deutz MWM 916-6	94	0.208
Deutz F 4L 1001F	59	0.187
Deutz F 3L 1011F	44	0.147
Permissible Engines		
CAT 3306	150	0.306
CAT 3304	100	0.297
Deutz MWM 916-6	94	0.271

MSHA, Mine Safety and Health Administration.

Table 11.6 8-Mode Test in ISO 8178

Mode	Engine Speed	Percent Load	Weight Factor
1	Rated r.p.m.	100%	0.15
2	Rated r.p.m.	75%	0.15
3	Rated r.p.m.	50%	0.15
4	Rated r.p.m.	10%	0.10
5	Intermediate[a]	100%	0.10
6	Intermediate	75%	0.10
7	Intermediate	50%	0.10
8	Idle	—	0.15

[a]This usually corresponds to the maximum torque. CO concentration in the exhaust in this mode is used to calculate the performance of diesel oxygen catalyst and activated soot filter.

50%–60% of the DPM by weight, reduction in sulfur content of fuel was paramount. Fortunately, now the ultralow sulfur fuel with only 15 ppm sulfur is generally available and is required for coal mine use in many states of the United States.

Many additives to further improve the quality of the old D_2 fuel were discovered. They are listed in Table 11.8 for reference only, but they may not be needed with ultralow sulfur fuels.

Table 11.7 8-Mode Test for a Typical Diesel Engine

Manufacturer: ABC Company
Serial Name: AB 123-45
H.P. 120 at 2200 r.p.m.
Test Engineer: John Doe
Fuel: Carbon, 87% (mass); H, 12.9%; S, 0.06%
Specific Gravity: 0.84
Air/Fuel Ratio (wt.): 14.46

Mode	1	2	3	4	5	6	7	8
Speed (r.p.m.)	2200	2200	2200	2200	1400	1400	1400	600
Torque (NM)	391	299	198	39	470	352	235	0.0
Corrected NO (ppm)	417	339	281	157	594	536	444	171
Corrected NO_2 (ppm)	6.3	21.6	21.8	30.4	14.1	6.2	12.6	19.2
Corrected CO (ppm)	61.5	50.7	66.3	175.5	104.7[a]	87	95.9	106.2
Corrected CO_2 (vol %)	6.62	5.26	4.40	2.40	6.19	7.39	5.91	1.36
Ventilation, CFM (NO based)	5167	4004	2798	1090	4259	3159	2271	266
Ventilation, CFM (NO_2 based)	389	1277	1082	1051	605	183	321	149
Ventilation, CFM (CO based)	380	299	329	607	374	266	245	82
Ventilation, CFM (CO_2 based)	4163	3106	2187	801	2934	2177	164	108

Maximum ventilation rate, CFM	5167
Maximum ventilation rate rounded, CFM	5500
CFM/hp	46
CO emission	0.5 g/bhp-h
NOX emission	3.96 g//bhp-h
HC emission	0.044 g/bhp-h
Particulate emission	0.07 g/bhp-h = 4.385 g/h (Average)
[b]Particulate index	2580 CFM
Particulate index rounded	3000 DFM or 25 CFM/hp

[a]This CO level is used to calculate the efficiency of CO reduction.
[b]Required ventilation to dilute DPM to 1 mg/m^3.

Table 11.8 Properties of Fuel Additives for Reducing DPM Emissions

Additive	Suggested Blend Ratio (Additive/Fuel)	Weight Percent Oxygen	Flash Point (°F)	DPM Reduction (Percent)
Diethylene glycol Methyl ether	10/90	40.0	189	25.3
Liquid DME 96% 1,2 dimethoxylethane 4% dimethoxyethane	11/89	36.5	71[a]	32.8
Triethylene glycol Dimethyl ether (triplyme)	11/89	36.0	185	12.1
Diethylene glycol Dimethyl ether (diglyme)	11/89	35.8	156	19.3
1,2-dimethoxy ethane (glyme)	11/89	35.6	78[a]	17.5
Methyl soyate	35/65	11.0	207	30.9

[a]These fuel additives have a flash point temperature lower than ASTM specification of 125°F.

11.3.3 Catalytic Converter

A well-designed catalytic converter placed in the exhaust system next to the engine can drastically reduce all gaseous emissions, including carbon monoxide, unburnt hydrocarbons, and soluble organic fractions (containing polynuclear aromatics). It is a compact, durable, reliable, and a low-cost component. It should, therefore, be an integral part of all diesel engines approved for underground coal mines. Table 11.9 shows the characteristics of a well-designed catalytic converter.

Table 11.9 Characteristics of a Well-Designed Catalytic Converter

CO	Reduced by 80%–95%
HC	Reduced by 85%–90%
DPM	Reduced by 25%–35%
Odor control	Very good
Influence on engine	Low-pressure drop; no fuel penalty
Reliability	Very good
Durability	≥5000 h

The core of a catalytic converter is an open-channel ceramic monolith or a metallic honeycomb substrate that provides support to the catalyst. Cordierite ($2MgO$-$2Al_2O_3$-$5SiO_2$) is the most popular ceramic material used for the substrate. It has many good characteristics, such as high surface area, large open frontal area, low heat capacity, low thermal expansion coefficient, and good mechanical strength [7].

Metallic substrates are made of metal, silica, iron, chromium, and aluminum alloys. They have higher surface areas and low-pressure drop but are more expensive. Both types of substrates are coated with catalysts formulated with noble metals, such as platinum, palladium, and rhodium. The catalyst application is a two-step process. The first step is a wash-coat of aluminum, silica, titanium, cerium, and other compounds. In the second step, the noble metal catalysts are impregnated on to the wash-coat.

The catalytic converter oxidizes all gases such as CO, NO, and HC to CO_2, NO_2, and CO_2 and water. The catalysts need to have a temperature above 300°C for efficient performance. This is the reason they are installed next to the engine. The catalytic converter usually lasts a long time. Five thousand hours of life is common when ultralow sulfur fuel is used. It is always designed to match the engine exhaust without creating too much back pressure. The current cost varies from \$5000 to \$10,000 depending on the size of the equipment. The surface of the catalytic converter is always coated with insulating material to keep the surface temperature below 302°F. For permissible engines, they are water-jacketed.

11.3.4 Diesel Particulate Filters

To collect the DPM and further oxidize the diesel exhaust components (in some cases), a diesel particulate filter is used on most diesel equipment except for very small engines working in outby areas that do not need it to meet legal requirements.

DPM filters can be broadly classified as (1) low-temperature filters and (2) high-temperature filters depending on the design.

The low-temperature filters are suitable for nonpermissible heavy-duty diesel equipment working outby in a mine or permissible equipment working in the face areas, such as shuttle cars, front-end loaders. The high-temperature filters have a universal appeal and it can be found on all diesel equipment except those that are permissible equipment. The latter can work safely even if the mine air contains 1% methane by volume. This is a legal limit for methane in most mines.

11.3.4.1 Low-Temperature Filters

These are basically designed to collect DPM and cool the exhaust to a safe level. Fig. 11.1 shows a schematic of a permissible filtration system. The hot exhaust from the engine goes to a water-jacketed catalytic converter. Next it goes through either a water bath where gases mix with water (as shown) or a water cooler without mixing with water that only cools the exhaust. The cooled gas goes through a water/flame trap before it enters the filtration housing. Mostly disposable paper or synthetic paper is used for DPM collection. The filter has to be replaced periodically to keep the intake air pressure generally below 15 in. of water gauge.

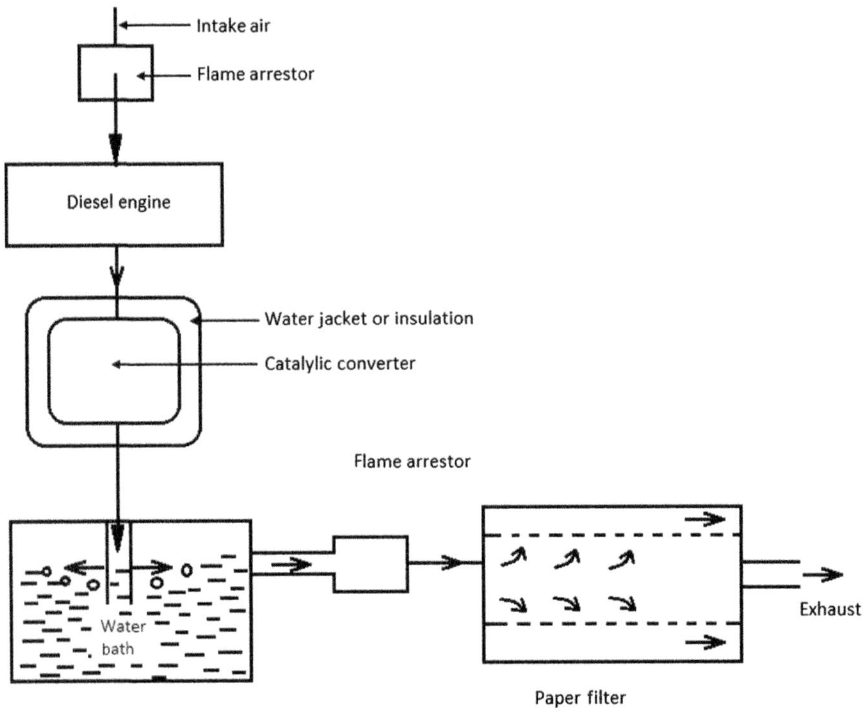

Figure 11.1 A schematic for a permissible filtration system.

The efficiency of the filtration system varies from 70% to 95% but most of them have 90% to 95% efficiency. Synthetic paper filter is thermally resistant and hence preferred but it is more expensive than the ordinary paper filter.

11.3.4.2 High-Temperature Filtration System

They are a bigger version of the catalytic converter but similar in principle. The exhaust gases pass through a filtration system made of a ceramic or silica body that are coated with catalytic metals such as platinum, rhodium, and palladium. They require that exhaust temperature be maintained above 350°C for a short duration each shift for proper operation. They are ideally suited for heavy-duty diesel equipment working out by but may not be suitable for light-duty equipment for they cannot raise the exhaust temperature high enough for the filter to regenerate. Gradual soot buildup in the filtration system often reduces the efficiency and increases the pressure drop across the unit. In most cases, it is desirable to remove the filtration system and regenerate (burn all soot) in an external regenerating kiln. A spare is needed for continuous operation of the equipment.

Table 11.10 compares the various performance characteristics and cost for the two main types of DPM filtration systems.

Table 11.10 Comparison of Low- and High-Temperature Diesel Particulate Matter Filters

Characteristics	Low-Temperature Filters	High-Temperature Filters
Heat exchanger	Yes	No
Cost	\$30–45,000	\$3–5000
Size	Very large	Small and compact
Maintenance needed	8 h	2000 h
Collection efficiency (in conjunction with a catalytic converter)	70%–95%	70%–90%

11.4 Diesel Exhaust Dilution

The ventilation air recommended by MSHA called the name plate ventilation (as shown in Table 11.7) is adequate for dilution of gases in the exhaust. By judicious selection of a catalytic converter and a filtration system, the DPM can also be diluted to less than 0.12 mg/m^3 with the name plate air. However, it is helpful to mathematically analyze diesel applications in the coal mines to confirm that such assumptions are correct. Thakur [8] did extensive work in this area. Only a summary is presented here. For details, the original work should be consulted.

There are four different applications of diesel equipment in coal mines as follows:

1. Single engine in a stationary mode.
2. Single engine moving in a straight airway.
3. Multiple engines (two to three) moving in a loop of straight airways.
4. Multiple engines (10–100 or so) working in a mine in different airways.

The last case was already analyzed in Chapter 5 of the book. The rest will be discussed here.

11.4.1 A Single Diesel Engine in a Single Roadway

Time-Independent Model:

The physical situation is represented by a stationary diesel engine (e.g., at loading or unloading points and spot cleanup situations) and schematically illustrated in Fig. 11.2.

The following assumptions are made for modeling:

1. Isotropic homogeneous turbulence, i.e., longitudinal coefficient of turbulent dispersions, (E_x) and transverse coefficient of turbulent dispersions (E_r) have same values.
2. Point source at the origin (0,0,0).
3. Concentration is being measured in the vicinity of the engine (i.e., no reflection from the walls is likely), and hence, the free space solution is applicable.
4. Density of air and coefficient of dispersions is constant.

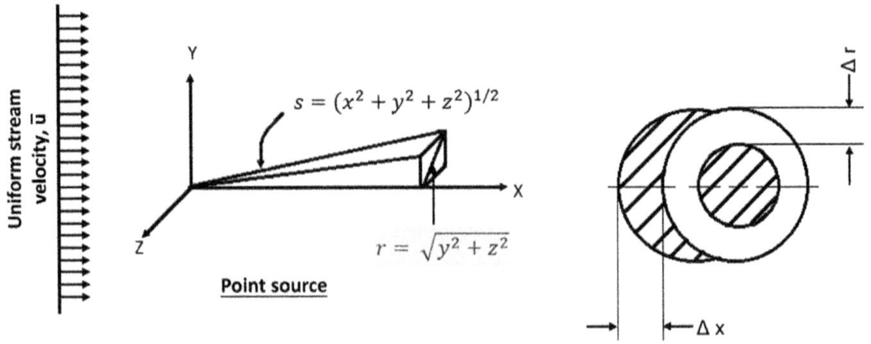

Figure 11.2 Diffusion from a point source in a uniform velocity stream.

The following differential equation is obtained by taking a mass balance over the cylindrical ring in Fig. 11.2.

$$\overline{u}\frac{\partial c}{\partial x} = E_r\left(\frac{1}{r}\frac{\partial}{\partial r}\left(r\frac{\partial c}{\partial r}\right) + \frac{\partial^2 c}{\partial x^2}\right) \tag{11.1}$$

where, $\overline{u}$ = average air velocity, E_r = the coefficient of transverse turbulent dispersion, $r = (Y^2 + Z^2)^{1/2}$, c = concentration of pollutant in the general body of air.

In Eq. (11.1), $\overline{u}\frac{\partial c}{\partial x}$ represents the dispersion of material by the velocity of air, $\overline{u}$, i.e., it is the convective term. $E_r\frac{\partial^2 c}{\partial x^2}$ gives turbulent dispersion of material in the x direction, whereas $\frac{E_r}{r}\frac{\partial}{\partial r}\left(r\frac{\partial c}{\partial r}\right)$ is an analogous term for dispersion in the radial direction. Eq. (11.1) is obtained by simply equating the input and output.

To solve Eq. (11.1), two boundary conditions are needed, which are obtained by considerations of the physical situations.

Boundary Condition 1: It is reasonable to assume that at an infinitely distant point, the concentration of diesel exhaust would be zero, i.e.,

$$c = 0 \text{ at } s = \infty \tag{11.2}$$

where, $s = (x^2 + y^2 + z^2)^{1/2}$.

Boundary Condition 2: It can also be assumed that at any cross section in the roadway, the concentration would be the highest at the center, i.e.,

$$\frac{\partial c}{\partial r} = 0 \text{ at } r = 0 \text{ for any } x \tag{11.3}$$

Finally, to relate the solution to the rate of emission of exhaust, one can consider conservation of mass at the source, i.e., at the origin. Mathematically, it is expressed as

$$-4\pi\, s^2 E_r \left(\frac{\partial c}{\partial s}\right)_x = q c_i \text{ at } s \rightarrow 0 \tag{11.4}$$

where, q, volume rate of engine exhaust discharge; c_i, concentration of pollutants in the engine exhaust.

The solution of Eq. (11.1) with the special boundary conditions has been derived elsewhere [9] and is given below:

$$C = \frac{qc_i}{4\pi E_r s}\exp\left(-\frac{\overline{u}(s-x)}{2E_r}\right) \tag{11.5}$$

By inspection, it can be easily seen that the concentration, c, is the maximum at $x = 0$, i.e., in the plane of the source itself. Hence, concentration at the diesel engine is given by

$$C|_{x=0} = \frac{qc_i}{4\pi E_r r}\exp\left(-\frac{\overline{u}r}{2E_r}\right) \tag{11.6}$$

Eq. (11.6) can be further simplified if it is assumed that the concentration is measured in a manner to obtain average concentration within a radius, a, from the engine. Designating this concentration as C_L, one obtains

$$C_L = \frac{qc_i}{4_\pi E_r} \frac{\int_0^a \frac{1}{r} e^{-\overline{u}r/2E_r} \cdot rdr}{\int_0^a rdr} \tag{11.7}$$

On integration, Eq. (11.7) reduces to

$$C_L = \frac{qc_i}{\pi \overline{u} a^2}\left(1 - e\frac{-\overline{u}a}{2E_r}\right) \tag{11.8}$$

For any finite value of $\frac{-\overline{u}a}{2E_r}$, the expression $e\frac{-\overline{u}a}{2E_r}$ is greater than zero. Hence, C_L is always smaller than $\frac{qc_i}{\pi\overline{u}a^2}$, which is the concentration predicted by the static dilution formula. Derivation of E_r will be discussed later in this chapter. E_r in coal mines is typically higher than smooth pipes, at approximately 0.8 m^2/s. Eq. (11.8) yields dilution air quantities that is 50%−100% of static dilution air quantities [8].

11.4.2 Time-Dependent Model

When a diesel engine is moving in a roadway with a velocity, v, the growth of concentration becomes time-dependent. If the velocity of air is u, the relative velocity would be $(v \pm u) = V_r$. The mass flow diagram is shown in Fig. 11.3.

Assuming $v > u$ and the plane $x = 0$ moving with the diesel engine, the equation of convection diffusion becomes

$$\frac{\partial c}{\partial t} + V_r \frac{\partial c}{\partial x} = E_x \frac{\partial^2 c}{\partial x^2} \tag{11.9}$$

where, E_x is the coefficient of longitudinal turbulent dispersion.

In Eq. (11.9), the term $\frac{\partial c}{\partial t}$ is the rate of growth of concentration in the differential element, whereas $V_r \frac{\partial c}{\partial x}$ is the net gain of material due to convective transfer. These two terms balance the total loss of material owing to turbulent dispersion, which is represented by $E_x \frac{\partial^2 c}{\partial x^2}$. To solve Eq. (11.9), three conditions:two boundaries, and one initial are needed.

Boundary Condition 1: This is obtained by assuming that mass is conserved at the origin, i.e., at $x = 0$. The total input of exhaust from the engine per unit area is $\left(\frac{qc_i}{F}\right)$ where F is the cross-sectional area. Net loss of material at $x = 0$ given by the algebraic sum of convection and diffusive terms. Mathematically,

$$V_r c \,|_{x=0} - E_x \frac{\partial c}{\partial x}\bigg|_{x=0} = \frac{qc_i}{F} \quad \text{for } t > 0 \tag{11.10}$$

where, q, volume rate of exhaust emission; c_i, concentration of species, i, in the exhaust.

Boundary Condition 2: It is reasonable to assume that at a point very far from the engine, the concentration of exhausts would be zero, i.e.,

$$c = 0 \text{ as } x \rightarrow \infty \quad \text{for } t > 0 \tag{11.11}$$

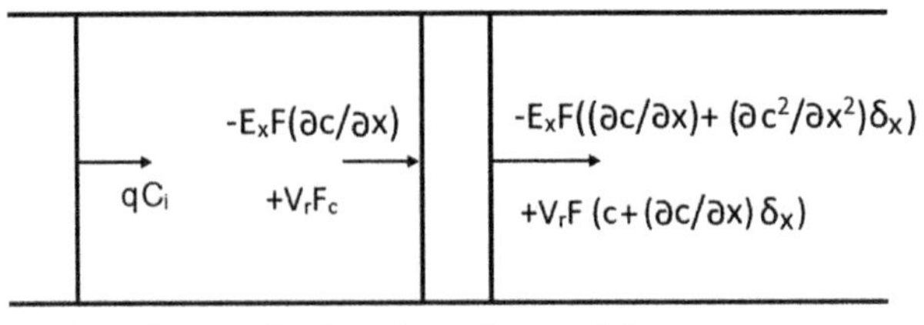

Figure 11.3 Mass flow diagram for time-dependent model.

Initial Condition 1: Just before the engine starts working, concentration of diesel exhaust in the mine air is zero, i.e.,

$$c = 0 \text{ at } t = 0 \text{ for all } x > 0 \tag{11.12}$$

The solution of Eq. (11.9) with the above boundary and initial condition is derived by Thakur [8] and is given below:

$$c = \frac{qc_i}{2FV_r}\left[\begin{array}{l} \text{erfc}\dfrac{x - V_r t}{2\sqrt{E_x t}} - \left(1 + \dfrac{V_r x}{E_x} + \dfrac{V^2 r^t}{E_x}\right) e\dfrac{V_r x}{E_x}\ \text{erfc} \\ \dfrac{x + V_r t}{2\sqrt{E_x t}} + 2\,V_r\left(\dfrac{t}{E_x \pi}\right)^{1/2} \exp -\left(\dfrac{x - V_r t}{2\sqrt{E_x t}}\right)^2 \end{array}\right] \tag{11.13}$$

Concentration, c, assumes its maximum value at $x = 0$ and is readily obtained from Eq. (11.13).

$$c\big|_{x=0} = \frac{qc_i}{2FV_r}\left[\begin{array}{l} \text{erfc}\dfrac{-V_r t}{2\sqrt{E_x t}} - \left(1 + \dfrac{V_r^2 t}{E_x}\right) \text{erfc} \\ \dfrac{V_r t}{2\sqrt{E_x t}} + 2\,V_r\left(\dfrac{t}{E_x \pi}\right)^{1/2} \exp -\left(\dfrac{V_r^2 t}{4\,E_x}\right) \end{array}\right] \tag{11.14}$$

Now, if we let $t \rightarrow \infty$, the limiting value of the maximum concentration at $x = 0$ is

$$C_L = \frac{qc_i}{2FV_r}(1 + \text{erf}(\infty)) = \frac{qc_i}{FV_r} \tag{11.15}$$

This is the same as the effective ventilation formula of Holtz and Dalzell [10]. Mathematically, the maximum concentration growth around the engine for any finite travel time will, of course, be less than that predicted by the effective ventilation formula. However, for large travel times the concentration around the engine is likely to reach the limiting value.

To evaluate Eq. (11.13), q and c_i are available from engine test data. E_x can be calculated from friction factors. For noncircular roadways, the radius, a, is substituted by the hydraulic radius defined as $R_h = (2[\text{Area}]/[\text{Perimeter}])$. To determine the concentration growth around the engine with time, several evaluations of Eq. (11.13) are needed for different x and t. A computer program was written to solve Eq. (11.13) and is available in Thakur's thesis [8].

A special case of Eq. (11.13) is when the diesel engine travels in the same direction as the air current with the same velocity. This is the worst possible situation. In this condition, the only convective term results from exhaust emissions, i.e., $\frac{q}{2F}\frac{\partial c}{\partial x}$ instead of $V_r\frac{\partial c}{\partial x}$ in Eq. (11.9). All other terms have the same meaning as before.

In this case, the exhaust emissions are free to disperse on either side of the diesel engine. The differential equation now becomes

$$\frac{\partial c}{\partial t} + \frac{q}{2F}\frac{\partial c}{\partial x} = E_x \frac{\partial^2 c}{\partial_x^2} \tag{11.16}$$

And the boundary and initial conditions are obtained exactly as in the case of Eq. (11.13). They are

$$\frac{q}{2}c\,\big|_{x=0} - E_x F \frac{\partial c}{\partial x}\bigg|_{x=0} = \frac{q}{2}\,c_i \quad \text{for } t > 0 \tag{11.17}$$

$$c \to 0 \text{ as } x \to \infty \quad \text{for } t > 0 \tag{11.18}$$

$$c = 0 \text{ at } t = 0; x > 0 \tag{11.19}$$

The physical meaning of these equations is the same as in the previous case. The solution of Eq. (11.16) with the given boundary and initial conditions is obtained by substituting $\frac{q}{2F}$ for V_r in Eq. (11.13) and dividing the right-hand side by 2 and is given below:

$$c = \frac{c_i}{2}\left[\operatorname{erfc}\frac{x - \frac{qt}{2F}}{2\sqrt{E_x t}} - \left(1 + \frac{qx}{2E_x F} + \frac{q^2 t}{4F^2 E_x}\right)\exp\frac{qx}{2E_x F}\operatorname{erfc}\frac{x + \frac{qt}{2F}}{2\sqrt{E_x t}} + \frac{q}{F}\left(\frac{t}{E_x \pi}\right)^{1/2}\exp - \left(\frac{x - \frac{qt}{2F}}{2\sqrt{E_x t}}\right)^2\right] \tag{11.20}$$

11.4.3 Multiple Diesel Engines in a Single Roadway

The case of multiple diesel engines in a single roadway is a natural outgrowth of the case of a single engine in a single roadway. Typical mining situations with exhaust type ventilation are shown in Figs. 11.4 and 11.5.

Fig. 11.4A shows the layout for the drivage of a long tunnel, whereas Fig. 11.4B represents a long entry. Fig. 11.4C is a simplified representation of an actual mining layout shown in Fig. 11.5. The length of haul in Fig. 11.5 is only 500 ft but the real advantage of diesel engine deployment in coal mines will accrue if this length could be increased to cover the entire panel length, which is typically 5000–10,000 ft. In that case, more than one diesel engine will be needed to insure a continuous removal of coal from the working face. Generally, diesel engines move faster than the air current in the face area and consequently, the air flowing in the haulage roadway will be

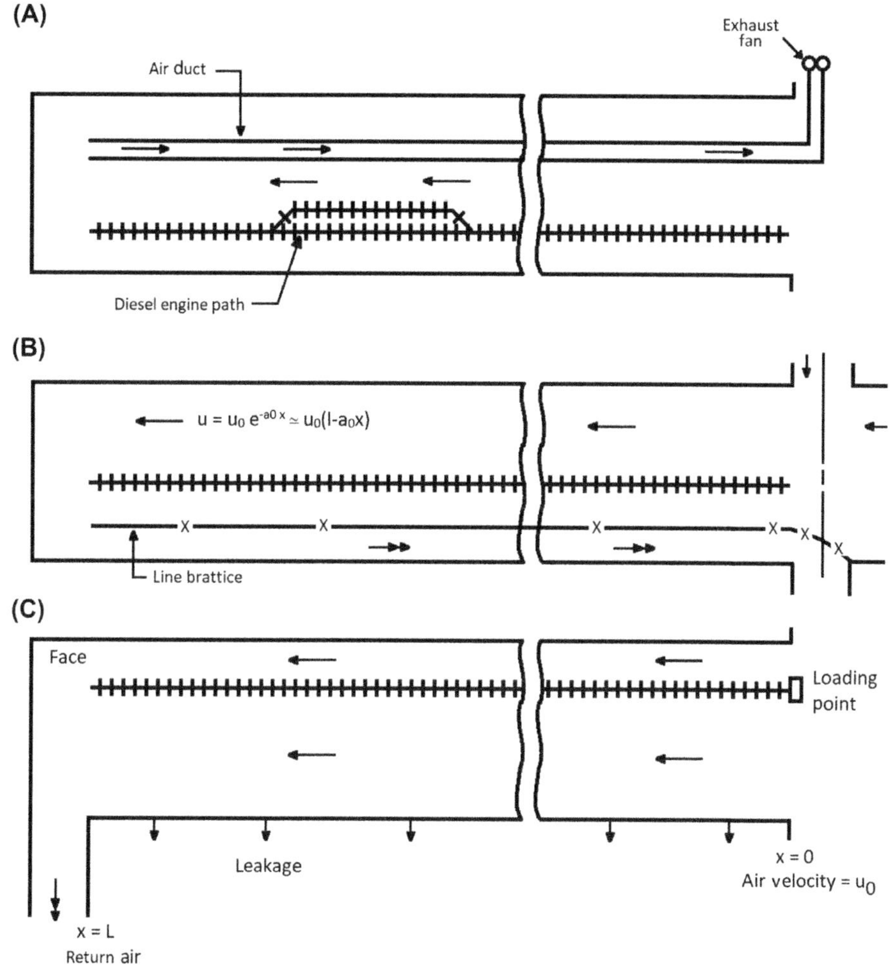

Figure 11.4 Layouts of various situations when multiple diesel engines work in a single roadway. (A) Layout of a long tunnel. (B) Layout of blind heading. (C) Simplified representation of the layout shown in Fig. 11.5.

contaminated several times by the diesel engines before it finally gets discharged into the return airways. This leads to a progressive rise in the concentration because of the superposition of contaminations. To facilitate the mathematical analysis of this situation, suppose a number of carbon monoxide monitors are installed equidistantly along the length of a roadway. The readings of CO are being continuously monitored, and the time-weighted exposure of persons at each sampling point is calculated by averaging the spot readings of concentration just as the calculation of TLV is supposed to be done

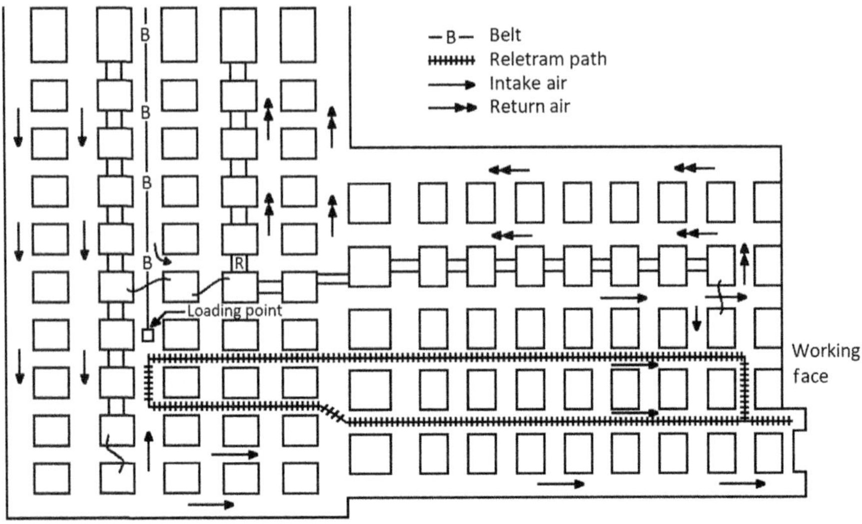

Figure 11.5 Layout of a typical development section.

over a complete cycle. Assuming continuous movement without any delay, the cycle time is given by

$$\frac{2L}{v} + \text{loading and unloading time}$$

where, L is the total length of haul, and V is the speed of the diesel engine.

If Δt is the time between arrivals of diesel engines at the face (i.e., the loading time), then for continuous removal of material:

$$\Delta \mathrm{t} = \frac{1}{n}\left(\frac{2L}{v} + \text{loading and unloading time}\right)$$

where, n is the number of diesel engines.

In subsequent mathematical development for multiple engines, it is assumed that the time-weighted concentration profile over the entire length, L, assumes a steady state with the highest concentration at the point $x = L$ when x is measured in the direction of air movement. Actual field studies [10] confirm the validity of this assumption.

Another factor that will influence concentration profile is air leakage through the ventilating devices. Because of leakage, which is a function of several factors such as the nature of stopping, ducting, the pressure differential, etc., the velocity is not uniform over the entire length. Velocity at any distance x can be represented by u_0 exp $(-a_0 x)$ where a_0 is a leakage coefficient and u_0 is the air velocity at $x = 0$. Usually a_0 is very small and velocity, u, can be approximated by a truncated Taylor series expansion of $[u_0 \exp(-a_0x)]$, i.e., $u = u_0(1 - a_0x)$.[1] Additionally, some of the components of diesel exhaust, e.g., oxides of nitrogen, sulfur dioxide, etc., are likely to be absorbed by the walls of roadways, especially if they are wet. Let γ be the coefficient of absorption. Using a concept similar to previous cases, it is seen that in a differential element Δx of the roadway, the input of the material is represented by the term $[v + u_0(1 - a_0x)]\frac{\partial c}{\partial x}$, whereas the loss of material from that same element Δx is given by the sum of $E_x\frac{\partial^2 c}{\partial_x 2}$ and γc, which are due to turbulent dispersion and absorption, respectively. It is assumed here that walls of roadways are perfect sinks, i.e., concentration of gases at the wall is zero. These terms are identical in the first two models to be developed here.

11.4.4 Diesel Engines Moving Continuously in a Cycle in a Roadway With Considerable Leakage

Making a mass balance on a small element Δx of the roadway, the equation of convective diffusion is obtained as

$$E_x\frac{\partial^2 c}{\partial x^2} - [v \pm u_0 \pm (1 - a_0x)]\frac{\partial c}{\partial x} + c = 0 \tag{11.21}$$

The + and − signs in the second term of the above equation to opposite and concurrent movement of diesel engine with respect to the air current. The value of E_x for this situation can be approximated by substituting the velocity v for u, and $v\Delta t$ for the characteristic length d; i.e., $E_x = \mathrm{A}v^2\Delta t$ where A is a constant whose value is given by 14.4 $\sqrt{\lambda/\lambda_r}$. The value of λ is obtained using Reynold's analogy [11] as

$$\lambda = \frac{\lambda}{8a}u_0(1 - a_0x) \tag{11.22}$$

Define

$$\frac{u_0}{v} = U_0; \frac{u_0}{v}\left[1 - \frac{a_0x}{2}\right] = U_m; \text{ and } \frac{u_0}{v}(1 - a_0x) = U_L.$$

With the above definitions and substituting for E_x and λ, Eq. (11.21) reduces to

$$\frac{\partial^2 c}{\partial x^2} - \frac{1 \pm U_L}{Av\Delta t}\frac{\partial c}{\partial x} + \frac{\lambda U_L}{8aAv\Delta t}c = 0 \tag{11.23}$$

[1] For blowing type ventilation, $x = 0$ plane is at the face and $u = u_0$ $(1 + a_0x)$. The value of a_0 is not necessarily same in both cases and is best obtained by actual observations.

Eq. (11.23) is a linear differential equation with variable coefficients. A solution is obtained by reducing it to the canonical form, i.e., by substituting:

$$c(x) = z(x)\cdot\exp\left[\frac{1}{2}\cdot\frac{1 \pm U_m}{Av\Delta t}\right]x \tag{11.24}$$

where, $z(x)$ satisfies the differential equation.

$$z'' - f(x)z = 0 \tag{11.25}$$

the solution of Eq. (11.25) is

$$Z = B_1 \exp(K_1 x) + B_2 \exp(-K_2 x) \tag{11.26}$$

The particular solution $B_2 \exp(-K_2 x)$ has no physical meaning in this problem because the concentration, c, is an increasing function of x. To solve the equation $Z = B_1 \exp(K_1 x)$, the following boundary condition is used:

$$c|_{x=0} = c* = \frac{0.14\, c_i\, q*}{B[1 \pm U_L]U_L} \tag{11.27}$$

where, B is a constant and $q^* = (q/\pi a^2 v)$. The derivation of this condition is shown by Thakur [8].

Using this boundary condition and superposing the concentrations due to the movement of diesel engine parallel and opposite to air current, the solution for Eq. (11.23) is

$$c = \frac{0.14}{B}\left[\begin{array}{l} \dfrac{c_{ip}q*_p}{[1-U_L]U_L}\exp\left\{\dfrac{1}{2}\dfrac{1-U_m}{Av\Delta t} + \left\{\dfrac{1}{4}\left(\dfrac{1-U_m}{Av\Delta t}\right)^2 - \dfrac{U_L}{8aAv\Delta t}\right\}^{\frac{1}{2}}\right\}x \\ + \dfrac{c_{iq}\, q*_q}{[1+U_L]\, U_L}\exp\left\{\dfrac{1}{2}\dfrac{1+U_m}{Av\Delta t} + \left\{\dfrac{1}{4}\left(\dfrac{1+U_m}{Av\Delta t}\right)^2 - \dfrac{U_L}{8aAv\Delta t}\right\}^{\frac{1}{2}}\right\}x \end{array}\right] \tag{11.28}$$

where the subscripts p and q refer to parallel and opposite movements of diesel engines.

Eq. (11.28) has been derived with a coefficient of absorption mainly to make the approach general. In practice, the absorption of diesel exhausts is not significant and can be discarded. Eq. (11.28) now becomes

$$c = \frac{0.14}{B}\left[\frac{c_{ip}\, q*_p}{(1-U_L)U_L}\exp\frac{1-U_m}{Av\Delta t}x + \frac{c_{iq}\, q*_q}{(1+U_L)U_L}\exp\frac{1+U_m}{Av\Delta t}x\right] \tag{11.29}$$

Data required for solution of Eq. (11.29) are easily available. Values for c_i and q are obtained from test data on the engine, whereas values for leakage coefficient (a_0) and interval between arrivals (Δt) are obtained from case studies. The velocity of air, u, and velocity of engine, v, are already known. A computer program was written to solve Eq. (11.29) in reference [8].

11.4.5 Diesel Engines Moving Continuously in a Cycle in a Roadway With Little Leakage

In this case, the leakage coefficient, a_0, is very small. Hence, the convective velocity can be reasonably well estimated by the average velocity over the entire length, L, of the roadway.

For exhaust type ventilation, the average velocity is given by

$$\overline{u} = u_0\left(1 - \frac{a_0 L}{2}\right) \tag{11.30}$$

Substituting $\overline{U}$ for U_L in Eq. (11.30):

$$c* = \frac{0.14\ c_i q*}{B\left[1 \pm \overline{U}\right]\overline{U}} \tag{11.31}$$

where,

$$\overline{U} = \frac{\overline{u}}{v} = U_o\left(1 - \frac{a_0 L}{2}\right) \tag{11.32}$$

The differential equation in this case is

$$\frac{\partial^2 c}{\partial x^2} - \frac{1 \pm \overline{U}}{Av\Delta t}\frac{\partial c}{\partial x} + \frac{\lambda \overline{U}}{8aAv\Delta t}c = 0 \tag{11.33}$$

The solution of Eq. (11.33) is obtained by proper substitution in Eq. (11.28) and is given below:

$$c = \frac{0.14}{B}\left[\begin{array}{l} \dfrac{c_{ip} q*_p}{\left(1-\overline{U}\right)\overline{U}}\exp\left\{\dfrac{1}{2}\dfrac{1-\overline{U}}{Av\Delta t} + \left\{\dfrac{1}{4}\left(\dfrac{1-\overline{U}}{Av\Delta t}\right)^2 - \dfrac{\lambda\,\overline{U}}{8aAv\Delta t}\right\}^{\frac{1}{2}}\right\}x \\ + \dfrac{c_{iq}\ q*_q}{\left(1+\overline{U}\right)\overline{U}}\exp\left\{\dfrac{1}{2}\dfrac{1+\overline{U}}{Av\Delta t} + \left\{\dfrac{1}{4}\left(\dfrac{1+\overline{U}}{Av\Delta t}\right)^2 - \dfrac{\overline{U}}{8aAv\Delta t}\right\}^{\frac{1}{2}}\right\}x \end{array}\right] \tag{11.34}$$

For the same reasons as before, the absorption terms can be dropped from Eq. (11.34), which then reduces to

$$c = \frac{0.14}{B}\left[\frac{c_{ip}\,q{*}_p}{(1-\overline{U})U}\exp\frac{1-\overline{U}}{Av\Delta t}x + \frac{c_{iq}\,q{*}_q}{(1+\overline{U})U}\exp\frac{1+\overline{U}}{Av\Delta t}x\right] \tag{11.35}$$

If leakage is neglected, i.e., $a_0 = 0$, then $U_m = U_L = \overline{U} = U_0$ and solution for both cases is the same. In either case,

$$c = \frac{0.14}{B}\left[\frac{c_{ip}\,q{*}_p}{(1-U_0)U_0}\exp\left(\frac{1-U_0}{Av\Delta t}\right)x + \frac{c_{ip}\,q{*}_q}{(1+U_0)U_0}\exp\left(\frac{1-U_0}{Av\Delta t}\right)x\right] \tag{11.36}$$

Solutions as obtained in Eqs. (11.29) and (11.35), of course, correspond to the highest concentration that can arise in a roadway.

Thus, these solutions constitute a limit. Data required to solve Eq. (11.35) are the same as those for Eq. (11.29).

11.4.6 Diesel Engines Moving Randomly in a Roadway With Uniform Air Velocity

When diesel engines move in a roadway without any definite pattern, they are defined as performing randomly. This situation can be best analyzed by assuming that the diesel exhaust source is uniformly distributed over the entire length of the roadway. Assuming further that a steady-state concentration is developed very soon, the required differential equation is

$$E_x\frac{\partial^2 c}{\partial x^2} + \overline{u}\frac{\partial c}{\partial x} = \frac{q\,c_i}{FL} \tag{11.37}$$

Physical situations are best represented by the following boundary conditions:

$$\frac{\partial c}{\partial x} = 0 \text{ at } x = 0 \tag{11.38}$$

i.e., there is no transfer of material at $x = 0$.

$$\overline{u}c|_{x=L} + E_x\frac{\partial c}{\partial x}\bigg|_{x=L} = \frac{qc_i}{F} \tag{11.39}$$

i.e., matter is conserved in the roadway.

The solution of Eq. (11.37) with the above boundary conditions is given below [8]:

$$c = \frac{q\,c_i}{FL\bar{u}}\left[x - \frac{E_x}{\bar{u}}\left(1 - e\frac{-\bar{u}x}{E_x}\right)\right] \tag{11.40}$$

here, the term $\frac{q\,c_i}{FL\bar{u}}$ is the time averaged source strength in parts/sec/unit volume of the roadway.

Eq. (11.40) forms the lower limit of concentration profile because it gives zero concentration at $x = 0$. The growth of concentration would depend on the movement of the engine, which is not known. However, a reasonable estimate of E_x is obtain by K_3uL where K_3 is a constant, u the average velocity of air, and L the length of the roadway. The value of K_3 appears to vary between 0.2 and 1.0. Ideally, the value of E_x should be measured experimentally. Approximations by formula are at best indicative rather than actual. As before, values of q and c_i are obtained from engine test data.

11.4.7 Coefficients of Turbulent Dispersion

When a gaseous matter is injected into a pipe through which a fluid is flowing with turbulent motion, as is generally the case in mine roadways, it is dispersed, relative to a frame of reference, which moves with the mean velocity of flow. This dispersion is caused by the movement of eddies or velocity fluctuations relative to the movement of the main mass of the fluid and its magnitude depends, among other variables, on the size of the airways, the velocity of air, the kinematic viscosity of the fluid, and roughness of walls of airways. The concept of dispersion coefficients is very useful in analyzing these flow phenomena. Two such coefficients are associated with turbulent flow. The coefficient acting in the direction of flow is called the longitudinal coefficient of turbulent dispersion, E_x, while that in the direction normal to the direction of flow is called the radial or transverse coefficient of turbulent dispersion, E_r. While considerable work has been done on the experimental measurement of these coefficients in smooth pipes [12,13], only limited information is available on the values of E_x and E_r for mine roadways.

11.4.7.1 Coefficient of Longitudinal Turbulent Dispersion

For smooth, circular pipes, the longitudinal coefficient of turbulent dispersion for gases is given by Taylor [12] as

$$E_x = 3.57\,a\,u\sqrt{\lambda} \tag{11.41}$$

where, a is the radius of the pipe, u is the velocity of the stream, λ is the coefficient of friction.

Eq. (11.41) yields a much smaller value of E_x than actually observed because mine airways are seldom circular and smooth. Also, the flow in them is neither ideally turbulent nor laminar. Actual values, 7–25 times higher than those predicted by

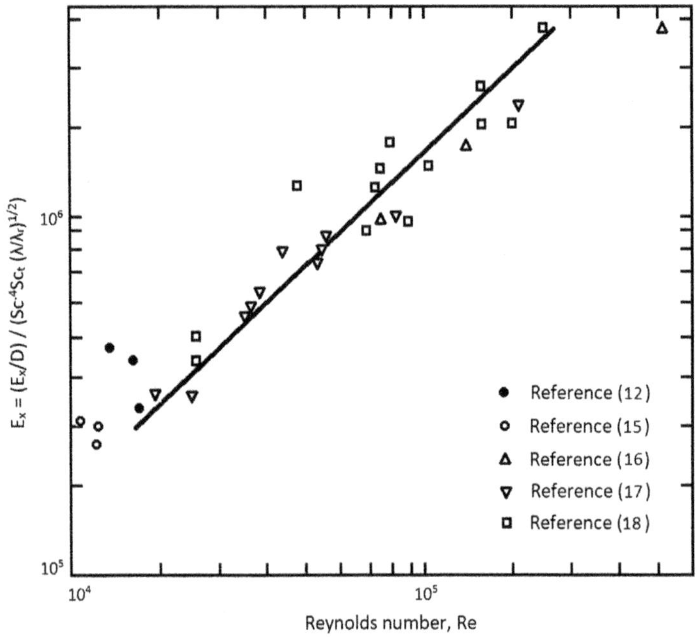

Figure 11.6 Plot of turbulent dispersion coefficient against Reynold's number.

Eq. (11.41), have been recorded [14]. A summary of the most relevant data available on this coefficient is given in Fig. 11.6 [12,15–18].

The results are plotted as a graph of normalized value of E_x as $\frac{E_x}{\left(D \cdot \mathrm{Sc}^4 \mathrm{Sc}_t \cdot \sqrt{\lambda/\lambda_r}\right)}$ against the Reynold's number, Re $= (ud/vv)$ both on logarithmic scales. Here, λ, λ_r are the coefficients of friction for smooth and rough pipes, respectively, D is the coefficient of molecular diffusion, ν is the coefficient of kinematic viscosity, $\mathrm{Sc} = (\nu/D)$ is the molecular Schmidt number, $\mathrm{Sc}_t = (E_u/E_c)$ is the turbulent Schmidt number, E_u is the coefficient of turbulent viscosity, E_c is the coefficient of turbulent diffusion, and d is a characteristic length; for pipes, it is the diameter. For values of Re between 30,000 and 200,000, E_x can be approximated within $\pm 10\%$ by

$$\begin{aligned} E_x &= 15.8\, D\, \mathrm{Re}\, \mathrm{Sc}^4 \mathrm{Sc}_t \sqrt{\lambda/\lambda_r} \\ &= 15.8\, ud\, \mathrm{Sc}^{-6} \mathrm{Sc}_t \sqrt{\lambda/\lambda_r} \end{aligned} \tag{11.42}$$

Under normal atmospheric conditions, $\mathrm{Sc} = 0.72$ and $\mathrm{Sc}_t = 0.75$. On substitution Eq. (11.42) reduces to

$$E_x = 14.4\, ud\, \sqrt{\lambda/\lambda_r} \tag{11.43}$$

11.4.7.2 Coefficients of Transverse Turbulent Dispersion

In turbulent systems, particularly in the face areas of mines, there is a strong tendency for isotropy (i.e., $E_x = E_r$). Consequently, a fairly good estimate of E_r is obtained from Eq. (11.41).

A generalized estimate of E_r can be obtained by analogy with the calculation of E_x as given below:

$$E_r = 3.57\, au\, Sc^{0.4}\, Sc_t \sqrt{\lambda} \tag{11.44}$$

For values of Sc and Sc_t as given earlier, Eq. (11.44) reduces to

$$E_r = 2.35\, au\, \sqrt{\lambda} \tag{11.45}$$

The value of E_r for rough pipes is obtained by substituting λ_r for λ in Eq. (11.45) where λ_r is the friction factor for rough pipes.

The relation of λ_r with Atkinson coefficient K was discussed earlier in Chapter 2.

11.4.8 Main Conclusion of Mathematical Modeling

Main conclusions and practical applications of mathematical models developed in the chapter and computer models of Chapter 5 are as follows:

1. To calculate air for a single engine in the stationary mode, Eq. (11.8) provides satisfactory results. For large values of E_r, these quantities are approximately 50%−100% of that provided by the static dilution formula which is currently used for certification.
2. For a single engine moving in a roadway for a long travel time, the results from the effective ventilation formula developed by Holtz and Dalzell [10] are in good agreement with Eq. (11.5) developed here as a limiting case for Eq. (11.13). At present, effective ventilation is not used for the certification of diesel engines.
3. When multiple diesel engines work in a roadway, it is difficult to generalize the air requirements. Eqs. (11.29) and (11.36) were used to determine the amount of air needed with multiple engines. The present certification requirements are that with a second and a third unit in the same roadway, the air quantities should be 200% and 300%, respectively, of that required with one unit. Under negligible leakage conditions (Eq. 11.36),[2] the model predicts air requirements of only 130% and 160% of that needed with one unit for two and three units, respectively. For high leakage conditions, no generalized conclusion is possible. In such cases, the actual leakage coefficient should be determined and air requirements calculated using Eq. (11.29).
4. Field studies conducted for multiple engines in a network of airways indicate good agreement with computed values of diesel exhaust emissions. Concentrations were computed by averaging emissions in the various modes of operation; e.g., idling, accelerating, cruising, and decelerating by suitable weight factors.

[2] Leakage coefficient less than 0.00005/m.

5. Operating a diesel engine in a roadway with zero relative velocity must be avoided. The rapid growth of concentration if this is done is shown by Eq. (11.20). Also, running an engine in a blind heading in the stationary mode should be avoided. Rebreathing of exhaust creates excessive rise in carbon monoxide.
6. Pollutant concentration is a complex function of engine size, number of engines, velocity of travel, air velocity, length of haul, and production. However, it pays to keep air velocities low but consistent with all safety statutes and run diesel engines at higher velocities.

11.5 Diesel Equipment Maintenance and Training of Personnel

The maintenance and training requirements stated in W.V. Diesel Rule (Title 196) are one of the best in the coal industry. The following is an excerpt from that.

11.5.1 Maintenance Plan

Following is a breakdown of the maintenance plan that will be utilized by a mine to insure compliance with the W.V. Diesel Rule (Title 196).

1. All maintenance, repair, and diagnostic testing of diesel-powered equipment will be performed by mechanics qualified under Section 196-1-24.
2. The maintenance of all equipment will begin with the operators pre-opt check list, which will be maintained on each piece of diesel-powered equipment until the next 100 h maintenance is performed at which time the old pre-opt check list book will be discarded and a new pre-opt record will begin. By maintaining the pre-opt check list on the equipment the qualified mechanic, who is to perform the 100 h maintenance, will be able to read over this document and define problem areas with this particular piece of equipment. This practice will allow the qualified mechanic to have an understanding of problem areas of each particular piece of equipment.
3. The 100 h maintenance required by Section 196-1-19 will be performed by a mechanic who has been qualified under Section 196-1-24 of this Act.
   a. The 100 h maintenance intervals will be tracked by a tag system. This tag system will require the qualified mechanic to check the hours of operation at the end of the required maintenance and add 100 h to the actual hours of operation. The qualified mechanic will then insert this number on the tag and attach it to the piece of diesel-powered equipment at a conspicuous place in the operator's compartment. This will allow the equipment operator easy reference as to when the next 100 h maintenance will need to be scheduled. We believe this tracking system will insure compliance of Section 196-1-19.
   b. A copy of the 100 h check list is attached to this Appendix as "100 h Required Maintenance Check List."
   c. The qualified mechanic will perform all checks and necessary repairs required by Section 196-1-19. All repairs of diesel-powered equipment will be recorded in the Diesel-Powered Equipment Maintenance and Repair Book.
   d. If the on-board diagnostics controls show that maintenance and/or repair is needed prior to the next 100 h maintenance interval, a qualified mechanic will perform the necessary maintenance and record the same in the Diesel-Powered Equipment Maintenance and Repair Book.

   e. All maintenance and repair of any diesel-powered equipment will be entered in the Diesel-Powered Equipment Maintenance and Repair Book, which will be kept on file at the mine as required by Section 196-1-17.

4. The record keeping at the mine will consist of

   a. Pre-opt Check List.
   b. 100 h Maintenance Check List.
   c. Baseline Sampling Form.
   d. 100 h CO Emissions Form.
   e. The Diesel-Powered Equipment Repair and Maintenance Form.
   f. All forms and check list will be made into book form and kept at the mine site as required by Section 196-1-17.

11.5.2 Training of Diesel Equipment Operators and Mechanics

It is ideal for would-be diesel mechanics to train in the schools offered by manufacturers. It is typically a 7-day short course. Diesel instructors are certified by the state and they hold classes at all mines to train diesel operators. It is an 8–16 h long course where all aspects of safe operation of diesel equipment are thoroughly explained.

11.6 West Virginia Diesel Regulations—A Model for Coal Industry

Realizing that diesel equipment can substantially improve underground mine safety, the Governor of West Virginia established a W.V. Diesel Equipment Commission[3] in July 1997 and provided funds to investigate how DPM emissions in the mines can be minimized. Diesel engines most commonly used for diesel equipment, such as locomotives, shuttle cars, and personnel transport were studied at the diesel laboratory of West Virginia University. Various after-treatment devices that were efficient, small in size, and cost-effective were studied for their collection efficiencies using the ISO 8178-1 8-mode test.

Based on these findings and the contents of federal and other state's diesel regulations, the Commission reached a consensus that the best strategy to control DPM in coal mines and tunneling was to take an integrated approach as described below.

Similar approaches to control diesel emissions were also advocated by others [5,6].

1. Reduce DPM generation: Use only MSHA approved clean engines and EPA-approved/ASTM D975 diesel fuels.
2. Collect/combust generated DPM: A properly designed catalytic converter on all diesel engines and, if needed, soot filters to minimize DPM emissions.
3. Dilute DPM: The minimum quantity of ventilation air provided to each engine would be MSHA approval—plate air based on gaseous components of engine exhaust.

[3] The author served on the commission for 18 years (1997–2015) until the commission was dissolved in 2015.

4. Monitor gaseous emissions from the engine periodically: Carry out "engine stall test" and check gaseous emissions, particularly carbon monoxide emissions, which should not exceed 100 ppm in the tail pipe.
5. Routine maintenance and record keeping.
6. Adequate training for safe operations and proper maintenance.

Based on the above premises, W.V. diesel regulations were drafted in 2003 and the final draft was approved unanimously by the Commission in March, 2004. The full text can be found in W.V. Underground Mining Laws, Rules, and Regulations under Title 196, Series 1. The main structure of the regulations is presented in Table 11.11.

11.6.1 Highlights of Diesel Regulations

It is not possible here to go into details but following are the essential highlights of these regulations.

1. Clean engines: All diesel engines will be MSHA certified and when tested with maximum fuel/air ratio, it will not require an MSHA Part 7 approval plate ventilation rate exceeding 75 CFM/rated brake horsepower. At this ventilation rate, all gaseous components in the exhaust are diluted to their respective TLVs.
2. The diesel fuel will meet the standards of the most recently approved EPA guidelines for over-the-road fuel. Additionally, the fuel shall also meet the ASTM D 975 fuel standards with a flash point of 100 degrees Fahrenheit or higher at standard temperature and pressure.
3. All underground diesel-powered equipment shall include an exhaust emissions control and conditioning system that has been laboratory tested with the diesel engine using ISO 8178-1 8-mode test and has resulted in DPM emissions that do not exceed an average concentration of 120 $\mu g/m^3$ when diluted with the MSHA approval plate ventilation.
4. Exhaust emission control system includes (1) a DPM filter with at least 75% efficiency, and (2) an oxidation catalyst that reduces carbon monoxide concentration in tail pipe to less than 100 ppm. The system also reduces the exhaust temperature below 320 degrees Fahrenheit. Automatic shut down, a spark arrestor, a flame arrestor, sampling ports and on-board performance and diagnostic systems are also required.
5. Ventilation requirements:
   a. Adequate ventilation shall be provided to ensure that the ambient concentration of exhaust gases in the mine atmosphere shall not exceed 35 ppm for carbon monoxide, 25 ppm for nitric oxide, and 3 ppm for nitrogen dioxide. Corrective actions will be taken when the gas concentrations reach 75% of their respective TLVs.
   b. Minimum ventilation required would be the MSHA approval plate ventilation for each engine.
   c. Where multiple engines are in operation, the minimum ventilation quantity will be the sum of MSHA Part 7 approval plate ventilation quantities. Air quantity measurements shall be made at the most downwind diesel unit in the air split.

11.6.2 Diesel-Powered Equipment Package Approval Process

A mining company intending to use diesel equipment in their mines must submit an application to the Director of the W.V. Office of Miner's Health, Safety, and Training.

Table 11.11 W.V. Diesel Regulation: Subsection

Subsection	Title
196-1-2	Definition
196-1-3	Underground Use
196-1-4	Diesel-Powered Equipment Package
196-1-5	Exhaust Emissions Control
196-1-6	Ventilation
196-1-7	Exhaust Gas Monitoring and Control
196-1-8	Fuel Storage Facilities
196-1-9	Transfer of Diesel Fuel
196-1-10	Containers
196-1-11	Fire Suppression for Equipment and Transportation
196-1-12	Fire Suppression for Storage Areas
196-1-13	Use of Certain Starting Aids Prohibited
196-1-14	Fueling
196-1-15	Fire and Safety Training
196-1-16	Maintenance
196-1-17	Records
196-1-18	Duties of Operator
196-1-19	Scheduled Maintenance
196-1-20	Emissions Monitoring and Control
196-1-21	Diagnostic Testing
196-1-22	Training and General Requirements
196-1-23	Equipment = Specific Training
196-1-24	Diesel Mechanic Training
196-1-25	Operation of Diesel-Powered Equipment
196-1-26	Diesel Inspectors; Employment; Training
196-1-27	Diesel Inspector—Training Course

The director will distribute copies of the application to all six Commissioners who generally meet at least once a month to review and approve all applications. A typical application is required to include the following information:

1. An inventory sheet listing all diesel-powered equipment packages to be used at a given mine. A copy of this inventory is also kept at the mine.

2. The inventory sheet for each diesel-powered equipment package is required to provide the following information:
   a. Name, address, and permit number of the mine.
   b. The name and phone number of the person responsible for maintenance and the testing of the diesel equipment.
   c. Manufacturer, model, and serial number of the equipment using the diesel power package.
   d. Manufacturer, model, and serial number of the diesel engine.
   e. MSHA 7E approval number.
   f. Rated horsepower and rpm.
   g. DPM emission in g/h; a copy of the ISO 8178-1 8-mode test should be attached.
   h. Ambient DPM concentration when diluted with MSHA approval plate ventilation. This must not exceed 120 $\mu g/m^3$.
   i. Type of fuel used.
   j. Ventilation rate.
   k. Manufacturer and model number of the after-treatment devices, i.e., the catalytic converter and the soot filter.
   l. Efficiency rating of the after-treatment devices as provided by MSHA or an acceptable third party testing facility.
   m. Manufacturer, type, and composition of an active or a passive regeneration system, if used (many ceramic soot filters are regenerated off-board using electrical systems).

A statement saying that they will strictly follow the requirements of the regulations with regard to the specifics of these subsections should be sufficient.

So far nearly 1100 diesel-powered equipment have been approved for use in West Virginia underground coal mines. It is hoped that their numbers will multiply and result in gradual replacement of all trolley wire equipment in West Virginia coal mines. Besides West Virginia, many other states have formulated diesel laws/regulations to supplement Federal laws. Pennsylvania diesel laws are very similar to West Virginia diesel regulations. It is hoped that other states and countries without diesel regulations can use West Virginia diesel regulations as a model to write their own [19].

References

[1] Johnson JH. An overview of monitoring and control methods for diesel pollutants in underground coal mines using diesel equipment. CIB Bulletin July 1980:73–87.
[2] Reinbold EO. Ambient pollutant concentration in two underground mines using diesel equipment. Mining Engineering 1981;33(1):57–67.
[3] Christie DGS, Brown AM, Taylor RJ, et al. Mortality in New South Wales coal industry. The Medical Journal of Australia 1995;163:19–21.
[4] Saverin R, Dahmann D. Diesel exhaust and lung cancer mortality in German potash mines. Atlanta, Georgia, USA: The Health Effects Institute Diesel Workshop; March 7–9, 1999.
[5] Thakur PC, Patts LD. An integrated approach to control diesel particulate matter in underground coal mines. In: The U.S. Mine Ventilation Symposium; 1998. p. 273–9.
[6] Schnackenberg GH. Estimate of technically feasible DPM levels in underground metal and non-metal mines. Mining Engineering September 2001:45–51.

[7] Diesel Net. In diesel net technology guide. 1997, 2000 and 2002. http://www.dieselnet.com.
[8] Thakur PC. Computer-aided analysis of diesel exhaust contamination of mine ventilation systems [Ph.D. thesis]. The Pennsylvania State University; 1974. p. 223.
[9] Sutton OG. Micrometeorology. New York: McGraw-Hill Book Company; 1953. p. 136–7.
[10] Holtz JC, Dalzell RW. Diesel exhaust contamination of tunnel air, USBM, RI 7074. 1968.
[11] Sherwood TK, Pigford RL. Absorption and extraction. New York: McGraw-Hill Book Company; 1954.
[12] Taylor GI. The dispersion of matter in turbulent flow through pipes. Proceedings of the Royal Society, London 1954;233(1155).
[13] Skobanov VV. Turbulent transport coefficients for mine airway. Fiz-Tekh Problemy Razrabotki Polez Iskop 1973;4.
[14] Aiery EM. Diffusion of firedamp in mine airway. In: The Mining Engineer; January 1969.
[15] Osipov SN, Grekor GG. Determination of the coefficient of mixing in a restricted current of gas. Fiz-tekh Problemy Razrabotki Polez Iskop 1968;4.
[16] Skobanov VV. Turbulent transport coefficients for mine workings. Fiz-Tekh Problemy Razrabotki Polez Iskop 1973;4.
[17] Fisher B. The mechanics of diffusion in turbulent streams. Journal of Hydraulics Division, Proceedings of the ASCE 1967;93(H46).
[18] Mass W, Wildschut H. La dispersion des bouchons de grisou dans une mine de charbon. Revue de L' Industrie Minerale 1966:824–34.
[19] Thakur PC. US Diesel Regulations: A Model for All Coal Mines. Proceedings of the 11th International Mine Ventilation Congress; 2018. p. 471–82.

Respirable Dust Sampling and Measurement

12

Chapter Outline

Over the past 60 years, respirable dust sampling and measurements in coal mines have undergone a great change. Many instruments were developed to measure the number of dust particles in air or the surface area of the airborne dust. It was assumed that the surface area of the dust deposited in human lungs is the best measure of health hazards. During 1950 to 1970, the British Pneumoconiosis Field Research (PFR) [1] conclusively proved that the mass of dust deposited in the lungs is most correlated with the growth of coal workers' pneumoconiosis (CWP). Hence instruments that can measure the dust concentration gravimetrically were developed. A further refinement was the sampling of dust on a "size-selective" basis just as the human nose/mouth would

Advanced Mine Ventilation. https://doi.org/10.1016/B978-0-08-100457-9.00012-2

do. These size-selection curves are already discussed in Chapter 7. Most health standards are based on these gravimetric instruments. However, these sampling instruments still suffered from some design defects and newer instruments, such as the tapered element oscillating microbalance (TEOM) instruments, were developed. This chapter will provide a brief introduction to prominent instruments of each era with their merits and demerits.

The second issue with sampling and measurement of a random variable, such as the dust concentration in air, is the statistical reliability. Like the measurement of any other random variable, a minimum number of samples are required to derive an average that has 95% confidence of being within ±20% of the true mean. A case study for silica measurement will be discussed.

12.1 Early Dust Measuring Instruments

Before 1960, there were three main instruments that collected and counted the number of particles in air. They are as follows:

- The Konimeter
- Midget Impinger
- Thermal Precipitator

In the light of current knowledge, they provided a poor measurement of true risk.

12.1.1 The Konimeter

This instrument was developed in 1916 by Kotze but is obsolete now. A spring-loaded piston pump draws 5 cc of dusty air and impinges it on a grease-coated glass plate. The glass plate can be rotated to collect up to 30 samples. The glass plate is put under a microscope. Each sample appears as a circle. Typically, only a 30° section is used to count the number of particles, and it is multiplied by 12 to get the total number of particles in 5 cc of air. The dust concentration was expressed as "dust index" that is equal to the number of dust particles per cc of air. It should be noted that 100 ppcc equals 2.83 million particles per cubic feet. Konimeter has a low collection efficiency for particles below one micron. Large aggregates of dust can be shattered by impact and give a much higher count. This instrument was mostly used in Australia, India, Canada, Germany, and Poland.

12.1.2 The Midget Impinger

This instrument was mainly used in the United States. The mine air sample was drawn through a 1 mm nozzle at a fixed rate for a fixed time. Typically, 0.5 ft^3 (14,163 cc) of air was collected in a flask containing about 30 cc of water. The dust particles are further collected in a 1 mm deep hemocytometer cell after letting the dust settle down for 30 min. The cells were put under a microscope and dust particles, smaller than 10 microns, were counted. There are 4000 fields per cc of water, but only two to three fields were counted to get an average. An average count of 50 per field was equal to:

$$58.8 \times 10^6 \text{ particles per cubic foot} = 2,079 \text{ pp cc.}$$

The instrument suffered from the same defects as the Konimeter and was eventually abandoned.

12.1.3 Thermal Precipitator

In this device, the sampled air is drawn vertically downward through a narrow channel between two microscope cover glasses as shown in Fig. 12.1A. An electrically heated wire is stretched horizontally across the center of the unit. If a cold body (the microscope slide) is inserted near the hot wire, airborne dust particles are deposited on the slide in the form of a strip because of the thermal gradient, hence the name *thermal precipitator*. After sampling, the cover glasses are evaluated microscopically, usually with a 2 mm oil immersion objective. The volume of air sampled is metered so that the dust concentration can be expressed as "number of particles per cc."

The sampling rate of the thermal precipitator is 7 cc/min. The instrument is provided with a water aspirator and powered with a miner's cap lamp battery, which makes it safe for gassy mines. Sampling time is from a few minutes to 24 h, depending on the dust concentration.

If the sample contains siliceous dust, the slides may be acid treated, as in a konimeter, to remove carbonaceous and acid-soluble particles.

The outstanding features of the thermal precipitator are the following:

1. It collects all particles less than 10 μm size with 100% efficiency (thermal deposition becomes more effective the smaller the particle size).
2. It avoids further breakage of aggregates.

Although the instrument provides a dust sample of high accuracy, the final result is still subject to error, as only a portion of the total sample is counted. A variant of this

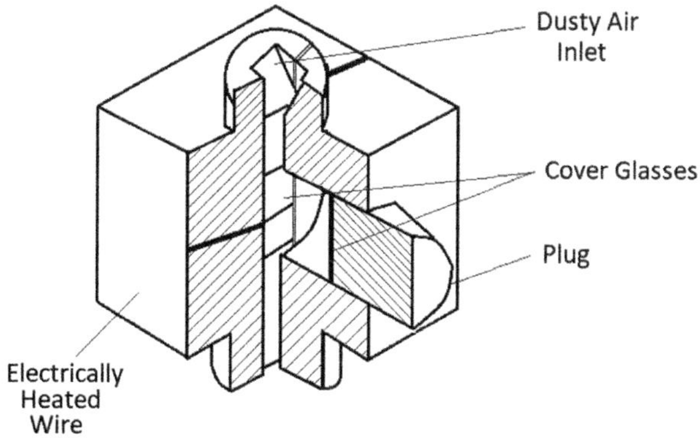

Figure 12.1A Standard thermal precipitator dust sampler.
Source: By Hartmann, John Wiley, 1982.

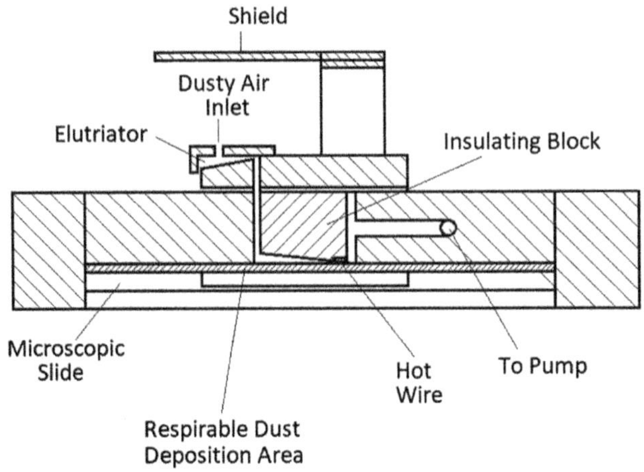

Figure 12.1B Long-running thermal precipitator dust sampler.
Source: By Hartmann, John Wiley, 1982.

instrument is the long-running thermal precipitator (Fig. 12.1B), with a size-selective entrance. The dust passes through an elutriator where gravitational deposition of particles occurs before the dust cloud reaches the thermal precipitator [2,3].

12.2 Gravimetric Personal Dust Samplers

Midway through the PFR study (1952–69), it was generally agreed that the mass concentration of dust in air is the best measure of health hazard. At the conclusion of the PFR study in 1969, several countries changed the respirable dust standard to the mass of respirable dust in unit volume of air (mg/m^3) as discussed in Chapter 9.

The United Kingdom and United States both developed gravimetric personal dust samplers and established their dust standards based on these instruments. They are the Mining Research Establishment (MRE) gravimetric dust sampler and the US personal dust sampler that used a cyclone for sampling the respirable fraction. Both instruments are described in greater detail in the following section.

12.2.1 Mining Research Establishment Gravimetric Dust Sampler

The MRE sampler as shown in Fig. 12.2 was developed by the National Coal Board's (NCB's) MRE and has been adopted as the standard dust sampling instrument for coal mines in Britain and the United States. Mine air is drawn at 2 L/min through an elutriator (which meets the British Mining Research Council [BMRC] sampling curve) and is collected on an internal glass filter made of fibers of 5-μm pore size membrane. The plus 7 μm particles are deposited on the elutriator plates. The filter is preweighed and measured again with dust collected over an 8-h period. The difference in weight divided by the volume of sampled air gives the dust concentration in mg/m^3.

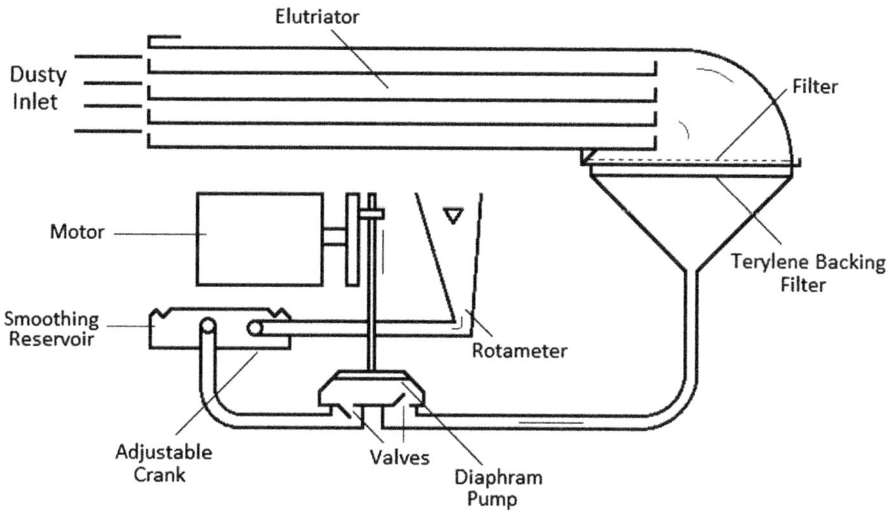

Figure 12.2 Mining research establishment gravimetric dust sampler.
Source: By Hartmann, John Wiley, 1982.

12.2.2 The US Personal Gravimetric Sampler

A 10 mm cyclone is used for size selection in this instrument. Fig. 12.3 shows the details.

The dusty air is drawn at 1.7 L/min through a cylindrical chamber tangentially and follows an inward spiral path with increasing velocity as it approaches the axis before escaping through a smaller diameter axial outlet tube. Because of the centrifugal force, large particles are removed from the airstream and collected in a receptacle at the bottom of the lower conical section of the cyclone chamber. Respirable dust penetrating the cyclone is collected on a filter. The preweighed filter is weighed again to determine the respirable mass per unit volume of air in mg/m^3 for an 8-h duration.

The personal gravimetric sampler consists of two basic components: (1) a rechargeable, battery-powered, diaphragm-type pump with calibrated flow rates and (2) the cyclone assembly with preweighed cassette or filter holder to collect the dust sample. The pump is worn on the miner's safety belt, and the cyclone assembly containing the filter is clipped onto the shirt lapel. The pump and cyclone are connected by a synthetic tubing.

In spite of the design differences, there is very good linear relationship between the two instruments' readings. In general, the mass concentration as measured by MRE instrument at 2 L/min flow is equal to 1.38 × mass concentration measured by the US personal sampler at 1.7 L/min. The personal sampler has an accuracy of ±20%, and several measurements on consecutive shifts are necessary to derive an average reliable respirable dust concentration. The law requires five samples, each taken over a consecutive 8-h shift, for averaging.

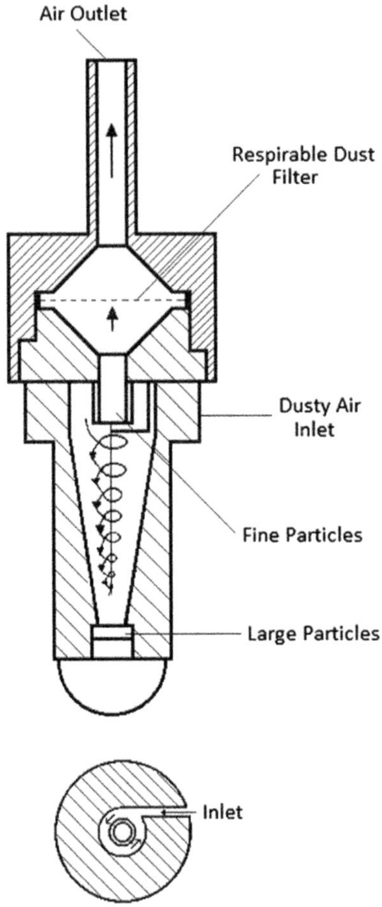

Figure 12.3 Cyclone dust sampler.
Source: By Hartmann, John Wiley, 1982.

12.2.3 GCA Beta-Ray Sampler

GCA developed a gravimetric sampler that used beta ray for mass measurement. Respirable dust is collected inside the instrument by impaction on a thin, grease-coated plastic [4]. The mass of dust collected over a short time period (usually 1 min) is measured by the change in attenuation of beta radiation caused by the buildup of dust on the disk. The instrument weighs 6.6 lb and is intrinsically safe for use in coal mines. It is battery-powered and completely self-contained. At the end of the sampling period, the average dust concentration over the period is read out on a digital display.

The source for beta ray was a piece of C_{14} with a half-life of 5700 years. This instrument, although very handy, did not receive a government approval as the previous two instruments did because of several drawbacks. The main objection was that it was not size selective. Second, it tended to give a lower than MRE dust concentration whenever the dust concentration was high. The instrument was also considerably heavier and more expensive than the personal samplers.

12.2.4 The Microorifice Uniform Deposit Impactor

Hering [5] describes a multistage impactor (Model 298) where the size separation was done by impaction. The incoming air was made to pass through a number of orifices decreasing in size, thus imparting higher velocities to the airstream. Larger particles were collected on the first few stages, and the finer particles were collected on successively lower stages. The deposits were made on substrates that were preweighed. The increase in weight was a measure of the dust deposited. It was an excellent research tool but could not be used to measure compliance with legal standards. A further refinement of this instrument was called "microorifice uniform deposit impactor" or MOUDI. It was also a good research tool but impractical for in-mine use to determine compliance. These instruments yielded a mass distribution of airborne dust from which the respirable fraction could be calculated.

12.3 Dust Concentration Measurement by Light-Scattering Instruments

In late 1970s, the US government made a proposal to monitor dust concentration at a mining machine continuously (in real time) and calculate the time when the limit of 16 mg/m^3-h (2 $mg/m^3 \times 8$ h exposure) dust exposure was reached. At this time, the power to the mining machine would be cut, and a new crew of workers will be called in to restart mining.

In the early 1980s, three instruments based on the principle of scattered light measurement came into the market. They are briefly described in the following article.

12.3.1 The British SIMSLIN Dust Monitor

The Research Division of NCB first developed the SIMSLIN II (safety in mines scattered light instrument) monitor. The mine air passed through an elutriator to conform to the BMRC sampling curve and next passed through an open chamber where it scattered a laser beam, produced by a gallium arsenide diode in the near-infrared (IR) range. With a mean wavelength of 0.9 μm, the scattered radiation over 12–20 degrees in forward direction was measured by photometers, and dust concentration was given in mg/m^3. The instrument could give both instantaneous and cumulative readings. It measured the surface area of dust particles. It had two ranges for dust concentration, 0–20 and 0–200 mg/m^3.

12.3.2 The US GCA RAM-1 (and RAM-1-2G)

GCA Corporation of the United States developed an instrument similar to SIMSLIN, but it used a 10 mm nylon cyclone for size separation. The instrument was intrinsically safe and could be used on mining machines at the working place. It could give instantaneous as well as cumulative dust concentration over the 8-h shift. The scattered light of 0.9 μm wavelength is collected over a forward angle of 45−95 degrees. It had three ranges for dust concentration, 0−2, 0−20, and 0−200 mg/m^3.

12.3.3 The German Tyndallometer

It was a modification of their old Tyndall beam reflector instrument and performed similar to SIMSLIN and GCA RAM-1 instruments. The laser light beam had a wavelength of 0.9 μm. The instrument could only give instantaneous readings, but it did not use any size-selective devices. The monitor had two dust concentration ranges: 0-2 and 0-99 mg/m^3. The instrument is calibrated using a light-scattering standard provided by the manufacturer.

12.3.4 Performance Evaluation of Light-Scattering Instruments

Thakur, Reister, and Hatch [6] carried out extensive field evaluation on all three light-scattering instruments, namely, SIMSLIN, GCA RAM-1, and the Tyndallometer. Main defects that disqualified them as a dust monitoring device for compliance with the law were as follows:

1. These instruments could not distinguish between dust particles and water particles used for dust suppression. They read a dust concentration of 4−5 mg/m^3 when no coal cutting was being done and only water sprays were turned on. This was the biggest shortcoming of these instruments. The GCA RAM-1 was least impacted by water and humidity.
2. Size distribution of coal dust particles impacted the readings. The instruments would give reliable readings only for the size distribution of the dust cloud that it is calibrated for. As discussed before, different ranks of coal and minerals in coal, such as silica, have different characteristic size distributions. Particle size, shape, and its refractive index also impacted the readings.
3. Light-scattering instruments failed to yield reliable readings when compared with personal gravimetric samplers. Under controlled laboratory conditions, the variation in calibration was 20%−30%, but under in-mine conditions the readings differed by a factor of 2.23.

12.4 The Tapered Element Oscillating Microbalance Instrument (A Personal Dust Monitor)

After the failure of the light-scattering instruments for continuous dust monitoring, the US government funded research to develop a more reliable instrument for personal dust monitoring. The TEOM is a promising development in this area, but it still lacks approval for compliance measurements. It is, however, a good tool for the engineering control of dust and related research.

The core component of the TEOM is a vibrating hollow tube called the tapered element. It is fixed on one end but free to oscillate at the other. The resonant frequency of the element (approximately 252 Hz) is monitored by an electronic system along with the oscillation amplitude. A small filter (16 mm diameter) assembly is mounted on its free end, which collects the dust particles in the sampled air. As the weight of the filter assembly changes owing to dust deposit, its frequency changes according to Eq. (12.1).

$$\Delta W = A\left[\frac{1}{f_t^2} - \frac{1}{f_o^2}\right] \tag{12.1}$$

where ΔW is the change in the weight of the filter, mg; f_o, initial frequency in Hz; f_t, measured frequency in Hz after dust deposition in time, t; A, a constant that converts change in frequency to weight gained.

The respirable dust concentration is calculated from the mass of dust collected, rate of air flow, and time of sample collection. The personal dust monitor (PDM) collects respirable dust through a Higgins–Dewell (HD) cyclone that follows the International Organization of Standardization (ISO) convention for respirable dust. The ISO standard is slightly different from the AEC (Atomic Energy Commission) curve used in the personal dust sampler [7]. PDM draws air at the rate of 2.2 L/min. The air intake is heated to 46°C to keep moisture out and minimize the error due to excessive moisture in mine air.

PDM readings are made unreliable by a variety of confounders. The main sources of error are as follows:

1. Moisture in air and the coal dust sample.
2. Atmospheric temperature fluctuations.
3. Mine air humidity.
4. Air flow fluctuations.
5. Pressure drop across the filter element.
6. Heat-induced loss of volatile matter in coal.

PDM is powered by the miner's lamp battery. An illuminated data display on the PDM continuously shows the previous 30 min dust concentration, cumulative mass concentration for the shift, and a projected end of 8-h shift concentration. A full description of the PDM is available in the literature [8,9].

A summary of all dust measuring instruments is presented in Table 12.1 for a ready reference.

12.4.1 Comparison of PDM with a Personal Dust Sampler and Marple Impactor

An extensive study by the National Institute for Occupational Safety and Health (NIOSH) [10] shows that the PDM does have a linear relationship with the standard personal dust monitor and the Marple impactors (MIs) with some qualifications.

Table 12.1 Classification of Dust Measuring Instruments

Type	Collection Method	Instrument Names and *Remarks*
Particle number counter	Impinging in water/glass plate	Konimeter, midget impinger; *All obsolete. Poor predictor of health hazards.*
Gravimetric measurement	**(a)** Size selective; elutriator **(b)** Size-selective cyclone collection on filters **(c)** Size selective; frequency change **(d)** Collection by impaction on glass plate	**(a)** MRE gravimetric sampler. **(b)** US personal dust sampler. *Currently used for compliance measurements.* **(c)** Tapered element oscillating microbalance. **(d)** *Proposed compliance tool.* Anderson/Marple Model 298 MOUDI (microorifice uniform deposit impactor). *Good research tools.*
Surface area measuring light scattering	**(a)** No size selection **(b)** Elutriator **(c)** Cyclone	**(a)** Tyndallometer **(b)** SIMSLIN **(c)** GCA RAM-1 *Good engineering control tools.*
Thermal collectors	Temperature gradient	*Difficult to use in mines.*
Electrostatic charges	Electrostatic charging and collection	*Not usable in mines.*

A systemic bias of 4.3% is indicated in comparison with personal dust samplers. Ignoring this error, the results of PDM must be multiplied by a factor of 1.059 to make it equivalent to the personal dust sampler. The 95% confidence interval for the multiplier is 1.031−1.087. The PDM tends to sample a lower weight compared with personal dust sampler.

A similar comparison with MI shows that PDM also has a linear relationship with MI readings. The linear regression shows an intercept of 0.013 mg/m^3, which is a systemic error. Ignoring this error, the PDM reading must be multiplied by 1.183 to make it comparable with MI readings. The 95% confidence interval for the multiplier is 1.111−1.256. The multiplying factor increases to 1.303 to make the PDM reading comparable with MRE readings. The 95% confidence interval for the multiplying factor is 1.223−1.383.

The PDM is very expensive compared with the legal personal dust sampler. It has the advantage of yielding dust concentration readings for shorter intervals, which can help in better engineering control of respirable dust in mines. In view of the great variation in dust collections during a shift and from one shift to another working shift, a minimum number of measurements must be taken to obtain a reliable estimate on dust exposure. The next article will highlight the importance of a minimum of five shift-long samples for obtaining a reliable average.

12.5 Respirable Dust Sampling Strategy

The enforcement of "respirable dust standard" at 1.5 mg/m^3 in the United States is based on the average of five samples collected on five consecutive shifts. This is based on very sound scientific basis because respirable dust concentration is a random variable and averaging is necessary for accuracy. There are some proposals to decide compliance with the legal standard based on a single sample. To illustrate its inaccuracy, a scientific study of variations in silica concentrations was conducted, and the minimum number of samples needed to arrive at a mean that is within ±20% of the true mean with 95% level of confidence was calculated. The number of samples needed varied from 5 to 86 depending on the randomness of silica concentrations in air and reemphasized the need for at least five samples for a better estimate on true dust concentration [11].

When the respirable dust in the atmosphere of any working place in a mine contains more than 5% quartz, the operator of the mine is required to reduce the concentration of respirable dust to a value lower than the standard 2.0 mg/m^3. The reduced standard, as published in CFR 30 Part 71.101 of the Federal Register, is determined by dividing the number 10 by the percent of quartz. Thus, the maximum allowable level of quartz in the mine atmosphere is 0.1 mg/m^3 (5% of 2 mg/m^3).

The respirable dust standard has been determined by this method since 1971. Some recent changes in the Mine Safety and Health Administration's (MSHA's) analytical procedure for measuring quartz and in the Federal regulations regarding sampling locations have resulted in reduced standards being applied to a greater number of mine sections. In 1980, only 130 mine sections were subject to a reduced standard; however, in 1981, the number of sections on a reduced standard increased by an order of magnitude to 1300 [12]. A part of the increase could be due to larger number of samples being analyzed, but the majority of this increase remains unexplained.

Before February 1981, the analytical procedure used by MSHA's Pittsburgh laboratory to determine the percentage of quartz in respirable dust required a sample, or a composite of samples, weighing at least 5.0 mg. In February 1981, the procedure was revised, and the quartz analysis can now be done on as little as 0.5 mg of dust [13]. Consequently, samples previously too small for quartz analysis, are being analyzed. In other words, before February 1981, respirable dust samples collected in a mine section would have to be added together over many shifts to obtain enough dust for quartz analysis. This effectively resulted in the quartz exposure being determined on a time-averaged basis.

With the revised analytical methods, MSHA can now analyze for quartz on a single shift dust sample, which is not indicative of the workers' long-term exposure to quartz. The other factor that contributes to a greater number of reduced standards being applied lies in the definition of a working place. The original definition of a working place in the Federal Mine Safety and Health Act of 1977 was changed to mean any place in a coal mine where miners are normally required to work or travel.

Previously a working place was defined as the area of a coal mine inbye the last open crosscut in a working section. The standard is applied to designated occupations in mechanized mining units in underground sections, designated areas in nonface areas, and designated surface work locations. Once a reduced standard is applied to a mechanized mining unit, the standard remains with the unit regardless of locations. To have the standard reevaluated, the operator must ask MSHA to return to the mining unit to collect samples for analysis. Reduced standards applied to designated areas or designated work places remain with the area or work place until the quartz level at the location is reduced (again, MSHA must be asked to return for evaluation) or until the area or work place is abandoned.

In summary, the current MSHA procedure for the determination and enforcement of quartz standard suffers from the following drawbacks:

1. Compared with world standards, the US quartz standard appears to be arbitrarily severe.
2. The number of samples (one) taken for quartz analysis appears to be inadequate.
3. The analytical technique used for silica analysis needs to be standardized.

12.5.1 The Quartz Standards

In the United States, and other countries, exposure limits for dust containing free silica have been based on the concept that the toxicity of the dust is directly proportional to the concentration of free silica in the dust. The exposure limits were developed from epidemiologic studies relating incidence of silicosis to the concentration of dust in the atmosphere. The studies of silica exposure were primarily in the granite industry, metal mines, and foundries. However, some researchers [14,15] do suggest that the presence of quartz in coal dust contributes to the progression of CWP.

The United States has taken a fundamentally different approach to the development of a quartz exposure standard compared with the European countries. In the United States, the standard was set at a value low enough to prevent the occurrence of silicosis in any worker, regardless of technical feasibility, the workers' personal susceptibility to the disease, and his personal habits. The European countries recognize that even at very low quartz concentrations, some individuals may contract silicosis and that the present state of the art in dust control may not be sufficient to achieve excessively low dust standards. They set their standards at a technically achievable value and carefully monitor the health of the workers. If any signs of silicosis or susceptibility to silicosis appear, the worker is transferred to a less dusty work area.

In Great Britain, the dust standard, when no quartz is present, was 7 mg/m^3 for a longwall face and 5 mg/m^3 for heading machines. These concentrations are measured in the returns, 210 ft from the face. If quartz is present in concentrations greater than 0.45 mg/m^3, then the respirable dust standard is reduced to 3.0 mg/m^3 [16].

In West Germany, dust exposure limits were divided into five categories for quartz concentrations less than 5% and five categories for quartz concentrations greater than 5% as shown in Table 12.2.

Table 12.2 Dust Concentration Limits in West Germany

	<5%	>5%	
Category	**Dust Concentration Limits (mg/m^3)**	**Quartz Concentration Limits (mg/m^3)**	**Factor***
0	2.5	0.125	0.8
1	2.5–5.0	0.125–0.25	1
2	5.0–7.5	0.25–0.375	2
3	7.5–9.5	0.375–0.475	4
4	9.5–12.0	0.475–0.60	5
Forbidden	12.0	0.60	—

* The factor is multiplied by the number of shifts for the cumulative dust exposure.

Workers are allowed to attain a predetermined cumulative dust exposure depending upon their medical classification, 2500 for workers with no pneumoconiosis. Points for the exposure limit are determined by multiplying the factor in Table 12.2 (last column) by the number of shifts. Thus workers can work in areas with high silica concentration for a period of time if this is offset by a corresponding period in an area with low silica concentrations. This method recognizes, unlike that in the United States, that the development of pneumoconiosis depends on long-term exposure to dust and not on a single day's exposure.

In the Russia, the safe exposure limits have been set even higher than those for other countries as shown in Table 12.3.

Table 12.3 Dust Concentration Limits in the Russia

Mineral and Organic Dust	**Maximum Permissible Concentration (mg/m^3)**
Over 70% crystalline silica	1
10%–70% free silica	2
Silicate dust with 10% free silica	4
Other mineral dust with 10% free silica	5
Minerals and mixtures with no silica	6
Coals with more than 10% free silica	2
Coals with less than 10% free silica	4
Coals with no silica	10

Thus, the US silica and respirable dust standards are already much lower than what other major mining countries have deemed necessary to protect the health of mine workers and maintain high productivity. In 1974, NIOSH published a report [17] recommending that workers should not be exposed to levels of quartz greater than 0.05 mg/m^3, i.e., one-half of the present limit. In April 1978, MSHA circulated a draft proposal of revisions to 30 CFR Part 71 in which it was recommended that the 2.0 mg/m^3 standard be reduced when the quartz present in respirable coal dust exceeded 0.05 mg/m^3.

There has been no evidence of silicosis in coal mining with the present standard in effect and therefore there is no apparent need to further reduce the standard. NIOSH's reasoning for reducing this standard was based primarily on epidemiologic studies in granite sheds. In one study [17], no indication of harmful effects to the workers was found for free silica concentrations in the range of 0.03–0.09 mg/m^3, with an average of 0.06 mg/m^3. In a separate study [17], no new cases of silicosis were found in men starting to work in Vermont granite sheds after 1937, where few exposures exceeded 0.05 mg/m^3. Thus, NIOSH apparently found it appropriate to recommend 0.05 mg/m^3 as the new quartz standard.

Also in reference [17], the authors offer evidence that silica in the forms of cristobalite and tridymite are much more dangerous than alpha quartz and recommend that the standard for these forms of free silica be one-half of that recommended for quartz. Then on the same page they offer evidence showing that the exposure of 0.05 mg/m^3 is safe for cristobalite and tridymite. These forms of silica have not even been found in coal mines, and reducing the current standard to 0.05 mg/m^3 would thus constitute an overkill.

12.5.2 Measurement of Quartz in Respirable Dust

Three principal methods are used for quantitative determination of silica in dust. These are the colorimetric chemical procedure, IR spectrophotometry, and X-ray diffraction. In comparative tests performed by NIOSH for determining quartz percentage in coal dust, all of the methods were found to have a maximum deviation of 5% from the overall mean value [17], thus showing that all three methods can be used to determine quartz percentage accurately. Owing to their sensitivity, speed, minimum sample preparation, and capabilities for automation, the IR and X-ray methods have been used far more often than the colorimetric method for routine quartz analysis.

The X-ray diffraction method has the capability of detecting the various forms of free silica, e.g., quartz, cristobalite, and tridymite. The X-ray method is highly recommended and routinely used by NIOSH. However, NIOSH recommends using IR spectrophotometry if interfering materials in the sample decrease the sensitivity of the X-ray diffraction or when the quantity of the total sample is very small. The IR method is used by MSHA for analysis of quartz in respirable coal dust, but it has restricted use

Table 12.4 The Variation of Quartz Percentages Between Laboratories in the 1980 NIOSH PAT Program

Method	Range of Variation
Overall	33.7%–47.4%
Colorimetric	35.4%–48.4%
Infrared	23.6%–44.0%
X-ray	31.2%–39.5%

for the analysis of dust from metal mines owing to interference. MSHA initially chose the IR method owing to its capability to detect quartz in very small sample sizes as found on the personal sampler filter [18].

Apparently, both the IR and X-ray diffraction methods are capable of accurately determining the percentage of quartz in a dust sample. However, the reported percentage error is dependent upon the primary quartz standard used for comparison. The latter is often a greater source of error. Human errors can also be significant. Since 1975, NIOSH has sponsored a Proficiency Analytical Testing (PAT) program to monitor the performance of various analytical laboratories. The NIOSH 1980 study [19] is summarized in Table 12.4. A total of 61 laboratories were surveyed; 28 laboratories used the X-ray diffraction, 21 laboratories used the calorimetric method, and 12 laboratories used the IR method.

NIOSH sent several dust samples to all of the laboratories, and they determined the percentage of quartz in the dust. NIOSH then determined the percentage variation (shown in Table 12.4) in the reported results. The large variation between laboratories emphasizes the need to standardize the analytical methods used to determine the quartz percentage. The weight of the respirable dust sample must be large enough to avoid large weighing errors. Different laboratories must use an approved primary quartz standard. At present, it is not uncommon for a laboratory to produce its own primary quartz standard. Duplicate analysis or analysis of more than one sample can reduce human errors considerably.

12.5.3 An Independent Study

The author conducted a survey of many of the working sections in mines, which were on a reduced dust standard. The purpose of the survey was (1) to identify the sources of quartz in each mine, (2) to determine the accuracy of MSHA's sampling, and (3) to determine the number of samples needed for reliable (true mean $\pm$ 20%) measurement of quartz concentration in the working place.

12.5.3.1 Sources of Quartz

The sources of quartz in each mine were identified by collecting channel samples of the roof, floor, and coal seam. These samples were analyzed by Materials Consultants and Laboratories Inc. (MCL), a subsidiary of Carnegie Mellon University, by the X-ray diffraction method. Quartz percentages are summarized in Table 12.5.

As can be seen, little quartz exists in the coal seam itself. The major sources of quartz are the roof and floor. Thus, excessive quartz percentages found in respirable dust samples are most probably due to mining practices, such as mining the roof or floor to provide adequate working height or to eliminate bad top or mining of coal seams with siliceous rock bands. The practices are often unavoidable and sometimes even necessary for the safety of workers.

12.5.3.2 Comparison of X-ray Diffraction and Infrared Techniques

A limited effort was made to compare the IR analytical technique with the X-ray technique. Mining sections in different mines put on reduced dust standards by MSHA were resampled immediately after MSHA sampling at the operator's canopy of the continuous miner. These samples were sent to MCL for quartz analyses by X-ray diffraction. Results of these analyses along with corresponding MSHA analyses are listed in Tables 12.6 and 12.7.

The samples were not collected on the same day as were MSHA's but were in the same section during the time the reduced standard was in effect. In all cases, the percentage quartz determined by MCL was lower than MSHA's. This difference could be due to two factors. First, MCL used X-ray diffraction and MSHA used the IR method; and as was shown previously, a large systematic difference could exist between the two methods. But, it is not likely that the difference could be as great as is shown

Table 12.5 Quartz Percentages in Coal Mines

Sample	Range of Quartz Percentage	Average Percent Quartz
Roof (16)	4.0–21.0	12.0
Floor (16)	3.2–32.0	12.9
Coal seam (20)	0.0–4.3	2.0
Rock dust (16)	0.0	0.0

( ) Number of samples.

Table 12.6 Comparison of Quartz Percentages as Measured by MCL to the Percentage Calculated From MSHA's Reduced Standard for Respirable Dust for Northern Appalachian Mines

Mine (Section)	% Quartz (MCL)	Reduced Standard (mg/m^3)	% Quartz (MSHA)
Mine A			
Section 1	0.0	1.1	9.09
Section 2	3.8	1.2	8.33
Mine B			
Section 1	—	1.2	8.33
Section 2	0.0	1.2	8.33

in many samples. The other factor is the possibility of spatial variation within the section from day to day.

12.5.4 Spatial Variation of Quartz

To investigate the spatial variation of quartz further, respirable dust samples were collected in seven mines, on a single section, on consecutive working days. The samples were analyzed by the IR method, following the same procedure as MSHA's. Quartz percentages obtained are listed in Table 12.8.

The results show that the spatial variation of quartz in respirable dust, within a given section, can be as high as (200 - 2000%). This clearly demonstrates that MSHA's method of determining a reduced standard based on a single sample may not be indicative of true quartz exposure of the workers.

The minimum number of samples needed to estimate the mean value of the quartz exposure can be calculated using the following formula [20]:

$$N = \left(\frac{Z_{1-\frac{1}{2}\alpha}\sigma}{d}\right)^2 \tag{12.2}$$

where σ is the population standard deviation; For a 90% confidence interval $\alpha = 0.10$ and $Z_{0.950} = 1.645$; N is the minimum number of samples needed; and the confidence interval is 2d.

The data in Table 12.8 were analyzed using the above formula, and the minimum number of samples needed to obtain a quartz percentage within plus or minus 20% of the true mean value was determined. The results are shown in Table 12.9. The number

Table 12.7 Comparison of MCL Quartz Percentages to MSHA Quartz Percentages in Respirable Dust for the Southern Appalachian Region

Mine (Section)	% Quartz (MCL)	% Quartz (MSHA)	Reduced Standard (mg/m^3)
Mine C			
Section 1	4.1	7.1	1.4
Mine D			
Section 1	5.2	14.3	0.7
Mine E			
Section 1	2.3	7.1	1.4
Mine F			
Section 1	5.7	9.1	1.1
Mine G			
Section 1	0.0	12.5	0.8
Mine H			
Section 1	4.5	9.1	1.1
Mine I			
Section 1	3.5	9.1	1.1
Mine J			
Section 1	0.0	8.3	1.2
Mine K			
Section 1	0.0	8.3	1.2
Mine L			
Section 1	0.0	8.3	1.2

of required samples varies between 5 and 86. Clearly the use of one sample for quartz determination is unjustified. However, it would also be impractical to collect, analyze, and average 86 samples.

A logical alternative would be to analyze the five samples collected for respirable dust concentration determination. The samples are already being taken periodically and sent to MSHA's Pittsburgh laboratory to determine the concentration of respirable dust. The same samples could then be analyzed for quartz and the mean exposure determined. The samples would be taken every other month ensuring a prompt follow-up.

Table 12.8 The Variation of Quartz Percentages in Respirable Dust Samples on Consecutive Working Days

Quartz Percentage						
Mine A	**Mine M**	**Mine C**	**Mine N**	**Mine G**	**Mine O**	**Mine K**
3.39	0.2	5.77	5.93	14.3	8.0	1.0
3.10	4.05	4.80	4.78	10.2	3.5	1.2
2.88	3.91	2.14	7.00	7.0	3.4	0.9
19.88	3.84	3.75	10.35	13.6	5.8	0.9
17.85	2.88	9.69	5.25	11.7	1.7	1.1
	3.75			16.2	0.0	2.1
					0.3	3.4
					0.0	3.0
					0.0	4.0
					0.8	0.6

Table 12.9 The Minimum Number of Samples Needed to Obtain a Quartz Percentage Within 20% of the Mean Value at a 90% Confidence Level

Mine	Number of Samples
A	46
M	13
C	16
N	6
G	5
O	86
K	28

12.6 Threshold Limits for Various Dusts Prevailing in Mines

Besides coal, silica, and DPM (diesel particulate matter), there are other dust clouds in mines that can create a health hazard. The American Congress of Governmental Industrial Hygienists (ACGIH) has set a threshold limit value (TLV) for each of them. A short list is provided in Table 12.10 to illustrate the relative pathogenicity of various dust particles.

Table 12.10 ACGIH Threshold Limit Values for Some In-Mine Dusts [21]

Substance	TLV—TWA	Comments
Arsenic, elemental	0.01 mg/m^3	Confirmed carcinogen
Asbestos, amosite	(0.5 fiber/cm^3)[a]	Confirmed carcinogen, applies to fibers >5 μ long
Asbestos, chrysotile	(2.0 fibers/cm^3)[a]	Confirmed carcinogen, applies to fibers >5 μ long
Asbestos, crocidolite	0.2 fiber/cm^3	Confirmed carcinogen, applies to fibers >5 μ long
Asbestos, other forms	(2.0 fibers/cm^3)[a]	Confirmed carcinogen, applies to fibers >5 μ long
Calcium carbonate	10 mg/m^3	If no asbestos and <1% silica present
Coal dust, respirable	2 mg/m^3	For coal dust with <5% silica; otherwise silica limit applies
DPM	0.12 mg/m^3	Only in West Virginia and Pennsylvania (United States)
Graphite (natural)	2 mg/m^3	All forms except fibers; respirable fraction only
Kaolin, respirable	2 mg/m^3	
Magnesite	10 mg/m^3	For total dust with no asbestos and <1% silica
Mica, respirable	3 mg/m^3	
Nuisance dusts	10 mg/m^3	For total dust with no asbestos and <1% silica
Oil mist, mineral	5 mg/m^3	Excludes vapor; TLV-STEL is 10 mg/m^3
Perlite	10 mg/m^3	For total dust with no asbestos and 1% silica
Portland cement	10 mg/m^3	For total dust with no asbestos and 1% silica
Silica, crystalline		
Cristobalite	0.05 mg/m^3	Respirable fraction
Quartz	0.1 mg/m^3	Respirable fraction
Tridymite	0.05 mg/m^3	Respirable fraction
Soapstone, respirable dust	3 mg/m^3	With no asbestos and <1% silica
Soapstone, total dust	6 mg/m^3	With no asbestos and <1% silica
Talc	2 mg/m^3	With no asbestos; otherwise use asbestos values
Welding fumes	5 mg/m^3	Composition of welding fumes varies; may be subject to other TLVs

ACGIH, American Congress of Governmental Industrial Hygienists; *STEL*, short-term exposure limit; *TLV*, threshold limit value; *TWA*, time-weighted average.
[a]Proposed for change to 0.2 fiber/cm^3.

Table 12.11 Specifications for Disel Particulate Monitor

System Weight	1.35 lbs (611 g) Excluding Battery
Power requirements	7.4 VDC (Li-ion battery) 100–240 VAC (wall charger)
Sensitivity	<15 μ/m^3 elemental carbon (EC)
Dynamic range	9–600 μ/m^3 (8 h. TWA EC)
Output	LCD display with user-controlled backlight User selectable 1, 5, 15 min averaging EC and TC data logged 8 h TWA DPM levels Mini-USB connection
Alarms	Low battery, filter change necessary Pump flow LED/LCD alerts
Battery life	>12 h
Dimensions	Approximate 6″(H) x 4.5″(W) x 2.75″(D)

12.7 Diesel Particulate Monitor

It has a potential application in metal mines but not in coal mines. The diesel particulate monitor (DPM) from ICx Technologies [22] displays elemental carbon levels in real time, taking the measurement out of the laboratory and placing it in the hands of the mine operator. Being sensitive, rugged, and easy to use, the DPM provides real-time results that are time- and space resolved. This capability enables rapid modification of vehicle use, personnel placement, and mine ventilation. The monitor uses technology developed by the diesel particulate group at the NIOSH Pittsburgh Research Laboratory and has been determined to precisely replicate results from their method 5040 test [22].

In addition to being compact and lightweight enough to be worn on a miner's belt, the DPM can be mounted in a vehicle cab, on a mine wall, or on ventilation equipment. The monitor operates on a Lithium-ion battery for more than a full shift or via AC power using an adapter. DPM readings are displayed on an LCD screen with a user-selectable backlight and may be downloaded via USB connection for review of extended monitoring operations. An integrated air pump and submicron particle size selector are included. Specifications of the instrument are shown in Table 12.11.

Rigorous field testing and comparison with NIOSH 5040 method is yet to be done.

References

[1] Jacobsen M, et al. New dust standards in British coal mines. Nature August 1, 1970;227: 445–7.

[2] Hartman HL, et al. Mine ventilation and air conditioning. 2nd ed. Wiley-Interscience Publication; 1982. p. 105–11.

[3] Dunmore JH, et al. An instrument for the sampling of respirable dust for subsequent gravimetric assessment. Journal of Scientific Instruments 1964;41:669–72.
[4] Lilienfeld MP, Dulchines J. Portable instaneous mass monitor for coal mine dust. American Industrial Hygienists Association 1972:136–45.
[5] Hering SV. In: Hering SV, editor. Inertial and gravitational collectors, air sampling instruments. Cincinnati, Ohio: ACGIH; 1989. p. 337–85.
[6] Thakur PC, Reister JB, Hatch R. Performance evaluation of machine mounted respirable dust monitors for US coal mines. In: Marple V, Liu B, editors. Aerosols in the mining and industrial work environments, vol. 3. Ann Arbor Science Publishers; 1983. p. 665–88.
[7] Soderholm SE. ISO respirable dust breathing curve. Annals Occupational Medicine 1989; 33:301–20.
[8] Rupprecht and Patashmick Co, Inc. Operating manual for TEOM, series 3600/3700, personal dust monitor, revision A. R&P part number 42–009904. September 2004.
[9] Volkwein JC, et al. Laboratory and field performance of a personal dust monitor. U.S. Department of Health and Human Services, CDC, National Institute of Occupational Safety and Health, Report number 7669; 2006.
[10] Page SJ, et al. Respirable dust sampler equivalency of a personal dust monitor, internal NIOSH report. 2008. p. 18.
[11] Thakur PC, Reister JB, Parisi WC. Determination of quartz in respirable coal dust. In: Karmis M, editor. Proceedings of 14th Annual Institute on coal mining health, safety and research. Blacksburg, VA: VPI; 1983. p. 127–37.
[12] Sutherland WH. Silica determination in coal dust, presented at the AMC coal convention. May 11, 1982.
[13] Holgate W.A.. The current quartz program and part 90 miner rules, 30 CFR 9. In: Proceedings of the 12th Annual Institute on Coal Mine Health, Safety and Research, VPI, Blacksburg, VA, August 25–27, 1981.
[14] Le Bouffant L, Martin JC, Daniel H. Harmfulness of dust in relation to its quartz content. In: Conference on technical measures of dust prevention and suppression in mines, Luxembourg; 1973. p. 127–8.
[15] Naeye RL, Dellinger WS. Lung diseases in Appalachian soft coal miners. American Journal of Pathology 1970;58:557–64.
[16] Saltsman RD, Constantine JP. Coal mine dust control and pneumoconiosis research in Europe and United States. In: Proceedings of the 12th Annual Institute on Coal Mine Health, Safety and Research, VPI, Blacksburg, VA; 1981.
[17] NIOSH report number 75-12 Criteria for a recommended standard - occupational exposure to crystalline silica. 1974.
[18] American Conference of Governmental Industrial Hygienists Proceedings of round table discussion on analytical techniques for quartz. December 6–7, 1972.
[19] Chung FH. Synthesis and analysis of crystalline silica. Environmental Science and Technology 1982;16(11):796–9.
[20] Dixon, Massey. Introduction to statistical analysis. McGraw Hill; 1969. p. 80.
[21] ACGIH. Threshold limit values for chemical substances and physical agents and biological exporter indices. Cincinnati, Ohio. 1996 [updated annually].
[22] Diesel Particulate Monitor, ICx Technologies. Air Hazard Monitoring Bulletin, Arlington, vol. A. www.icxt.com.

Section Three

Combustible Gas Control

Origin of Gases in Coal Mines

13

Chapter Outline

13.1 Introduction

Coal seams were formed over millions of years by the biochemical decay and metamorphic transformation of the original plant material. This process, known as coalification, produces large quantities of by-product gases increasing with the rank of coal and is the highest for anthracite at about 27,000 ft^3/t [1] for methane alone. Most of these gases escape to the atmosphere during the coalification process but a small fraction is retained in coal. The amount of gas retained in coal depends on a number of

Advanced Mine Ventilation. https://doi.org/10.1016/B978-0-08-100457-9.00013-4

factors such as the rank of coal, the depth of burial, the immediate roof and floor to the coal seam, geologic anomalies, tectonic pressures, and temperature prevailing at the end of the coalification process. In general, the higher the rank of coal and the greater the depth of coal seam, the higher is the gas content of coal. Gas content of coal seams vary from a few ft^3/t to 800 ft^3/ton for depth up to 3000 ft. Coal seams are the source as well as the reservoir for all gases.

Methane is the major component of gas in coal, comprising 80%–95% of the total gas content. The balance is made up of ethane, propane, butane, carbon dioxide, hydrogen, oxygen, and argon. All coal seams begin to oxidize when exposed to ventilation air in a mine and produce some carbon monoxide and hydrogen sulfide. Sulfur dioxide and nitrogen oxides are produced by diesel equipment in coal mines. A brief introduction to all these gases is presented below.

13.2 Properties of Gases in the Mine Atmosphere

13.2.1 Methane

It is a colorless, odorless, tasteless, nontoxic gas that is flammable in the range 5%–15% (by volume). It can be also fatal if a person walks into an atmosphere containing very high concentrations of methane. It is lighter than air with a specific gravity of 0.55 and, therefore, has a tendency to accumulate in cavities in the roof or layering against the roof. Measurement of methane concentrations in such cavities should be done remotely with extended probes.

Excess methane can slow down mining or even completely stop it creating great economic losses. Methods to drain and control methane in mine airways will be discussed later in the book. Table 13.1 shows the main characteristics of methane and other mine gases. The law requires that methane concentration in mine air be kept below 1% (except bleeders can have 2%).

13.2.2 Ethane

It is very similar to methane except its concentration in coal seam gases is 0.25%–2% only. Only deep, highly gassy coals have 2% ethane. Table 13.1 shows the main properties of ethane. It is usually not measured separately in mine air.

13.2.3 Propane and Butane

Coal seam gas has only a trace of these gases (a few ppm). Higher concentrations usually indicate a leakage from a natural gas well that may not have been plugged properly. The presence of helium also indicates a leakage of natural gas into the mine.

13.2.4 Carbon Monoxide

It is a colorless, odorless, tasteless, toxic gas. It is also flammable in the range 12.5%–74%. It is mostly created by spontaneous combustion of coal, explosions, blasting, and internal combustion of diesel engines.

Table 13.1 Properties of Common Mine Gases

Name	Symbol	Specific Gravity (Air = 1)	Density lb/ft^3	Harmful Effects	TLV-TWA ppm	TLV-STEL ppm	TLV-C ppm
Oxygen	O_2	1.1056	0.083	Nontoxic	—	—	—
Nitrogen	N_2	0.9673	0.073	Nontoxic, simple asphyxiant	—	—	—
Carbon dioxide	CO_2	1.5291	0.115	Asphyxiant, increased respiration	5,000	15,000	—
Methane	CH_4	0.5545	0.042	Asphyxiant, explosive	—	—	20,000
Ethane	C_2H_6	1.03	0.080	Asphyxiant, explosive	—	—	—
Carbon monoxide	CO	0.9672	0.073	Toxic, explosive	50	400	—
Hydrogen sulfide	H_2S	1.1912	0.890	Toxic, explosive	10	15	—
Sulfur dioxide	SO_2	2.2636	0.170	Toxic	2	5	—
Oxides of nitrogen	NO, NO_2	1.5895	0.119	Toxic	25	—	5
Hydrogen	H_2	0.0695	0.005	Explosive	—	—	—

Adapted from Anon: Threshold limit values for chemical substances and physical agents in the workroom environment with intended changes, American Conference of Governmental Industrial Hygienists, Cincinnati, Ohio, USA.

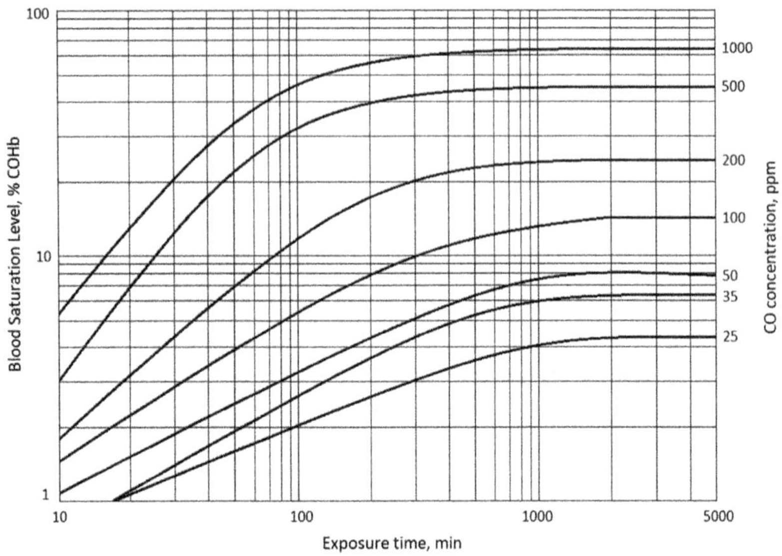

Figure 13.1 Carbon monoxide blood saturation [2].

The main danger from CO is due to its toxicity. It has a great affinity for the hemoglobin in the blood with which it forms a permanent bond —called carboxyhemoglobin (COHb). If the blood saturation by COHb increases, various symptoms become apparent, and they are listed in Table 13.2. The rate of saturation depends on CO level in mine air. If the CO concentration is 1% (10,000 ppm), death can occur in a minute. Death can also occur at 70%–80% blood saturation with COHb (Table 13.1).

Fig. 13.1 shows gradual blood saturation as a function of CO concentration and exposure time [2].

Table 13.2 Gradual CO Poisoning of Blood and Symptoms

Blood Saturation (% COHb)	Symptoms
5–10	First noticeable effect, loss of some cognitive function.
10–20	Tightness across forehead, possible headache.
20–30	Headache, throbbing in temples.
30–40	Severe headache, weakness, dizziness, dimness of vision, nausea and vomiting, and collapse
40–60	Increased likelihood of collapse and unconsciousness, coma with intermittent convulsions.
60–70	Coma, possible death.
70–80	Respiratory failure, death.

13.2.5 Carbon Dioxide (Oxygen Depletion)

It is a colorless, odorless gas with a mild acidic taste in high concentrations. It is heavier than air (specific gravity 1.5) and thus it accumulates near the mine floor. The coal seam as well as the mined out area can be, and usually is, the main source of CO_2. It is an asphyxiating gas. Increasing concentration of CO_2 in air makes breathing faster and exacerbates the health effects of oxygen depletion. Table 13.3 shows the health effects of reduced oxygen in air.

Properties of CO_2 are listed in Table 13.1.

13.2.6 Hydrogen

It is a colorless, odorless, tasteless, and nontoxic gas. It is explosive in the range 4%–74% and can explode with as low as 5% of oxygen [3]. Main source of hydrogen in mines is the battery charging station and mine fire.

Table 13.1 shows the properties of hydrogen.

13.2.7 Hydrogen Sulphide

It is a colorless, toxic gas with a smell like rotten eggs. It is therefore called "stink damp." It is also explosive in the range 4%–44%. It is extremely toxic and the threshold limit value for it is 10 ppm [2] with a short-term exposure not to exceed 15 ppm. Concentration above these values creates eye and throat irritations. At 1000 ppm, a person may become unconscious and death follows in a few minutes.

Table 13.1 shows properties of H_2S.

It is often smelled in coal mines but seldom detected in measurable concentrations. It is highly soluble in water.

13.2.8 Oxides of Nitrogen

The main two oxides of nitrogen, NO and NO_2, are produced mainly by diesel equipment in the mines. NO gets quickly oxidized to NO_2 which is soluble in water. NO_2 has a ceiling limit of 5 ppm that should not be exceeded in mine air.

Table 13.3 Health Effects of Reduced O_2 Concentration

O_2% (by Volume) in Air	Effects
20.8 (normal air)	None
17 (safety lamp extinguishes)	Deep breathing.
15	Dizziness; buzzing in ear.
13	May lose consciousness.
9	Unconscious.
7	Life threatened.
6	Convulsions, death.

In higher concentration, it creates throat irritation and coughing but if inhaled for long time, it can cause death the next day by pulmonary edema (retention of water in lungs). All diesel engine exhausts are continually monitored with hand-held instruments to make sure no one is overexposed.

The properties of NO and NO_2 are listed in Table 13.1. The threshold limit values (TLV) for NO is 25 ppm.

13.2.9 Sulfur Dioxide

It is colorless, nonflammable and a highly toxic gas. Mine fires (in coal seams containing sulfur), blasting, and some internal combustion engines produce SO_2. It irritates the eyes, nose, and throat. The TLV is set at 2 ppm but the STEL is at 5 ppm [2].

Properties of SO2 are listed in Table 13.1.

13.2.10 Variations in TLV

It will be appropriate here to define the variations in the TLV for gases. They are three in kind.

1. TLV—Time-weighted average (TWA)
   It is an average value over an exposure period of 8 hours per work-day or 40 h per work-week that does not create any adverse health effects. In short, it is just called the TLV.
2. TLV−TWA—Short-term limit (STEL)
   A higher concentration of gas above its TLV−TWA can be tolerated but only for a very short period. Thus, the TLV for CO is only 50 ppm but the TLV−TWA−STEL is at 400 ppm for 15 min provided no more than four such exposures occurs in a day with at least 60 minutes of normal air between two high CO exposures. The TLV−TWA will not be exceeded in spite of short durations of higher exposures.
3. TVA—Ceiling
   Some gases are so toxic that their TLV has a ceiling limit that cannot be exceeded at any time. For example, the TLV for NO_2 at 5 ppm is a ceiling limit or TLV-C. For methane, it is 2%.

13.3 Characteristics of Coal

Coal is a heterogeneous and anisotropic material. As far as the gas storage in coal is concerned, only the rank of the coal is of significance. To determine the rank, two types of laboratory tests are done on all coal:

Proximate analysis and
Ultimate analysis.

The data thus obtained is used for estimating gas contents, respirable dust index, and spontaneous heating characteristics of a particular coal seam. These subjects are discussed in detail under appropriate headings in this text.

13.3.1 Proximate Analysis

This analysis basically measures four constituents of coal as follows:

1. Moisture
2. Volatile matter
3. Fixed carbon
4. Ash

American Standards for Testing Material (ASTM) uses the ash-free−fixed carbon and calorific value of coal to rank coal seams. Table 13.4 shows a complete list.

International coal classification standards were developed by the Coal Committee of the Economic Commission for Europe. Table 13.5 shows both ASTM and European classifications side by side for ease in comparison.

International system uses a three digit number to describe three different properties of the coal. The first digit shows the rank as shown in Table 13.5. The second digit shows the Roga index or free swelling index when coal is heated rapidly. The third digit shows the coking properties of metallurgical coal when it is heated slowly and is measured by Gray-King Coke index [5].

13.3.2 Ultimate Analysis

Ultimate analysis of coal breaks it into four elements: O_2, H_2, C, and N_2. Usually carbon content increases with the rank of coal but H_2 and O_2 contents go down. Oxygen on dry, ash-free basis is used to determine the liability of coal to spontaneous combustion (discussed in Chapter 21).

13.3.3 Rank of Coal Versus Vitrinite Reflectance, R_o

Vitrinite reflectance is a measure of percentage of incident light reflected from the surface of vitrinite particle/bands in coal seams. R_o is usually an average of several readings. A typical coal seam has four megascopic bands known as vitrain, clarain, durain and fusain. Only vitrain is used for the ranking of coal (Table 13.6).

13.4 Characterization of Methane from Coal

Most coalbed methane (CBM) can be classified in two groups: (a) biogenic and (b) thermogenic. Low-rank coal usually contains methane of biogenic origin, whereas more mature, high-rank coals have thermogenic methane. Besides rank (or R_o), there are other indices that characterize CBM as either biogenic or thermogenic. Most important indices are listed below.

13.4.1 Hydrocarbon Index

It is defined as the ratio of CH_4 to (C_2H_6 + C_3H_8).

Table 13.4 Classification of Coals by Rank

Rank	Group	Fixed Carbon Limits, % (Dry, Mineral-Matter—Free Basis)		Volatile Matter Limits, % (Dry, Mineral-Matter—Free Basis)		Calorific Value Limits, Btu per Pound (Moist[a] Mineral-Matter—Free Basis)	
		Equal or Greater Than	Less Than	Greater Than	Equal or Less Than	Equal or Greater Than	Less Than
I. Anthracitic	1. Metaanthracite	98	—	—	2	—	—
	2. Anthracite	92	98	2	8	—	—
	3. Semianthracite	86	92	8	14	—	—
II. Bituminous	1. Low-volatile coal	78	86	14	22	—	—
	2. Medium-volatile coal	69	78	22	31	—	—
	3. High-volatile bituminous A coal	—	69	31	—	14,000[b]	—
	4. High-volatile bituminous B coal	—	—	—	—	13,000[b]	14,000
	5. High-volatile bituminous C coal	—	—	—	—	{11,500 {10,500	13,000 11,500
III.Subbituminous	1. Subbituminous A coal	—	—	—	—	10,500	11,500
	2. Subbituminous B coal	—	—	—	—	9,500	10,500
	3. Subbituminous C coal	—	—	—	—	8,300	9,500
IV. Lignitic	1. Lignite A	—	—	—	—	6,300	8,300
	2. Lignite B	—	—	—	—	—	6,300

[a]Moist refers to coal containing its natural inherent moisture but not including visible water on the surface of the coal.
[b]Coals having 69% or more fixed carbon on the dry, mineral-matter—free basis shall be classified according to fixed carbon, regardless of calorific value.
Adapted from Leonard JW, Mitchell DR. Coal preparation. The American Institute of Mining, Metallurgical and Petroleum Engineers, Inc.; 1968. 1-27-29.

Table 13.5 Parameters of International Coal Classification Compared to American Standards for Testing Material (ASTM) Rank

International Class	ASTM Rank[a]	Volatile Matter, Dry, Ash-free Percent or Btu, Moist, Ash-Free
1A	Anthracite	3–6.5
1B	Anthracite	6.5–10
2	Semianthracite	10–14
3	Lvb	14–20
4	Mvb	20–28
5	Mvb	28–33
6	Hvab	>13,950
7	Hvbb	12,960–13,950
8	Hvcb or Sba	10,980–12,960
9	Sbb	10,260–10,980

Report of Investigations 5435 and Ref. [1] of this chapter.
[a]*Lvb*, low-volatile bituminous; *Mvb*, medium-volatile bituminous; *Hvab*, high-volatile A bituminous; *Hvbb*; high-volatile B bituminous, *Sba*, subbituminous A, *Sbb*, subbituminous B.

Table 13.6 Shows the Dependence of Rank on Ro Values [4]

Coal Rank	R_o (%)	Remarks
Peat/Lignite	<0.2	As the rank of coal increases from lignite to anthracite, generally the carbon content and calorific value increases but volatile matter and moisture contents decrease.
Subbituminous	0.2–0.6	
High-volatile bituminous	0.6–1.1	
Mid-volatile bituminous	1.1–1.5	
Low-volatile bituminous	1.5–2.0	
Semianthracite	2.0–2.5	
Anthracite	2.5–6.0	
Met-anthracite	6.0–10.0	
Graphite	>10.0	

13.4.2 CO_2 Percent

It is not very reliable but biogenic methane has less than 5% CO_2, whereas thermogenic coal can contain 2%–15% CO_2. In some coals in Australia, it can be as high as 50% by volume.

13.4.3 Gas Wetness Index

It is defined on a ratio of

$$\frac{(C_2H_6 + C_3H_8 + C_4H_{10} + C_5H_{12})}{(CH_4 + C_2H_6 + C_3H_8 + C_4H_{10} + C_5H_{12})}$$

Typically the index is above 3% for thermogenic methane. If less then 3%, the methane is of biogentic origin.

13.4.4 The δ ^{13}C and δD Isotropic Ratios

Isotopes are different forms of the same element. Carbon, for example, has three naturally occurring isotopes: ^{12}C (carbon-12), ^{13}C (carbon-13) and ^{14}C (carbon-14). ^{14}C is radioactive and gives out beta ray that has been used for respirable dust measurement, but its concentration in coal is low, on the order of 1×10^{-10} percent in atmospheric carbon dioxide. Most carbon is bout 99% ^{12}C and 1% ^{13}C.

Similarly, there are three different forms of hydrogen. Tritium or hydrogen-3 is radioactive and is formed in the upper atmosphere by cosmic rays. The stable isotopes are ^{1}H (hydrogen-1) and ^{2}H (hydrogen-2).

In a compound like methane, there can be three kinds of carbon and three kinds of hydrogen present. By measuring the relative concentration of these isotopes, we can characterize CBM from different sources.

Stable isotope compositions are generally expressed using the δ (delta) notation. Thus,

$$\delta^{13}C_{sa} = \frac{(^{13}C/^{12}C)_{sa} - (^{13}C/^{12}C)_{st}}{(^{13}C/^{12}C)_{st}} \times 1,000 \qquad (13.1)$$

and

$$\delta D_{sa} = \frac{(^{2}H/^{1}H)_{sa} - (^{2}H/^{1}H)_{st}}{(^{2}H/^{1}H)_{st}} \times 1,000 \qquad (13.2)$$

sa—means sample of methane.

st—standard.

For carbon, the standard is Vienna Pee Dee Belemnite.

For hydrogen, the standard is Vienna Standard Mean Ocean Water.

Table 13.7 shows the range of all the above indices for biogenic and thermogenic methane [5,6].

13.4.5 Differentiating Between Thermogenic Coalbed Methane and Natural Gas

The geochemical fingerprint of a thermogenic gas is dependent on the type of organic material from which the gas was originally formed [7]. Most CBM originated from

Table 13.7 Indices for Distinction Between Thermogenic and Biogenic Coalbed Gases

Indices	Origin of Gas	
	Thermogenic	**Biogenic**
Vitrinite reflectance (R_o in %)	0.6%–3.0%	0.3%–0.8%
Hydrocarbon index $[CH_4/(C_2H_6 + C_3H_8)]$	<20	>1000
Gas wetness index $C_{2+} = [C_2H_6 + C_3H_8 + C_4H_{10} + C_5H_{12})/(CH_4 + C_2H_6 + C_3H_8 + C_4H_{10} + C_5H_{12})]$	>3%	<3%
CO_2 content	2–15 vol%	<5 vol%
$\delta\ ^{13}C$ of methane (in %) vs. Vienna Pee Dee Belemnite	>–50%	<–55%
δD of methane (in %) vs. Vienna Standard Mean Ocean Water	–275 to –100%	–400 to –150%
$\Delta\ ^{13}C_{CO_2 - CH4}$	<40%	>60%

Adapted from Thakur PC, et al. Coalbed methane from Prospect to pipeline. Elsevier; 2014. p. 7–29.

plant material whereas natural gas (from deeper sandstone/limestone bedrocks) is of marine origin.

Most prominent differences are the following:

1. CBM has a higher methane/ethane ratio because the ethane concentration is low.
2. Natural gas contains significant amount of butane and helium, whereas CBM has only traces of these gases.
3. When δD of CH_4 is plotted against $\delta\ ^{13}C$ of CH_4, natural gas has usually higher values of $\delta\ ^{13}C$ (typically –40) and δD (–150 to –100). The $\delta\ ^{13}C$ for CBM is lower than –50 and δD is also lower than –150. The difference is quite distinct (Fig. 13.2).

13.5 Coalbed Methane—An Energy Source

Coal mine degasification started in United States in 1970s. The vast amount of gas recovered was discharged into the atmosphere in the beginning. This was liable to make global warming worse because methane is 23 times more effective in trapping infra-red radiations (radiative forcing) than CO_2. Hence, the gas was processed to meet the gas pipeline specifications and was marketed for additional profit. Shortly after that CBM production from deeper coal seams (that were not mined) started in western United states. The CBM production peaked at 1.8 TCF, about 10% of the total gas production in the United States in 2008. It is, therefore, appropriate to highlight this commercial potential of gas in coal.

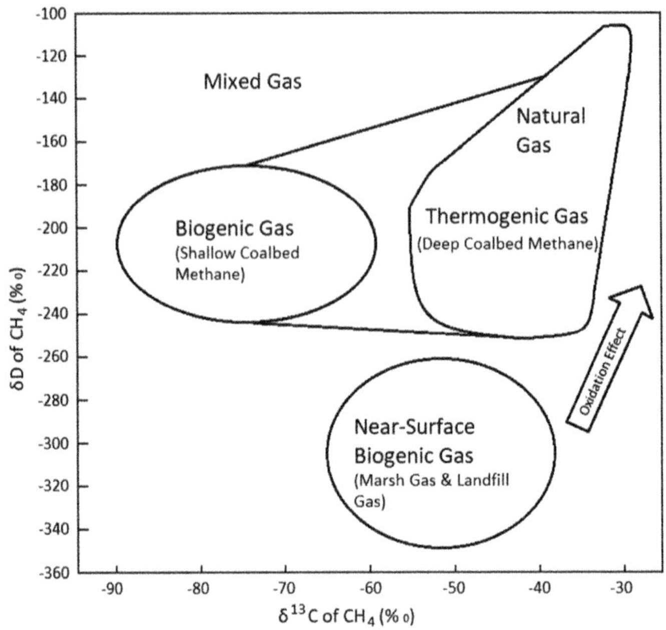

Figure 13.2 Carbon and hydrogen isotopic compositional ranges of methane from different sources.
Chemical Geology, Ed. M Schoell, Vol 71, 1988, pp. 1-10.

13.5.1 Global Reserve of Coalbed Methane

Coal is found in 70 countries around the world and currently provides 41% of all electricity consumed [8]. Table 13.8 shows the estimated CBM reserve for selected countries.

It is a vast reserve of gas that can supplement natural gas reserves globally. Reservoir and production engineering for commercial CBM production is available in literature [8].

Application of horizontal drilling technology with hydrofracturing can easily double the US CBM production. For working mines, the benefits of CBM production are many as listed below:

1. Safety in mines and prevention of disasters.
2. Increased productivity for coal resulting in reduced cost of mining.
3. Additional revenue from gas sales.

13.5.2 The United States Reserves of Coalbed Methane

There are currently 12 coal basins in the United States producing CBM. Detailed descriptions are available in recently published books [5] [8]. Table 13.9 lists the main basins with their location, CBM production, and reserves.

Table 13.8 Estimates of World Coalbed Methane Reserve

Country	Estimated Coal Reserve (10^9 tons)	1992 Estimated TCF	1987 Estimated TCF
United States	3,000	388	30–41
Russia	5,000	700–5,860	118–790
China	4,000	700–875	31
Canada	300	212–2682	92
Australia	200	282–494	N.A.
Germany	300	106	2.83
India	200	35	0.7
South Africa	100	35	N.A.
Poland	100	106	0.4–1.5
Other Countries	200	177–353	N.A.
Total gas in place		30,958–33,853	275–11,296

Table 13.9 US Coal Basins and Coalbed Methane (CBM) Reserves

Coal Depth	Basins	CBM Production (BCF/Yr)	CBM (TCF) Reserves
I. Shallow (100–1500 ft)	**(a)** Powder River Basin	280	100[a]
	(b) Cherokee Basin	5	20[a]
	(c) Illinois Basin	1	21
	(d) Northern Appalachian Basin	10	61
II. Medium depth (1500–3000 ft)	**(a)** Central Appalachian	94	21
	Basin	52	21–22
	(b) Warrior Basin	105	11
	(c) Raton Basin	100	3
	(d) Arkoma Basin		
III. Deep (+3000 ft)	**(a)** San Juan Basin	650	84
	(b) Uinta Basin	40	42
	(c) Piceance Basin	5	84
	(d) Green River Basin	20	83
Total		**1362**	**571[b]**

[a]Estimated from coal tonnage in the basin.
[b]Does not include Alaska.

References

[1] Hargraves AJ. Planning and operation of gaseous mines, CIM Bulletin. March 1973.
[2] Anon: Threshold limit values for chemical substances and physical agents in the workroom environment with intended changes, American Conference of Governmental Industrial Hygienists, Cincinnati, Ohio, USA.
[3] Coward HF, Jones GW. Limits of flammability of gases and vapors. USBM Bulletin No 1952;503.
[4] Leonard JW, Mitchell DR. Coal preparation. The American Institute of Mining, Metallurgical and Petroleum Engineers, Inc.; 1968. 1-27-29.
[5] Thakur PC, et al. Coalbed methane from Prospect to pipeline. Elsevier; 2014. p. 7–29.
[6] Schoell M. Multiple origin of methane in the Earth. In: Schoell M, editor. Origins of methane in the earth. Chemical Geology, vol. 71; 1988. p. 1–10.
[7] Coleman DD. Advances in the use of geochemical fingerprinting for gas identification. In: American Gas Operations Conference, San Francisco, CA; May 9–11, 1994. p. 9.
[8] Thakur PC. Advanced reservoir and production engineering for coalbed methane. Elsevier; 2017. p. 210.

Reservoir Properties of Coal Seams

14

Chapter Outline

The most important reservoir properties that dictate not only the gas production and mine degasification rates but also the correct degasification techniques are

- Gas content of coal and its gas isotherm;
- Coal matrix permeability;
- Diffusivity of coal or sorption time and;
- Reservoir pressures and ground stress.

These subjects have been discussed in great detail in a recently published monograph [1]. A summary is provided here of only those properties that influence the drainage of methane from coal prior to mining and postmining.

14.1 Gas Content of Coal

The volume of gas contained in coal at standard temperature and pressure (STP)* is termed the gas content of the coal and is expressed in cubic feet per ton. It is generally accepted that gas is stored in a monolayer on the microscopic particles of coal that are

Advanced Mine Ventilation. https://doi.org/10.1016/B978-0-08-100457-9.00014-6

smaller than the micropores in the coal matrix. At greater depth, the gas may be in a "condensed, liquid-like state" [2]. The volume of gas retained in coal is dependent on the rank, temperature and pressure, and the depth of the coal seam. The microscopic surface of coal is large; a ton of coal has a surface area of approximately 2218 million ft^2 (200 Mm^2). Thus, one cubic foot of coal can store two to three times the amount of gas contained in a typical sandstone reservoir for natural gas of the same volume but at higher pressure.

Gas content measurement methods are classified as (1) conventional and (2) pressurized desorption techniques. In the conventional technique, coal cores or drill cuttings are retrieved from the core holes and immediately put in a sealed container to measure the desorbed gas. This method suffers from uncertainty in the estimate of gas lost during sample retrieval and handling. To eliminate this problem, the pressurized core desorption technique has been developed. In this technique, gas loss is minimized by sealing the coal samples while they are in the core hole. Both methods provide positive proof of gas presence. Desorbed gases are chemically analyzed to determine the composition and calorific value of coalbed methane.

14.1.1 The Direct Method of Gas Content Measurement

This technique was originally developed by Bertard and Kissell [3,4]. It was further improved by Diamond and Schatzel [5] and it became the "ASTM standard practice for determination of gas content of coal" [6]. In this technique, the desorbed gas from the coal sample is measured first. Next, the cumulative gas production is plotted against the square root of time to determine the lost gas. Finally, a small, weighted portion of coal sample is crushed in a hermetically sealed mill to get the residual gas. The total gas content is the sum of the three components: (1) desorbed gas, (2) estimated lost gas, and (3) residual gas.

14.1.2 Desorbed Gas

After coal cores or drill cuttings are put in a hermetically sealed container, called a desorption canister, the desorbed gas is measured periodically. In the first few days, readings may be taken every hour, but later a measurement once a day is sufficient. The general layout of the experimental setup is as shown in Fig. 14.1 [5].

The desorption canister is about 18 in. tall, with a 4 in. internal diameter. It is equipped with a pressure gage and a valve to let the desorbed gas out. The desorbed gas is measured by water displacement in a graduated glass cylinder 4 in. in diameter and 12 in. high. The glass cylinder is connected to a leveling water reservoir, and the gas volume measurement is taken when the water levels in the cylinder and leveling reservoir are the same. The precision of the measurement is about $\pm 4\%$ [7].

The desorption process typically extends to 4–6 weeks. It is stopped when gas desorption is less than 10 cm^3/day. The cumulative gas production is plotted on a graph paper against $(time)^{0.5}$ to determine the lost gas component of the total gas content (discussed in the next paragraph). The desorbed gas is periodically analyzed using a gas chromatograph to determine its composition and calorific value.

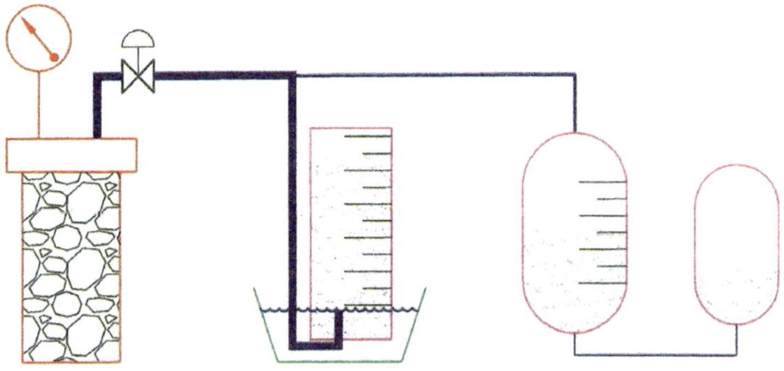

Figure 14.1 Gas content measuring apparatus.

14.1.3 Lost Gas

This portion of the total gas content is the gas that escapes from the sample during its collection and retrieval, prior to being sealed in an airtight canister. It is estimated indirectly. Most gas desorption processes from coal or shale follow a power law [8,9].

$$Q = At^n \tag{14.1}$$

where Q is the cumulative volume of gas desorbed in ft^3; A is a characteristic of the coal (equals initial production in gas wells); t is time in days or minutes; and n is a characteristic of the coal or shale.

Eq. (14.1) can be expressed in its logarithmic form as

$$\ln Q = \ln A + n \ln t \tag{14.2}$$

The value of "n" for most coal is 0.8−1.00. Hence, a plot of ln Q against ln t yields a straight line. The intercept on the "y" axis is equal to ln A.

In a simplified version of Eq. (14.1), (refer to Equation (14.16))

$$Q = B\, t^{\frac{1}{2}} \tag{14.3}$$

Hence, a plot of cumulative desorbed gas, Q, against $(t)^{0.5}$ yields a straight line. Here, B is the intercept on the y axis and is a measure of the lost gas as shown in Fig. 14.2.

14.1.4 Residual Gas

Even when the coal sample in the desorption container has stopped producing gas, a significant volume of gas is still left in the sample. It can only be retrieved and measured by crushing the sample to very fine sizes.

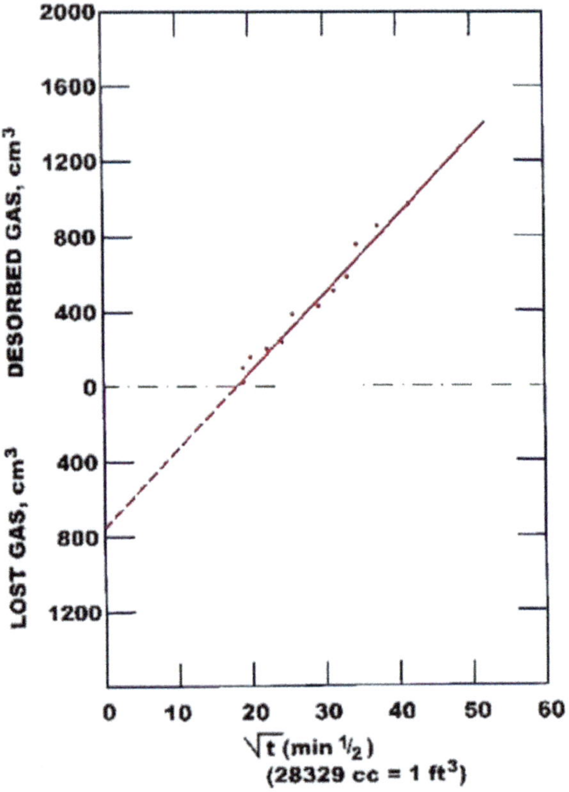

Figure 14.2 Lost gas estimation graph.

A hermetically sealed modified ball mill (Bleuler Mill) [10] is used for this purpose. A measured quantity of the coal core or drill cutting is put in the mill and crushed. The released gas is measured by the same setup that was used for desorbed gas measurements.

The total gas content of the coal sample is obtained by adding the three components, i.e., desorbed gas, lost gas, and residual gas. The coal sample is next weighed and sent to a laboratory for a proximate analysis which yields the moisture, ash, volatile matter, and fixed carbon contents of coal. The weight of coal is calculated on a dry, ash-free basis. The total gas content of the coal sample is divided by the weight of the coal sample (dry, ash-free) to get the final gas content of coal in ft^3/t ($1\ cm^3/gm = 32\ ft^3/t$).

Table 14.1 shows the gas content and gas composition data for some typical US coal seams.

14.1.5 Gas Isotherms and Indirect Methods of Gas Content Determination

At constant temperature, each coal seam shows a measurable relationship between the total gas adsorbed (or desorbed) and the confining pressure. Fig. 14.3 shows typical gas isotherms for five US coal seams.

Table 14.1 Coalbed Methane Content and Composition of US Coal Seams

Coal Seam	Rank	Gas Content (ft^3/t)	Composition[a] (%)					Calorific Value BTU/ft^3
			CH_4	C_2H_6	C_3H_8	H_2	CO_2	
Pocahontas #3 (VA)	LV	450–650	97–98	1–2	Trace	0.02	0.2–0.5	949–1058
Hartshorne (OK)	LV	200–500	99.20	0.01	–	–	–	900–1058
Kittanning (PA)	LV	200–300	95–98	0.02	Trace	–	0.1–0.2	1020
Mary Lee (AL)	LV	200–500	96	0.01	–	–	–	1024
Pittsburgh #8 (WV)	HVA	100–250	89–95	0.25–0.5	Trace	–	2–11	949–1000
Mesaverde Formation (NM)	Sub bituminous	100–300	88	–	–	–	12	938

[a]N_2 , LV - Low Volume, HVA - High Volume Anthracite, and argon contents are not listed but are needed to make the total 100%.

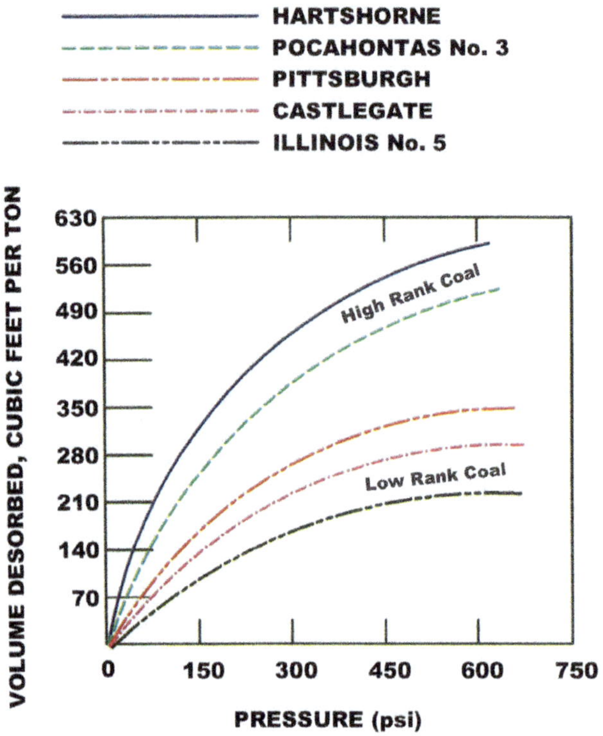

Figure 14.3 Gas isotherms for US coal seams.

It is to be noted that high-rank (low-volatile bituminous) coals contain more gas than the low-rank (high-volatile bituminous) coals (HVA, HVB, HVC) at the same confining pressure. It is also clear that the sorption capacity of all coal increases with pressure, but the increase occurs at an ever-decreasing rate as the sorption capacity reaches an asymptotic limit—the saturation limit.

A mathematical representation for these isotherms was given by Langmuir [11] and is expressed as

$$V = V_m \frac{bP}{1 + bP} \tag{14.4}$$

where V is the volume of gas contained at pressure P, ft^3/t; V_m is the maximum sorption capacity of coal, ft^3/t; P is the pressure, psi; and b is the Langmuir constant, $psi.^{-1}$

For indirect determination of the gas content of coal at a given pressure, Eq. (14.4) can be rewritten as

$$\frac{P}{V} = \frac{P}{V_m} + \frac{1}{b\,V_m} \tag{14.5}$$

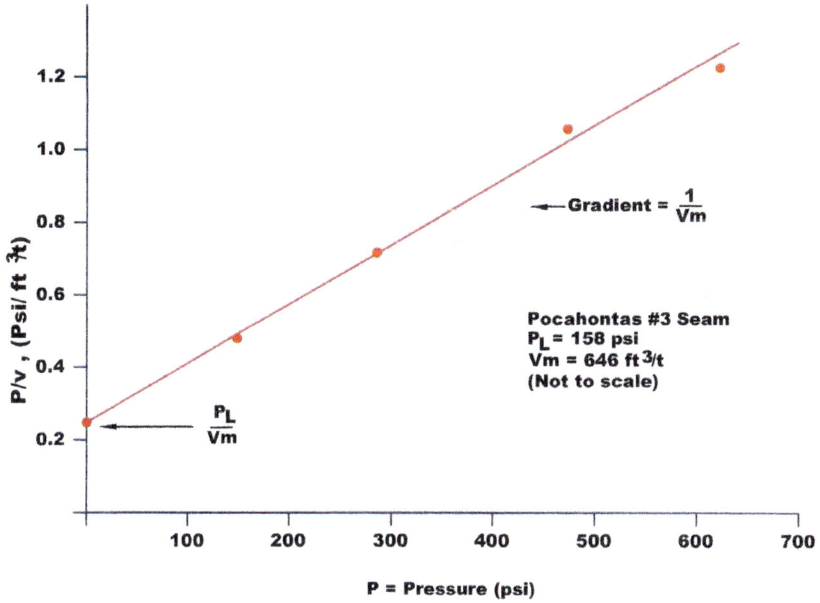

Figure 14.4 A plot of P/V against P.

The term "b" is experimentally found to be equal to $1/P_L$, where P_L is the characteristic pressure that corresponds with $V_m/2$ on the gas isotherm.

Thus, Eq. (14.5) can be written as

$$\frac{P}{V} = \frac{P_L}{V_m} + \frac{P}{V_m} \tag{14.6}$$

If the isotherms shown in Fig. 14.3 are replotted with P/V on the y axis and P on the x axis, a straight line is obtained. The plot for the Pocahontas #3 seam is shown in Fig. 14.4. The slope of the line is $1/V_m$, from which V_m can be determined. The intercept on y axis is P_L/V_m, from which P_L can be determined.

Calculated values of P_L and V_L for all gas isotherms shown in Fig. 14.3 are shown in Table 14.2.

Table 14.2 Calculated V_L and P_L Values for US Coal Seams

Coal Seam	V_L (ft^3/t)	P_L (psi)
Hartshorne	788	205
Pocahontas #3	646	158
Pittsburgh	443	170
Castlegate	409	229
Illinois #6	353	273

Some interesting conclusions can be drawn from data in Table 14.2 and the existing gas content and reservoir pressures of these coal seams. The Hartshorne and Pocahontas coal seams are deep (1500–2500 ft depth). Their gas contents are 550–650 ft^3/t and reservoir pressures are 500–650 psi. This indicates that these coal seams are still near their saturation points and have not lost much gas. They are potentially good reserves and will yield high rates of gas production as they have in the field. The Pittsburgh and Illinois coal seams, on the other hand, are relatively shallow (at 1000 ft depth). The measured gas content is typically 100–200 ft^3/t, and the reservoir pressure is less than 200 psi. This shows that a considerable amount (50%–60%) of their original gas content has been lost and, therefore, these reservoirs would be low producers. These observations are, in fact, confirmed by actual gas production data.

14.2 Coal Matrix Permeability

Permeability is a property of a porous rock such as coal and is a measure of the capacity of the medium to transmit fluids. It depends on the driving pressure differential, the area of the specimen, and the viscosity of the fluid.

Mathematically, it can be written as

$$u = \frac{Q}{A} = -\frac{k}{\mu}\frac{dp}{dx} \tag{14.7}$$

where u is the average fluid velocity in cm/s; A is the cross-sectional area in cm^2; k is the permeability of the medium in darcy; μ is the viscosity of gas/liquid in centipoise; and $\frac{dp}{dx}$ is the pressure gradient in atm/cm.

A negative sign indicates that fluid flows in the direction of the declining pressure gradient. As most mineable coal seams are shallow (less than 3000 ft in depth), the fluid can be assumed to be noncompressible. Integrating Eq. (14.7) for the length of the specimen, L,

$$\frac{Q}{A}\int_{o}^{L} dx = \frac{-k}{\mu}\int_{P_2}^{P_1} dp \tag{14.8}$$

or

$$Q = \frac{kA}{\mu L}(P_1 - P_2) \tag{14.9}$$

In an experiment to measure k, all the parameters in Eq. (14.9) are known, and hence permeability can be easily determined.

A cube of coal 1 cm on a side will have a permeability of 1 darcy, if a fluid of 1 cp viscosity flows between the back and front faces of the cube at a rate of 1 cc/s under a pressure differential of 1 atm at 68°F. Converted to SI units, 1 darcy is equivalent to

9.869233×10^{-13} m^2 or roughly 1 mm^2. As a darcy is a very large unit, the permeability is mostly expressed in 1/1000 of a darcy or millidarcy (md), and it has a dimension of L^2.

The above equation is valid for liquids. For gases, the volume q is introduced as defined by Eq. (14.10):

$$q = Q \cdot \frac{P_1 + P_2}{2P_b} \tag{14.10}$$

Substituting in Eq. (14.9) and expressing K in md, the equation for gas flow can be written as

$$k = \frac{2000\, qL\, \mu P_b}{A\left(P_1^2 - P_2^2\right)} \tag{14.11}$$

where k = permeability in millidarcy; q = gas flow rate in cm^3/s; L = length of the specimen in cm; μ = gas viscosity in centipoise; P = absolute pressure in atm; subscript 1 = upstream core; subscript 2 = downstream core; b = base pressure of gas measurement.

For example, let us assume the following:

$q = 2\ cm^3/s$
$P_1 = 2$ atm
$P_2 = 1$ atm
$L = 2$ cm; $A = 3\ cm^2$
$P_b = 1.00$ atm
$\mu = 0.018$ cp at 68°F

$$k = 1\left(\frac{2000 \times 2 \times 0.018}{3(4-1)}\right) \times \frac{1.0}{1.0} = 8 \text{ md}$$

14.2.1 Measurement of Permeability

There are numerous, theoretical, laboratory-based, and field techniques available for the measurement of coal matrix permeability. Reliable data are obtained by only field measurements.

Field measurements of permeability can be done in a variety of ways as listed below:

1. Drill stem testing.
2. Slug testing.
3. Injection fall-off testing.
4. Pressure-buildup and drawdown testing.
5. Minifrac(ture) of the coal seam.

The first three tests are described by Rogers et al. [12], and the pressure build up and draw down are explained by Thakur in detail [1]. The minifrac technique is described here that gives accurate permeability data with minimal expenses. It also yields the coal reservoir pressure and fracture extension pressure for the design of hydraulic fracturing [1].

14.2.2 Minifrac Injection Testing

In this test, a small volume (1000–2000 gallons) of 2% KCl water is injected into the coal formation at a low rate of 3–5 bbl/min. Normal fracking of coal is done at a much higher rate of injection, 30–35 bbl/min. Bottom-hole pressure (BHP) is continuously measured as the minifrac progresses. The buildup of pressure until the coal minimally fracks is recorded. The injection is stopped as soon as about 1000–2000 gallons have been pumped in. The BHP at this point is immediately recorded. This pressure is called the instantaneous shut-in pressure (ISIP). The sum of the ISIP and the hydrostatic head divided by the depth of the borehole is called the "frac gradient". It can be used to predict permeability by history matching. A better estimate of matrix permeability is obtained by plotting the BHP against square root of time, $\sqrt{t}$. The point where the two straight lines intersect is the closure pressure, P_C, as shown in Fig. 14.5. Like the frac gradient, closure pressure can be used to estimate permeability.

Fig. 14.6 shows a relationship between the permeability and the closure pressure for a US coal field. In general, the higher the closure pressure, the lower the coal permeability.

The ISIP is 1250 psi. The depth of gas well is 1781 ft.

$$\text{Frac gradient,}\quad \text{F.G.} = \frac{\text{ISIP}}{\text{Depth}} + \frac{\text{Hydrostatic head}}{\text{Depth}}$$

$$\text{or}\quad \text{F.G.} = \frac{1250}{1781} + 0.434$$

$$= 1.14\ \text{psi/ft}$$

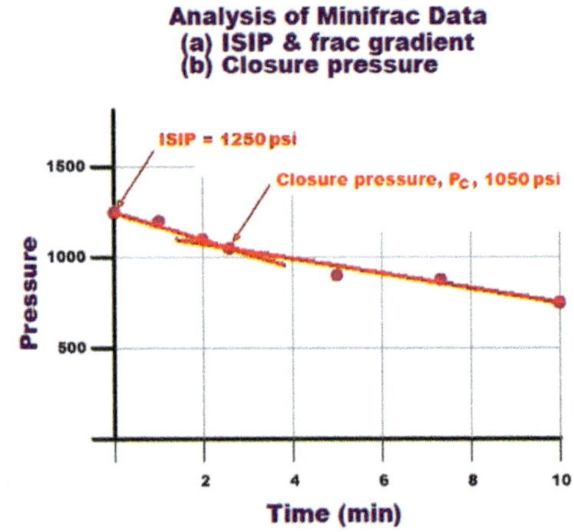

Figure 14.5 A plot of bottom-hole pressure against time. *ISIP*, instantaneous shut-in pressure.

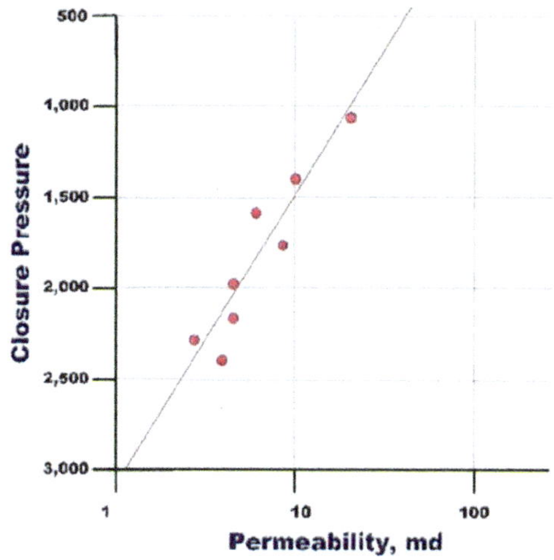

Figure 14.6 Permeability versus closure pressure.

The closure pressure, P_C, is 1050 psi. It is roughly equal to the lower principal, horizontal stress (σ_h) and is linearly related to the permeability of the coal matrix as shown in Fig. 14.6 for US coal seams.

Mathematically,

$$\ln\ k = a + bP_C \tag{14.12}$$

where k is the permeability in md; a and b are constants for the coal seam; and P_C is the closure pressure in psi.

Analyzing the data in Fig. 14.6,

$$\ln\ k = 5.1 - 0.001485\ P_C \tag{14.13}$$

For example, let us assume, $P_C = 1500$ psi.

Hence,

$$\ln\ k = 5.1 - 0.001485 \times 1500$$

$$\text{or} \quad k = 17.6\ \text{md}$$

Permeability of coal decreases with depth. Thakur [1] calculated the relationship between permeability and the depth of coal seam as

$$k = k_O\ \bar{e}\ \frac{D}{a} \tag{14.14}$$

where k_O is the assumed permeability of 100 md at 100 ft depth; a is a characteristic for the coal basin with a value of 700–1000 ft.

For example, calculate the permeability at 7000 ft depth using Eq. (14.14).

1. if a = 700

$$k = 100\ \bar{e}\frac{7000}{700} = 0.0045\ \text{md or } 4.5\ \mu d$$

2. if a = 1000

$$k = 100\ e\ \frac{-7000}{1000} = 0.09\ \text{md} \simeq 90\ \mu d$$

where μd means a microdarcy.

Eq. (14.14) gives only an approximate value of permeability. It always needs to be confirmed by a field test.

14.3 Diffusivity of Methane in Coal

Methane is held in adsorption on the surface of coal particles in a monolayer.

The flow of gases adsorbed on the coal matrix surfaces starts as soon as the confining pressure is reduced. The process goes through the following steps:

- Diffusion of gas from coal following Fick's law, i.e., concentration-dependent flow.
- Laminar flow of gases through the fractures in coal matrix. This follows Darcy's law, i.e., pressure-dependent flow. It is controlled by permeability.
- Turbulent gas flow in horizontal boreholes and vertical wells. This is controlled by the pressure gradient and by borehole/pipeline characteristics.

The net flow of gases is controlled by the first two factors: rate of diffusion and permeability-controlled flow. The sizes of horizontal boreholes and casings are designed to be so large that they do not impede the gas flow. Fig. 14.7 illustrates the flow sequence [1,9].

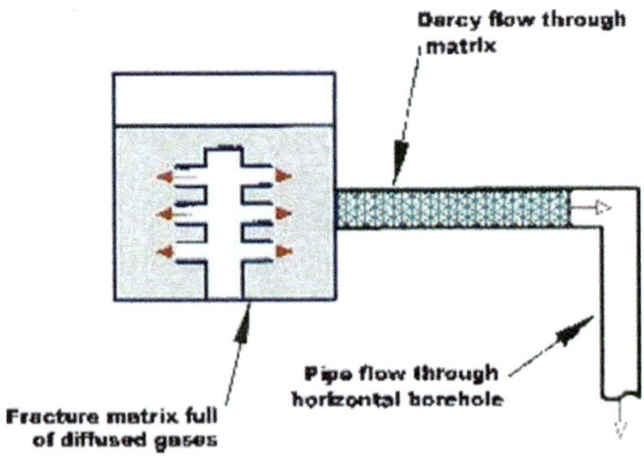

Figure 14.7 A model of methane flow in coalbeds.

The two processes work in series, and the one with a lower rate will control the net flow. Thus, in a shallow reservoir with high permeability and a low diffusion coefficient, the diffusivity determines the flow rate. In a deeper coal of high rank, the diffusivity is one or two orders of magnitude higher, and the permeability is lower. Hence, permeability determines the flow rate. It is, therefore, important to analyze the diffusion of gas from coal. It also determines how much methane can be ultimately drained from a coal seam.

14.3.1 The Diffusion Process

When a gas composed of molecule A (methane) comes in contact with gas composed of molecule B (air), the contact will cause diffusion of A into B and B into A. The process tends to produce a mixture of a uniform composition. Many enhanced methane production techniques from coal use a gas driver, such as carbon dioxide (CO_2) or helium (He), to increase production. Gas-to-gas diffusion is an important part of the study to predict the diffusion process. Similarly, when gas is passed through a porous medium wet with liquid (oil), the rate of attaining equilibrium between the gas and liquid phases depends on the diffusion process.

Fig. 14.8 shows a simple diffusion process. Container A has CO_2 at 100% concentration. Container B has 99% methane with 1% carbon dioxide. If the two vessels are connected by a conduit 1 × 1 cm and 1 cm long, CO_2 will try to go into container B and likewise methane will try to go into container A.

Assuming the containers are large in relation to the diffusion rate, the process is expressed mathematically as

$$\frac{dc}{dt} = -DA\frac{dc}{dx} \tag{14.15}$$

where c = number of molecules diffusing; t = time; D = diffusivity coefficient; A = area; dc/dx = concentration gradient.

14.3.2 Determination of Sorption Time

The rate of gas diffusion from coal matrix to bore spaces in coal is, of course, determined by the diffusivity coefficient, D. Because it is not certain what the coal "molecule" diameter, a, is, the diffusivity is often expressed as D/a^2 that has a unit of sec^{-1}.

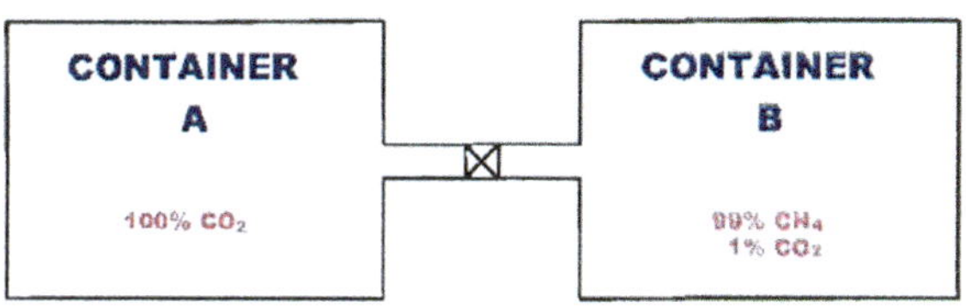

Figure 14.8 Diffusion process for gases.

Thakur [1] has shown mathematically that

$$\frac{M_t}{M_\infty} = \frac{6}{\sqrt{\pi}} \left(\frac{D}{a^2}\right)^{\frac{1}{2}} t^{\frac{1}{2}} \tag{14.16}$$

where M_t is the volume of gas diffused in time t; M_∞ is the Langmuir volume, V_m, discussed earlier, and t is the time.

A plot of $\frac{M}{M_\infty}$ against $t^{1/2}$ will yield a straight line. The gradient of the straight line is equal to $\frac{6}{\sqrt{\pi}} \left(\frac{D}{a^2}\right)^{\frac{1}{2}}$.

The time taken for a piece of coal to desorb (1 − 1/e) or 63.21% of gas is called its "sorption time" or τ. This expresses the rate of desorption in mining parlance better than the absolute value of D or (D/a^2).

Modifying Eq. (14.16), we can write

$$\frac{M}{M_\infty} = \left(1 - \frac{1}{e}\right) = \frac{6}{\sqrt{\pi}} \left(\frac{D}{a^2}\right)^{\frac{1}{2}} \tau^{\frac{1}{2}} \tag{14.17}$$

Rearranging and solving for τ, we get

$$\tau = \frac{3.49 \times 10^{-2}}{(D/a^2)} \tag{14.18}$$

Another way to determine τ is to solve the equation.

$$\frac{M}{M_\infty} = 1 - \exp\left(\frac{-t}{\tau}\right)^n \tag{14.19}$$

To illustrate and compare Eqs. (14.17) and (14.19), a typical gas desorption curve shown in Fig. 14.9 was plotted both ways and results were compared.

The plot of curve in Fig. 14.9 using Eq. (14.17) results in a straight line with an equation:

$$\frac{M}{M_\infty} = 0.000436\ \tau^{\frac{1}{2}} \tag{14.20}$$

This gives (D/a^2) equal to $1.6589 \times 10^{-8}\ s^{-1}$ and $\tau = 24.35$ days.

Eq. (14.19) was rewritten as

$$\ln \ln \left(\frac{1}{1 - \frac{M_t}{M_\infty}}\right) = n \ln t - n \ln \tau \tag{14.21}$$

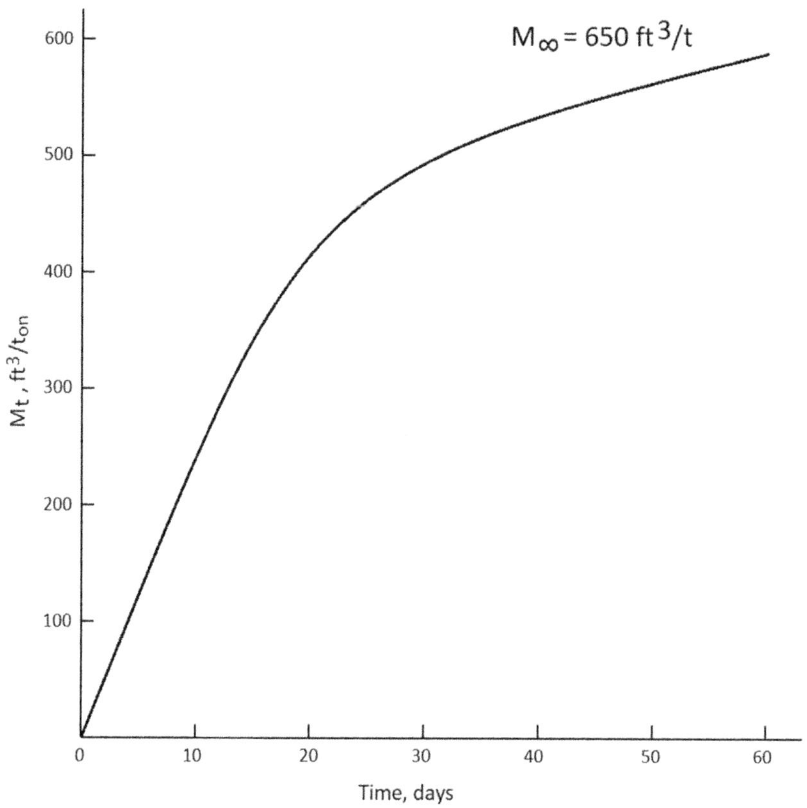

Figure 14.9 A hypothetical desorption curve for a low-volatile bituminous coal.

Hence, a plot of $\ell n\ \ell n\left(\frac{1}{1-\frac{M}{M_\infty}}\right)$ against ℓn t will yield a straight line with an equation:

$$\ln \ln\left(\frac{1}{1-\frac{M}{M_\infty}}\right) = 0.9545 \ln t - 3.0646 \qquad (14.22)$$

By substituting 1/e for M/M_∞, we get $\tau = 24.65$ days which is comparable with the result obtained by Eq. (14.20). The value of "n" is 0.9545. Table 14.3 shows the sorption time and (D/a^2) values for some US coal seams.

Table 14.3 Sorption Time and Diffusivity for Some US Coal Seams

Coal Seam	Sorption Time (days)	(D/a_2) (s^{-1})
Pittsburgh	100–900	4.0×10^{-9} to 4.4×10^{-10}
Pocahontas #3	1–3	4.0×10^{-7} to 1.34×10^{-7}
Mary Lee/Blue Creek	3–5	1.34×10^{-7} to 8.0×10^{-8}
San Juan Basin	1	4.0×10^{-7}

14.4 Reservoir Pressure

All coal seams have a gas pressure that keeps methane adsorbed in coal. This is called "reservoir or pore pressure." For design of degasification and assessment of the effect of degasification, it is essential to know it in advance of mining. It is primarily a factor of the depth of burial and the rank of coal. The greater the depth of the coal seam and the higher the rank of coal, the higher the pore pressure. However, there are a few exceptions.

Actual measured pore pressures at various depths from the United States, Canada, Australia, and South Africa are plotted in Fig. 14.10.

A linear relationship with depth appears to exist with a gradient of 0.33 psi/ft. For comparison, the hydrostatic head has a gradient of 0.434 psi/ft. Numerous readings of reservoir pressures in German coal seams showed that it is also highly correlated with the rank of coal. The maximum pressure observed in anthracite seams was 700 psi, whereas that in steam coal was only 250 psi [13]. Some coal seams display a higher pressure gradient than the hydrostatic head gradient of 0.454 psi/ft. Such coal seams

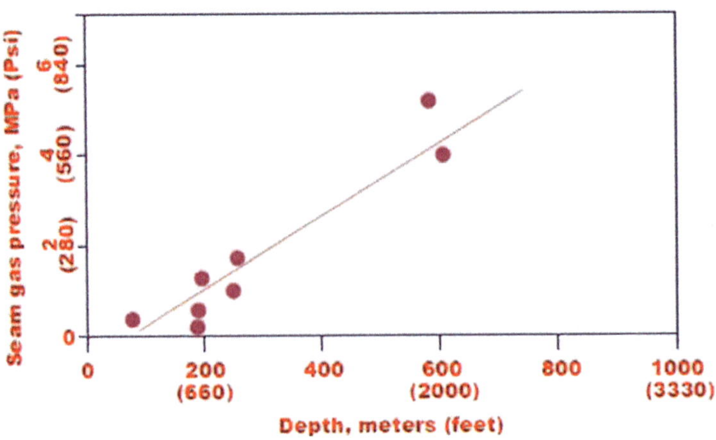

Figure 14.10 Reservoir pressure versus depth.

are overpressurized and highly productive. The Fairway region in the San Juan basin of the United States is a good example. Many vertical wells completed in thick, overpressurized coal seams have had a production of 2–10 MMCFD. The coal seam thickness is 40–60 ft. Similarly, there are a few coal seams that are seriously underpressurized and are poor producers.

14.4.1 Measurement of Reservoir Pressure

The simplest and perhaps the most reliable technique is to use a pressure gage, such as an RPG gage from Halliburton Services.

A vertical well is drilled into the coal seam and extended into the floor for 100–200 ft. A 4.5 in. casing is set in the well just above the coal seam using a formation packer shoe and cemented to the top. The coal seam is hydrojetted with high pressure water at about 3000 psi. Next, an RPG gage is lowered into the coal seam and a packer is set just above the coal seam. The well is kept shut for 72–96 h. The gage shows the pressure buildup on a graph paper. The asymptotic pressure on the graph is the reservoir pressure of the coal seam. Depending on depth, most of the world's coal seams have pressures in the range of 100–800 psi.

14.4.2 The Vertical Pressure, σv

It is a commonly accepted fact that

$$\sigma_V = 1.1\ D\ \text{psi} \tag{14.23}$$

where D is the depth in feet.

This is valid at least to a depth of 10,000 ft. Most coal seams occur above this depth.

14.4.3 Horner's Plot for Reservoir Pressure Measurement

As discussed earlier, the minifrac is done to determine coal seam permeability, but it can also be used to determine the reservoir pressure, P*. A plot of BHP against log $\frac{t_o+t_{si}}{t_{si}}$ gives a straight line, and its intercept of "y" axis is the reservoir pressure, as shown in Fig. 14.11.

Let us assume the minifrac was accomplished in 21 min of fluid pumping. The well is shut and ISIP is noted. The BHP is recorded at regular time intervals of 1 min or so. A typical data set is shown in Table 14.4.

Here $t_o = 21$ min; t_{si} is the time after shut-in.

It is popularly called Horner's plot.

Ground stress has three components: vertical stress, major horizontal stress, and minor horizontal stress. A reference should be made to a book published by Thakur [1] for details.

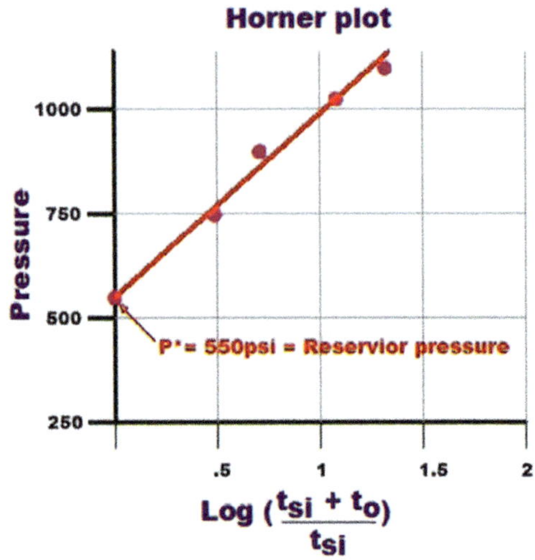

Figure 14.11 Horner's plot of minifrac data.

Table 14.4 Calculation of Data for Horner's Plot

T_{si}	$\frac{t_o+t_{si}}{t_{si}}$	Bottom-Hole Pressure (psi)	Log $\frac{t_o+t_{si}}{t_{si}}$
1	2.2	1100	1.34
2	11.5	1050	1.06
5	5.2	760	0.71
10	3.1	750	0.49

Problems

1. Calculate the value of "b" in Eq. (14.5) by plotting the gas isotherm of Fig. 14.3 as P/V, versus P. How does it relate to Langmuir volume P_L?
2. Using the data provided below:
   a. Plot BHP versus square root of time and determine closure pressure. If the ISIP was 4230 psi, what is the frac gradient (assume depth = 5000 ft).
   b. Plot the data to create a "Horner's plot" of BHP versus $\ln\left(\frac{t_o+t_{si}}{t_{si}}\right)$ and estimate the reservoir pressure P*. A plot of BHP against log $\frac{t_o+t_{si}}{t_{si}}$ gives a straight line, and its intercept of "y" axis is the reservoir pressure, as shown in Fig. 14.11.

      $t_o = 2.5$ min.

t (min)	Bottom-Hole Pressure (psi)
0	4030
0.25	3918
0.5	3870
0.75	3830
1.0	3803
1.5	3750
2.0	3710
3.0	3640
4.0	3570
5.0	3519
6.0	3470
7.0	3410
8.0	3335
9.0	3265
10.0	3200
11.0	3140
12.0	3080
13.0	3020

References

[1] Thakur PC. Advanced reservoir and production engineering. Elsevier; 2017. p. 210.
[2] Yu D, et al. In: Law RE, Rice DD, editors. Gas sorption on coal and measurement of gas content in hydrocarbons from coal. Tulsa, OK: AA&G; 1993. p. 159–84.
[3] Bertard C, et al. Determination of desorbable gas concentration of coal. International Journal of Rock Mechanics and Mining Sciences 1970:43–65.
[4] Kissell FN. The direct method of determining methane content of coalbeds for ventilation design. RI 7767: US Bureau of Mines; 1973. p. 17.
[5] Diamond WP, Schatzel SJ. Measuring the gas content of coal: a review. Coal Geology 1998;35:311–31.
[6] ASTM Designation: D7569–10. Standard practice for gas content of coal – direct desorption method. 2010.
[7] TRW. Desorbed gas measurement system design and application, US Department of Energy, Contract No. DE-AC21-78MCO8089. Morgantown, WV: METC; 1981.

[8] Thakur PC. Methane control on longwall gobs: longwall-shortwall mining, state of the art, AIME. 1981. p. 81–6.
[9] Thakur PC. Methane flow in Pittsburgh coal seam, the 3rd international mine ventilation congress. Harrogate, U.K. 1984. p. 177–82.
[10] Thakur PC. Mass distribution, percent yield, non-settling sizes and aerodynamic shape factors of respirable coal dust particles, MS Thesis. PA: The Penn State University; 1971. p. 133.
[11] Langmuir I. The adsorption of gases on plane surfaces of glass, mica and platinum. Journal of American Chemical Society 1918;40:1361–403.
[12] Rogers R, et al. Coalbed methane: principle and practice. Oktibbeha Publishing LLC; 2007. p. 504.
[13] Muche G. Methane desorption within the area of influence of workings, 1st international mine ventilation congress. Johannesburg, South Africa. 1975.

Premining Degasification of Coal Seams

15

Chapter Outline

Majority of global coal production comes from underground mining which is done by two methods: (1) room and pillar and (2) longwall mining. As shallow coal seams are mined out and mining reaches greater depths, longwall mining becomes the preferred method of mining. In United States, more than 50% of all underground mined coal is mined by the longwall method. The trend for panel sizes and mining equipment in the coal industry is to continue to go upward pushing production capacities and productivity to higher levels. Another reason for increasing the longwall panel size is that development sections cannot keep up with the rate of longwall advance. Increasing the width of the longwall panel slows down the rate of advance. Today it is quite realistic to plan longwall panels in mildly gassy coal seams that are

Advanced Mine Ventilation. https://doi.org/10.1016/B978-0-08-100457-9.00015-8

1500 ft wide and 15,000 ft long containing nearly 5 to 7 million tons of raw coal. Such large longwall panels offer many benefits as follows:

- Improved safety and reduced injury rate
- Improved recovery of in situ coal reserve
- Improved productivity and reduced cost/ton

On the other hand, large longwall panels cause some concerns, such as follows:

- Ventilation and methane control
- Respirable dust control
- Increased likelihood of geological anomalies, e.g., faults, washouts, etc.

These problems become more serious when the gas content of coal seams increases with the increase in mining depths. Costs of mine ventilation and coal seam degasification go up, and it becomes necessary to optimize the two processes to keep the mine environment safe and the combined cost at the minimum. Drilling process for degasification can also locate any geological anomalies in the panel prior to mining and helps in respirable dust control as discussed earlier.

The optimized plan for mine degasification and ventilation with added benefit of coal seam exploration can be summarized as follows:

1. Recover as much coal bed methane (CBM) as possible prior to mining and postmining to minimize ventilation requirements.
2. Optimize longwall panel sizes and the size of the district depending on the gas content of the coal seams to keep the combined cost of degasification and ventilation down.
3. Plan mine ventilation to maximize ventilation efficiency measured as the ratio of air horsepower delivered to face to the total air horsepower created at the fans.
4. Seal all mined out areas with approved design of stoppings and recover as much CBM as possible without making the atmosphere in the sealed area explosive.
5. Measure coal seam thickness while drilling and locate geological anomalies if any.
6. Gather all produced gases and market them after processing to defray the cost of degasification and ventilation.

15.1 Coal Seam Reservoir Parameters

In addition to the reservoir properties discussed in Chapter 14, the following parameters are the most important for planning a successful mine degasification.

15.1.1 Degree of Gassiness

It is safe to assume that all coal seams are gassy. They only vary in their degree of gassiness, i.e., gas contained per ton of coal. For planning, all coal seams can be classified as shown in Table 15.1.

The depth of a coal seam is only a rough indication of its gas content. Direct measurement of gas contents is highly recommended. The technique for direct measurement of gas contents was discussed in Chapter 14.

Table 15.1 Gassiness of Coal Seams

Category	Gas Content (ft^3/t)	Depth (ft)
Mildly gassy	<100	<600
Moderately gassy	100–300	600–1500
Highly gassy	300–700	1500–3000

15.1.2 Specific Gas Production

The second reservoir parameter that is equally important is the "specific gas production rate" of the coal seam. It is a measure of how much gas will be produced per day if an opening (a horizontal borehole or a vertical slit) is created in the coal seam by a given degasification technique. It is measured in MCFD/100 ft (e.g., 15 MCFD/100 ft for Pittsburgh coal seam). This parameter is a combination of reservoir pressure, permeability, porosity, gas content, and diffusivity of the coal seam.

15.1.3 Specific Gob Gas Production

The third reservoir parameter of importance is the "specific gob gas production" for the mine which is a measure of total gas production per unit area of the gob and is measured in MMCF/acre. It represents the total influx of gases into the gob from all overlying and underlying coal seams that are disturbed by longwall mining. For moderately and highly gassy mines in the United States, the specific gob gas production ranges from 8 to 30 MMCF/acre.

15.2 Premining Degasification

All premining degasification techniques can be classified in three broad categories.

- In-mine horizontal drilling
- Vertical wells with hydraulic fracturing
- Horizontal wells drilled from surface

In very deep coal seams (deeper than 3000 ft), the horizontal boreholes can also be hydraulically fractured to enhance gas production and expedite degasification. Selection of a particular technique depends primarily on the gas contents and other reservoir properties. Intensity of drilling is dictated by the specific gas production rate.

These three degasification techniques will be briefly described here. They are discussed in detail by Thakur [1] and a reference can be made for additional information.

15.2.1 In-Mine Horizontal Drilling

This is by far the cheapest and yet the most effective way of degasifying a coal seam prior to mining. The author [2] developed this technique, which can drill a 3- to 4-in.

diameter borehole to a depth of 3000–5000 ft. The drill rig is manufactured in the United States by J. H. Fletcher Company in Huntington, WV. Nearly 100 drill rigs are in use in all major coal mining countries, including the United States, China, India, Australia, and South Africa. Besides coal mine degasification, horizontal boreholes can be used for water drainage and advance exploration for faults, washouts, and other geological anomalies [3,4].

The equipment used to drill long horizontal boreholes can be divided into four major groups: the drill unit, the auxiliary unit, the bit guidance system, and the downhole drill monitor (DDM).

The drill rig provides the thrust and torque necessary to drill 3- to 4-in. diameter boreholes to a depth of 3000–5000 ft. The auxiliary unit provides the high-pressure water to drive a drill motor and flush the cuttings out. It also holds a gas and drill cutting separation system. The bit guidance system guides the drill bit up, down, left, and right as desired to keep the borehole in the coal seam. The DDM measures the pitch, roll, and azimuth of the borehole assembly. In addition, it indicates the approximate thickness of coal between the borehole and the roof and floor of the coal seam by using a gamma ray sensor that measures radiation from the roof or floor. The half-depth of gamma rays in coal is typically 8 inches. In recent years, many other uses of in-mine horizontal boreholes have come into practice, such as in situ gasification of coal, improved auger mining, and oil and gas production from shallow deposits [4].

15.2.1.1 The Drill Rig

Fig. 15.1 shows the drill unit. It is mounted on a four-wheel drive chassis driven by Staffa hydraulic motors with chains or torque hubs. The tires are 15 by 18 in. in size and provide a ground clearance of 12 in. The prime mover is a 50 hp explosion-proof electric motor which is used only for tramming. Once the unit is trammed to the drill site, electric power is disconnected and hydraulic power from the auxiliary unit is turned on. Four floor jacks are used to level the machine and raise

Figure 15.1 The drill unit for in-mine horizontal drilling.

the drill head to the desired level. Two 5-in. telescopic hydraulic props, one on each side, anchor the drill unit to the roof.

The drill unit houses the feed carriage and the drilling console. The feed carriage is mounted more or less centrally, has a feed of 12 ft, and can swing laterally by ±17 degrees. It can also sump forward by 4 ft. The drill head has a through chuck such that drill pipes can be fed from the side or back end. The general specifications of the feed carriage are as follows:

High Speed	RPM = 850	Torque 5000 lb-in
Low speed	RPM = 470	Torque 11,000 lb-in
Thrust	30,000 lbs (40,000 lbs pulling out)	
Maximum feed rate	10–20 ft/min	
Overall dimensions	Length = 16 ft Width = 8 ft Height = 4 ft	
Maximum tram speed	1.2 mph	

15.2.1.2 The Auxiliary Unit

The chassis for the auxiliary unit is identical to the drill unit but the prime movers are two 50 hp explosion-proof electric motors. It is equipped with a methane detector–activated switch so that power will be cut off at a preset methane concentration in the air. No anchoring props are needed for this unit. The auxiliary unit houses the hydraulic power pack, the water (mud) circulating pump, control boxes for electric motors, a trailing cable spool, and a steel tank which serves for water storage and closed-loop separation of drill cuttings and gas.

Fig. 15.2 shows a view of the auxiliary unit.

Fig. 15.3 shows a cross-sectional view of the separation system.

The tank is 10 × 3.5 × 3 ft in size and has two compartments. The inner compartment has sufficient capacity to hold drill cuttings from a 200 ft long hole of 4-in. diameter. Coal fines have a tendency to froth, but this is cured with suitable surfactants. At the end of the drilling shift, the vehicle is trammed to a crosscut and the cuttings are discharged by means of a screw feeder. Baffles in the tank collect the large cuttings, while fines were initially collected by the plate separator. The latter, however, did not perform entirely satisfactorily and was replaced by a cyclone. Clean water flows to the outer compartment which serves as the storage for fresh water. Float controls in this part of the tank ensure that the correct level of water is always maintained. The low-level float control opens a make-up water valve.

Gas is drawn from the tank via an outlet connected to the underground methane pipeline system. The tank works under slight positive pressure and is designed to withstand a gage pressure of 20 psi.

Figure 15.2 The auxiliary unit for in-mine horizontal drilling.

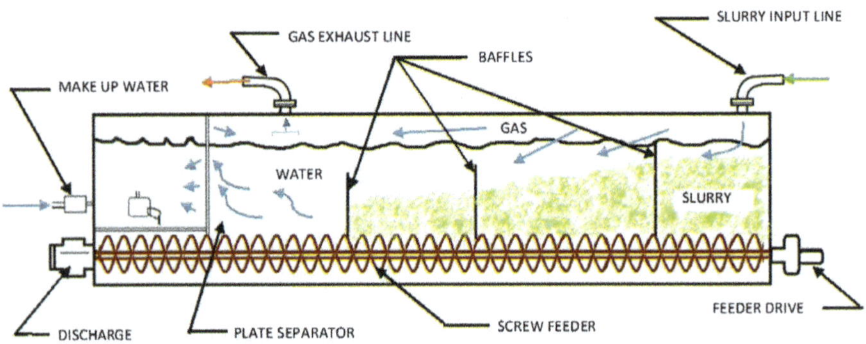

Figure 15.3 The gas and water separation system.

The water (mud) circulating pump is a triplex, reciprocating pump with a capacity of 70 gpm at 900 psi. In the rotary mode, an annulus fluid velocity of 3 ft/s is usually sufficient, but in nonrotary mode, the annulus velocity must be increased to 5 ft/s. The pump is driven by a 50 hp electric motor.

The hydraulic power pack consists of a number of hydraulic gear motors capable of delivering 80 gpm of hydraulic fluid at 2500 psi. The working pressure in the system seldom exceeds 2000 psi. Petroleum oil is the recommended fluid for the entire hydraulic system.

15.2.1.3 The Guidance Systems

When a horizontal hole is started in the middle of a relatively flat, 5–6 ft thick seam the drill bit usually ends up in the roof or floor before reaching 200 ft. To drill a deeper hole, it is imperative to guide the bit up and down as needed. In most cases, it is also necessary to guide the bit in the horizontal plane.

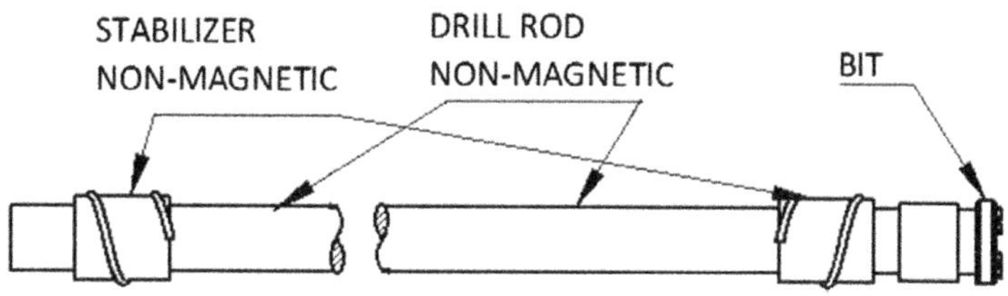

Figure 15.4 Rotary borehole assembly.

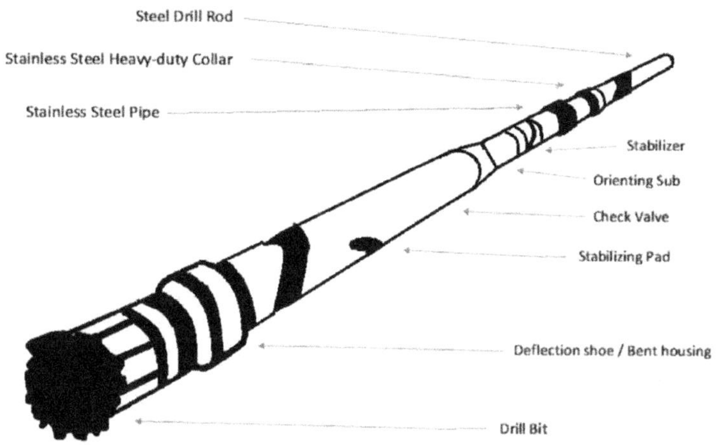

Figure 15.5 Nonrotary borehole assembly.

To achieve these goals two different modes of drilling, the rotary and the nonrotary modes, were employed. The design of the borehole assembly, i.e., the bit and the first 30 ft of drill column, in either case largely determines the rate of angle build. Figs. 15.4 and 15.5 show the borehole assembly design for the rotary and nonrotary modes of drilling, respectively.

15.2.1.3.1 Guidance of Rotary Borehole Assembly

In the rotary mode, the drill pipes rotate and all the torque and thrust are provided at the rotary head on the rig. As shown in Fig. 15.4, one stabilizer is used immediately behind the bit and a second is used 10–20 ft behind the first. The first stabilizer also has an internal orienting device for the borehole survey equipment. This stabilizer and 20–30 ft of drill column next to the bit are made of nonmagnetic material so that the borehole survey instruments will not be magnetically affected. Surveying is done with a pumpable tool that measures the pitch, roll, and azimuth of the borehole.

The guidance of the drill bit or, more precisely, the rate of angle built by the bit is actually a factor of two groups of variables: the design of the borehole assembly and the interaction between the bit and the material being drilled. As coal seams are not uniform, homogeneous strata and bits continuously change their characteristics with wear, and it is very difficult to forecast the rate of angle build precisely. For a given type of bit, usually a reasonable rotation speed is selected to yield a penetration rate of 3–5 ft/min, and the thrust is varied to make the bit go up and down. At low-thrust values the bit pitches down, but at high thrust it will go up. Thrust values and corresponding rates of angle build for a 4-in. diameter drag bit collected for a typical 500 ft of drilling were analyzed using a computer program. A straight-line relationship between the rate of angle build, $\Delta\theta$, and thrust, T, exists as given below:

$$\Delta\theta = 6 \times 10^{-5}\mathrm{T} - 0.30121 \tag{15.1}$$

where $\Delta\theta$ is in degrees per 10 ft and T is thrust in lb. In this particular case, the rotary speed was kept steady at 250 rpm and thrust varied from 1000 to 8000 lb.

The three-cone roller and Stratapax bits were also used. They showed similar trends, but the actual rate of angle build varied from bit to bit. With careful selection of drilling parameters, such as the rotary speed and thrust, different kinds of bits can be guided successfully. The drag bit is the easiest to guide but cannot drill through hard rock inclusions in coal. Three-cone roller bits are a little more difficult to guide but will cut through most materials. The life of roller bits is generally less than 1000 ft. Even if the teeth remain sharp, the bearings develop some play and guidance of the bit becomes very poor. Stratapax bits (made of poly crystalline diamonds) need higher torque but appear to be most suitable for drilling holes deeper than 3000 ft. The biggest drawback of rotary borehole assembly is that it cannot be guided in a horizontal plane. It therefore has a very limited use.

15.2.1.3.2 Guidance of Nonrotary Borehole Assembly

To overcome the deficiencies of a rotary drilling assembly, a nonrotary assembly was designed. It basically consists of a bit, a deflection device immediately behind the bit, and a downhole motor which runs on the drilling water or mud, as shown in Fig. 15.5. The deflection device was a spring-loaded eccentric sub which exerts a constant force on the side of the bit. The direction of this applied force depends on the orientation of the device and determines whether the bit will be deflected up, down, left, or right. The magnitude of this force and hence the rate of angle build are controlled by the size of the spring. Ideally, the rate of angle build is kept below 0.5 degrees per 10 ft. In coal seams, a side force of 50–100 lb is generally adequate. This device had a tendency to get plugged with coal fines, and it was replaced by a "bent housing" of one degree. The drill bit is forced to go up, down, left, or right depending on the orientation of the bent housing.

15.2.1.4 The Downhole Drill Monitor

To guide the drill bit successfully and contain it in the coal seam, it is essential to know both the position of the bit in relation to the roof and the floor of the coal seam and the pitch of the bit. In the case of nonrotary columns, the roll of the bit and azimuth must

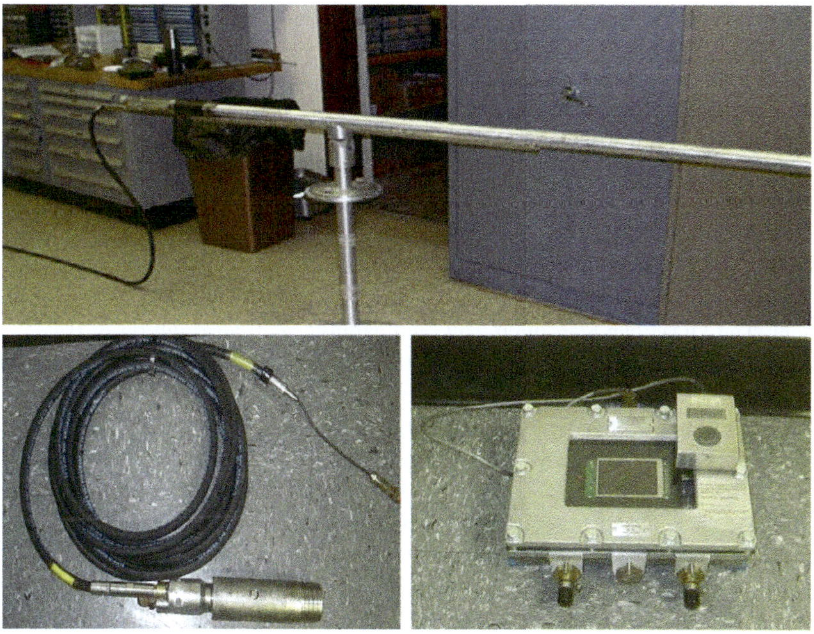

Figure 15.6 The downhole drill monitor.

also be known so that the deflection device can be properly oriented. Also, the Mine Safety and Health Administration (MSHA)[1] requires that the azimuth of degasification boreholes be plotted on mine maps to prevent inadvertent mining through such holes.

Borehole survey instruments incorporate sensors for the azimuth, pitch, and roll and a coal thickness indicator. The latter indicates the thickness of the coal between the borehole and the floor or the roof, depending on the orientation of the surveying tool. Fig. 15.6 shows the basic components of the survey instrument system, namely the DDM and the readout unit.

The DDM system consists of a downhole survey probe and a portable data collection and display unit situated outside the borehole.

The downhole survey probe is a battery-powered microprocessor-controlled data acquisition system contained in a 12 ft. long copper–beryllium tube. It is located just behind the downhole motor. The DDM remains downhole until the target depth is reached or until a battery change is needed. A triaxial magnetometer is used to measure the magnetic azimuth. Three accelerometers are used to measure pitch and roll of the drill bit. A solid-state gamma detector is used to monitor small amounts of natural gamma radiation emitted from the overlying and underlying shale deposits.

An approximation of roof and floor coal thickness can be made from the observed gamma ray count and the known half-depth value for gamma rays in coal. A built-in computer program controls collection and transmission of data to the collection and

[1] Mine Safety and Health Administration (MSHA) is the US certification agency for all mine equipment.

display unit. The collected data are digitized and transmitted acoustically through the drill string. Outside the borehole, a magnetic pickup located on the borehole wellhead (or the drill string) receives the signal and displays data sequentially on the display unit. This system has a depth limit of 3000 ft. Recently, a hard-wired communication system was put into use. All drill rods have an insert.

When put together, it provides a solid conductor to transmit data, with a range of well over 5000 ft. It works so well that it has totally replaced acoustic transmission-type instruments.

All downhole electronic parts are housed in approved explosion-proof aluminum tubing that is watertight and rugged enough to withstand the rigorous downhole environment. The DDM can read pitch, roll, and azimuth with a resolution of 0.1, 1, and 1 degrees, respectively. The ranges for pitch, roll, and azimuth are 0–90, 0–360, and 0–360 degrees, respectively.

The portable data collection and display unit is a battery-powered, intrinsically safe, MSHA-approved unit for use in return airways of underground mines. The display unit functions as a real-time analyzer to serve the operator in deciding how to orient the bent housing for subsequent drilling and to store various parameters of the borehole being drilled. The storage section of the display unit consists of solid-state memory components with the capability of retaining borehole data which can be taken to the surface and transferred to a larger and more powerful computer. This data can then be used to plot horizontal and vertical profiles of the boreholes. The horizontal profile (plan view) is plotted on mine maps for later use during mine development. The display unit can also be used by the operator to check vertical deviation, horizontal deviation, and drilling parameters such as water pressure and rotary speed if the drilling is done in the rotary mode (i.e., the drilling string is rotated from outside). Data are received by the display unit via a magnetically coupled piezoelectric crystal attached to the wellhead which converts small acoustic signals into electrical signals that are stored in the display unit memory or a hard disc. Each data set received includes the pitch, roll, azimuth, and gamma ray counts per minute. After the operator enters a value corresponding to the depth of the borehole, other parameters can be calculated, such as vertical deviation and horizontal deviation with respect to the wellhead. The internal memory of the display unit can store up to 200 sets of borehole data. Any particular data set can be recalled for the operator's review.

15.2.2 Gas Production From Horizontal and Vertical Wells

The initial time-dependent flow, steady-state flow, and eventual production decline from vertical and horizontal wells have been discussed in details by Thakur [1]. Only steady-state flow will be discussed here.

15.2.2.1 Production From Horizontal Boreholes

The most reliable production estimation for horizontal boreholes/wells is done by the "specific gas production" rates as defined earlier. The data for some important coal seams are presented in Table 15.2.

Table 15.2 Specific Gas Emissions for Coal Seams

Coal Seam	Depth (ft)	Rank	Specific Gas Production (MCFD/100 ft)
Pittsburgh	500-1000	High Vol. Bituminous	15.00
Pocahontas No. 3	1400–2000	Low Vol. Bituminous	8.00
Blue Creek/Mary Lee	1400–2000	Low Vol. Bituminous	9.00
Pocahontas No. 4	800–1200	Medium Vol. Bituminous	5.00
Sunnyside	1400–2000	High Vol. Bituminous	9.00

The specific production goes down with time and is about 3–4 MCFD/100 ft at the time of plugging just prior to mining through the wells.

Thakur [1] also derived mathematical expressions for gas flow from horizontal boreholes. It is given by Eq. (15.2).

$$Q = At^n \tag{15.2}$$

where Q is the cumulative production in MCF; A is the initial production in MCFD (months); t is the time in day (months); n is a characteristic of the coal seam.

Eq. (15.2) can also be written as follows:

$$\ln Q = \ln A + n \ln t \tag{15.3}$$

A plot of ln Q against ln t yields a straight line. The characteristic "n" varies from 0.8 to 1.00.

15.2.2.2 Gas Production From Vertical Wells

Production from a vertical well can be estimated by Eq. (15.2) but another estimate is provided by Eq. (15.4).

$$q = \frac{707.8\, kh\left(p_e^2 - p_w^2\right)}{\bar{\mu}\, \bar{z}\, T \ln(r_e/r_w)} \tag{15.4}$$

Where q = cubic ft/day at 60°F and 14.67 psia; k = permeability in darcy; h = thickness in ft; p_e = pressure at external radius, r_e; p_w = pressure at the well radius, r_w; $\bar{\mu}$ = average viscosity; $\bar{z}$ = average compressibility factor; T = temperature in degree Rankine (Fahrenheit + 460).

For liquid flow, Eq. (15.4) becomes:

$$Q = \frac{0.03976\, kh\, (p_e - p_w)}{\mu\, \ln(r_e/r_w)} \tag{15.5}$$

where Q is in CF/day and μ is liquid viscosity.

For example,

Calculate gas and water flow from a well producing steadily under the following conditions:

$k = 0.003$ darcy (3 md)
$h = 40$ ft
$\mu = 0.02$ cp
$z = 0.90$
$T = 60°F\ (+460)$
$r_e = 1000$ ft
$r_w = 0.25$ ft
$p_e = 500$ psi
$p_w = 50$ psi

Using Eq. (15.4),

$$Q = \frac{707.8 \times (0.003)40\left(500^2 - 50^2\right)}{0.9 \times 520 \times (0.02) \times \ln\left(\dfrac{1,000}{0.25}\right)}$$

$$= 270.9 \text{ MCFD}$$

The above conditions describe a typical well drilled into a thick seam with good permeability. The well is produced at a constant pressure of 50 psi. Similarly, using Eq. (15.5), the water flow can be calculated as 46.2 bbl/day.

15.3 Application of In-Mine Horizontal Drilling

First application of this technique was reported by Thakur [3,4] in 1978. A 1200 ft long borehole was drilled ahead of a development section from the return side of the section as shown in Fig. 15.7. It produced about 500 MCFD of methane.

The borehole was connected to a vertical borehole with pipes, and gas was safely discharged on surface. The impact of degasification was noted as follows:

1. The greatest impact of degasification was in the face area where methane concentration dropped to 0.25% in course of two to 3 months from an initial value of 0.95%.
2. The methane concentration in the section return (at the last open crosscut) fell to 50% of its original value indicating a methane capture ratio of 50%.
3. The immediate influence of those boreholes was felt up to a radius of 400 ft. This indicated that only one borehole drilled in the outermost airway of a section can adequately degas the

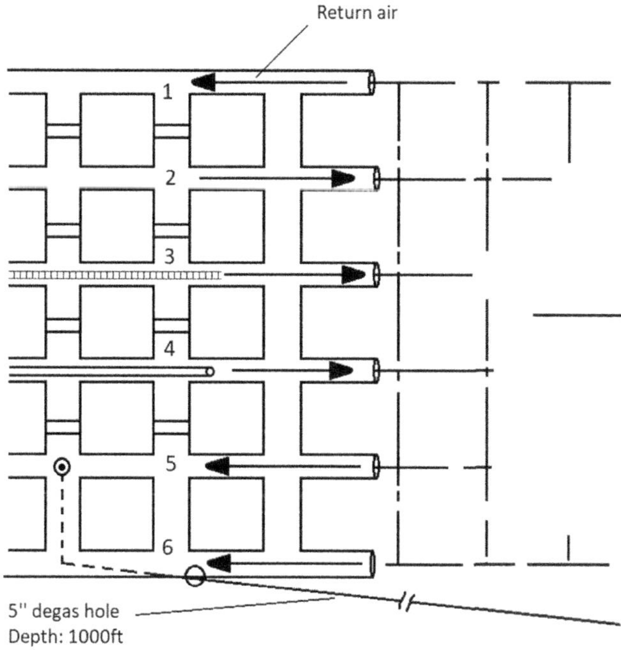

Figure 15.7 Layout of a development section with a long degas hole.

section if the width of the development section did not exceed 400 ft and the coal seam had good permeability.

Application of in-mine horizontal drilling for simultaneous degasification of the development headings and the longwall panel is shown in Fig. 15.8 [5].

In moderately gassy coal seams, the drilling is done at 1000 ft intervals for adequate degasification of the longwall panel. For highly gassy coal seams the first phase of degasification is done by vertical drilling and hydrofracturing as described in the next sections. The supplementary horizontal boreholes are drilled at closer intervals of 100–200 ft to remove 50%–70% in situ gas prior to mining.

15.4 Application of Vertical Wells With Hydraulic Fracturing

This is currently the most popular method of coal bed gas production, but it is also used for degasification of deeper coal seams (deeper than 1500 ft) where in-mine horizontal drilling is not enough. Deeper coal seams with gas contents of 300–700 ft^3/t must be drilled vertically, and the coal seams should be hydraulically fractured to drain about 50% of the gas in situ.

Fig. 15.9 shows the vertical section of a typical well. Details of well completion are given by Thakur [1] elsewhere.

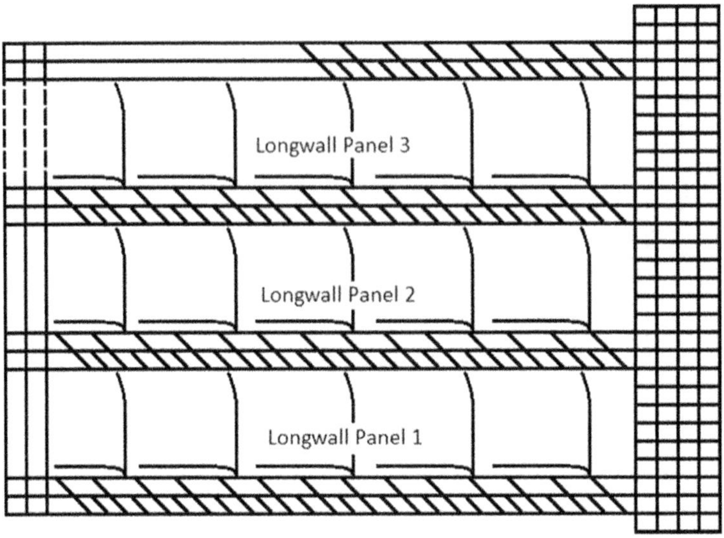

Figure 15.8 Coal seam degasification with in-mine horizontal drilling.

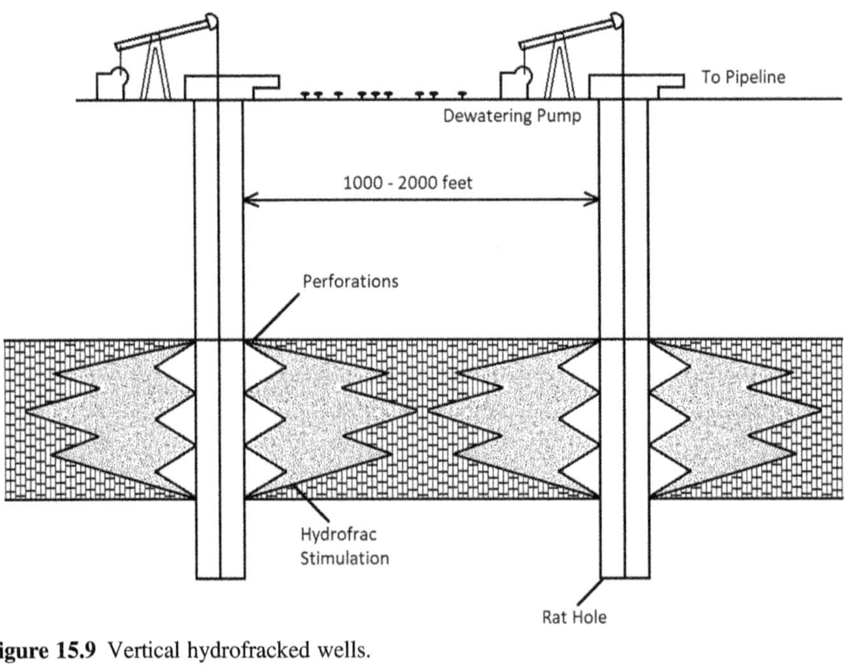

Figure 15.9 Vertical hydrofracked wells.

In summary, a mixture of sand and water is pumped into the coal formation at 30 bbl/min or a higher rate (1 bbl = 42 gallons). It creates a vertical fracture in coal seams that is 1000 to 2000 ft long, $^1/_2$ to $^3/_4$ in. wide, and about 20 ft high at the well bore. The width and height reduces to almost nil at the tip of the fracture at 1000 ft or so. Drilling is usually done three to 5 years ahead of mining to achieve 50% drainage of in situ gases.

A typical longwall face 1000′ × 15,000′ in area may need 10 to 12 vertical frac wells for adequate degasification. When the reservoir pressure is reduced to less than 200 psi, supplementary degasification is done with horizontal drilling to remove additional 25%–30% of in situ gas content. Hydrofracking is typically done with plain water or nitrogen foam, but gelled water and cross-linked gels have been also used depending on the specifics of the coal seam. Refer to Chapter 4, Figure 4.7 for the layout of frac wells on a typical longwall face. Average spacing is at 20–25 acres per well.

15.5 Application of Horizontal Boreholes Drilled From Surface

Figs. 15.10 and 15.11 show the common design and application of horizontal boreholes drilled from surface for coal seam degasification.

For a successful operation, water produced from the horizontal laterals must be removed. A production well with a sump (drilled below the coal seam to be degassed) is first drilled. Nearly 300 ft away, a vertical (access) well is drilled (Fig. 15.10). When it approaches the target coal seam, it is deviated by 90 degrees to intersect the coal seam horizontally passing through the production well.

The borehole can be laterally extended to 3000 to 5000 ft depending on the depth of the coal seam. A water pump is installed in the production well to remove water and

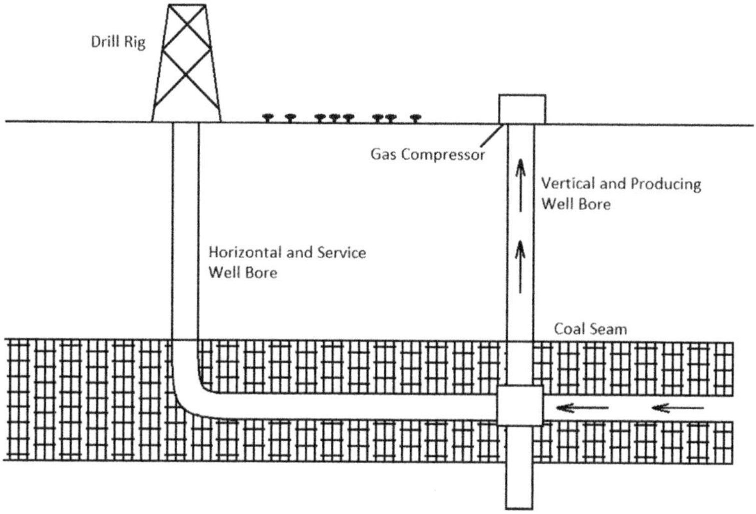

Figure 15.10 Horizontal boreholes drilled from the surface.

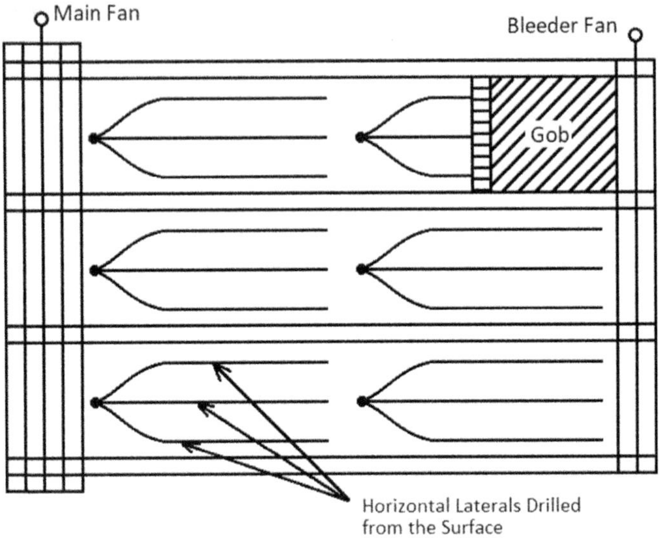

Figure 15.11 Coal seam degasification with horizontal boreholes drilled from the surface.

coal fines, and gas production is maintained. Very expensive drill rigs and instruments are needed to articulate a horizontal borehole assembly to stay in a thin coal seam and reach a depth of 3000 to 5000 ft. The cost of drilling horizontal boreholes from surface is, therefore, much higher ($100/ft) than the cost of drilling in-mine horizontal boreholes (usually less than $20/ft). The degree of degasification achieved is the same, about 50% of gases in situ.

Premining degasification of all moderately and highly gassy coal seams is highly recommended. Besides making mining safer, it improves coal productivity significantly cutting down the cost of mining. In one very gassy mine, the coal productivity jumped from 14 tons/man-day to 40 tons/man-day after about 70% of in situ gas was drained. The produced gas can be processed to meet the pipeline gas specifics and marketed to defray the cost of degasification [6].

15.6 Optimum Widths of Longwall Panels

Even after proper premining degasification, the coal seam contains 100–200 ft^3/t of gas. This gas is released when the coal is mined by mining machines and broken into small pieces. The total gas emissions reaching the tail end of the longwall face increases with the width of the longwall panel. Statutory requirements limit methane concentration at the tailgate to less than 1% (preferably 0.8%). It is also true that ventilation air reaching the tail end of the longwall face is dependent on the ventilation air quantities at the head gate end of the face and air leak-off on the face. The wider the longwall panel, the higher is the total leak-off as shown in Chapter 4, Figure 4.8. It is, therefore, logical to conclude that for a given set of conditions, an optimum limit on the width of the longwall panel is reached when it will not be possible to keep the methane concentrations below the statutory limits.

Various methods to estimate the total methane emission rate at the tailgate have been discussed by Thakur [7]. The optimum width of a longwall panel actually depends on the following three variables:

1. The rate of mining. For a 1,000 ft wide face, a 70–100 ft/day rate of mining is needed for economic reasons.
2. The total gas emissions on the longwall face.
3. The specific gob gas emission.

The rate of mining controls the other two variables to a large extent. Only longwall face methane emissions will be discussed here. The impact of specific gob emissions will be discussed in Chapter 16.

15.6.1 Estimation of Total Methane Emissions at the Longwall Tailgate by Direct Measurements

An estimate of methane emissions at the tail end of a longwall face can be derived by the following equation:

$$Q = Qo + V(A - B) - C(X) \tag{15.6}$$

where Q = total methane emissions at the tail end of a longwall face, ft^3/min; Qo = total methane emissions when no mining is being done, ft^3/min; V = the rate of mining in tons/min; A and B are the gas contents of coal prior to mining and after mining respectively; C(X) is the methane lost in the gob with air leakage.

An example:

In a properly degassed, moderately gassy coal seam, the following data were measured.

Qo = 100 CFM
V = 8 t/min
A = 120 ft^3/ton; B = 40 ft^3/ton
C(X) = averaged for the entire face at 250 CFM (50,000 CFM lost in gob containing an average of 0.5% methane).

Hence,

$$Q = 100 + 8(120-40)-250$$
$$= 490 \text{ CFM.}$$

To dilute it to 1%, the ventilation air should be 49,000 CFM with a gas layering index greater than five.

15.6.2 Mathematical Derivation of Limiting Methane Concentration at the Tailgate

Assuming that q is the net methane emission ([emission from solid coal + emissions from broken coal] minus methane lost with leak-off air) in a differential element, dx, the mathematical equation for methane concentration is (refer to Chapter 3) given by Equation (15.7). u is the ventilation rate, ft^3/min.

$$\frac{\partial^2 c}{\partial x^2} + u\frac{\partial c}{\partial x} - q = 0 \tag{15.7}$$

The boundary conditions are as follows:

$$\frac{\partial c}{\partial x} = 0 \text{ at } x = 0 \quad \text{(methane concentration is constant)}$$

$$\varepsilon_x \frac{\partial c}{\partial x}\bigg|_{x=L} + uc|_{x=L} = qL \quad \text{(from mass conservation)}$$

The solution of Eq. (15.7) is as follows:

$$c(x) = \frac{q}{u}\left[x - \frac{\varepsilon_x}{u}\left(1 - e^{-ux/\varepsilon_x}\right)\right] \tag{15.8}$$

At x = L,

$$C_L = \frac{q}{u}\left[L - \frac{\varepsilon_x}{u}\left(1 - e^{-uL/\varepsilon_x}\right)\right] \tag{15.9}$$

If L is large, we can discard e^{-uL/ε_x} term as zero, and

$$C_L = \frac{qL}{u}\left(L - \frac{\varepsilon_x}{u}\right) \tag{15.10}$$

Thus, the limiting value of C_L is $\frac{qL}{u}$.
An example:
Assume L = 1000 ft, q = 0.3 ft^3/ft-min
C_L = 0.8% (or 0.008).
Hence, $0.008 = \frac{0.3 \times 1{,}000}{u}$
or u = 37,500 CFM.

A check of gas layering index should be made to make sure it exceeds 5.00 (refer to Chapter 4).

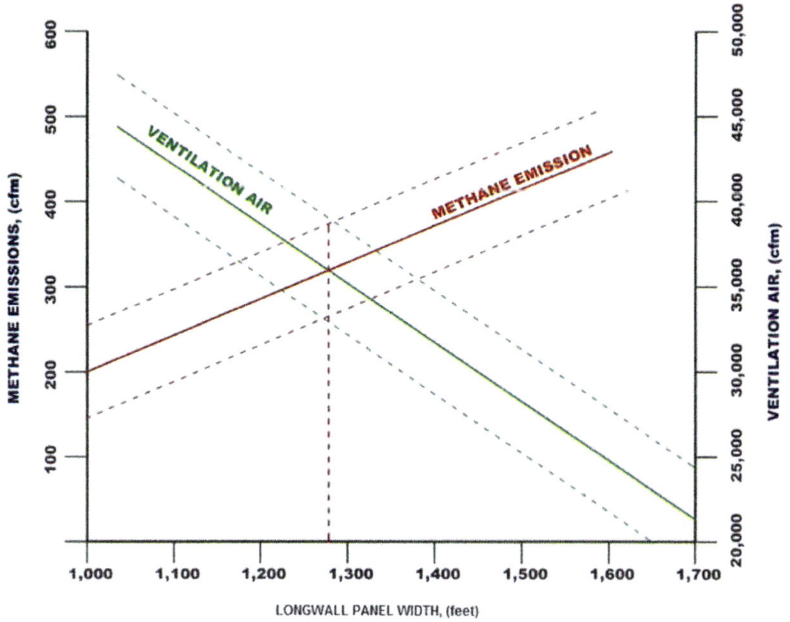

Figure 15.12 Optimum width of longwall panels in pittsburgh seam.

15.7 Field Observations of Optimum Longwall Panel Width

The Pittsburgh coal seam of Pennsylvania and Northern West Virginia is a moderately gassy coal seam with a gas content of 100–250 ft^3/ton. Premining degasification is done with in-seam horizontal drilling and about 50% of the gas is drained—prior to mining. The rate of mining varies from 70 to 100 ft/day. The total methane emissions and ventilation air at the tailgate of longwalls 1000 to 1600 ft wide panels are plotted against the panel width in Fig. 15.12.

An optimum width of 1300 ± 50 ft is indicated.

Longwall panel width can also be controlled by high specific gob emissions. It will be discussed in Chapter 16. For highly gassy mines with specific gas emissions of 30 MMCF/acre or more, the optimum width is reduced to 750 ± 50 ft.

Problems

15.1. Using Eq. (15.6), calculate total methane emissions at the tailgate, Q, of a highly gassy coal seam; given:

Qo = 200 CFM, A = 200 ft^3/ton, B = 60 ft^3/ton

Ave C(X) for a 1000 ft long face = 200 CFM.

15.2. Calculate the volume of air/min needed to dilute the above Q to 0.8%, and verify if the gas layering index exceeds 5.00.

15.3. Calculate the steady-state gas production from a vertical well for the following given conditions:

$k = 0.01$ darcy
$h = 6$ ft
$\mu = 0.02$ cp
$z = 1.0$
$T = 520$ degrees R
$r_e = 1000$ ft
$r_w = 0.25$ ft
$p_e = 1000$ psi
$p_w = 50$ psi

If the gas content of coal is 600 ft^3/ton, how many frac wells will be needed to degas a longwall panel, 1000 × 10,000 ft in 5 years?

Hint: the total gas production declines by a power law (Eq. 15.2) with the value of n equal to 0.8.

References

[1] Thakur PC. Advanced reservoir and production engineering for coal bed methane. Elsevier; 2017. p. 210.
[2] Thakur PC, Poundstone WN. Horizontal drilling technology for advance degasification. Mining Engineering 1980:676–80.
[3] Thakur PC, Davis JG. How to plan for methane control in underground coal mines. Mining Engineering 1977:41–5.
[4] Thakur PC, Dahl HD. Horizontal drilling – a tool for improved productivity. Mining Engineering 1982:301–4.
[5] Thakur PC. In: Darling P, editor. Gas and dust control, SME mining engineering handbook, Chapter 15; 2011. p. 1595–609.
[6] Thakur PC, et al. Coalbed methane from prospect to pipeline. Elsevier; 2014. p. 420.
[7] Thakur PC. Optimum widths of longwall panels in highly gassy mines – Part I, 11th U.S. Mine Ventilation Symposium. In: Mutmansky J, editor. The Penn State University. PA, USA: University Park; 2006. p. 433–7.

Postmining Degasification of Coal Mines

16

Chapter Outline

When coal is extracted either by the room and pillar technique or the longwall method, it causes the overlying strata (containing several coal seams) to subside and the underlying strata to heave and release gas. The mine atmosphere acts as a pressure sink drawing all gases to the mine workings. Thus, in the postmining phase, the gas emission reservoir is considerably expanded. The coal seam that is being mined does not make significant contributions to the gob gas emissions. The postmining gas emissions can overwhelm mine ventilation if it is not captured and controlled.

Depending on the number of gas-bearing zones in the gas emission space and their gas contents, the total methane emission from longwall gobs could vary from a few hundred to more than several thousand cubic feet of gas per ton of mined coal. Hence, the ventilation of longwall faces demands a large quantity of air. This ventilation need is further enhanced by high air losses specially on caving faces (with no stowing of the gobs). The old system of longwall ventilation, the "U" pattern, where all the air was brought down one gate road and exited through the other gate road, loses large quantities of air through gobs giving a high methane concentration at the return end of the face. The "Y" ventilation pattern where intake air is brought down the headgate and return air leaves via two or more tailgate roads relieves the situation to some extent but, on most longwalls, some kind of methane control is still needed. (Refer to Chapter 4 in the book.)

Advanced Mine Ventilation. https://doi.org/10.1016/B978-0-08-100457-9.00016-X

With the onset of longwall mining in Europe in 1940s, the initial work on methane control on longwall was naturally done there. The basic principle of methane control has always been, and still remains, some means of "bypassing methane from the gas emission space without letting it mix with the mine air". In most cases, the bypassing mechanism is either a strategically located borehole or a roadway. A successful methane control program depends on the following basic premises:

1. Determination of the geometry of the gas emission space and location of gas-bearing horizons therein.
2. Estimation of the rate of methane influx into the longwall gob.
3. A scheme to bypass the gas in the most economic and efficient manner and thus prevent it from entering the mine atmosphere.

In successful methane drainage programs a high proportion, usually between 50% and 70%, of the total gas emissions in a working district is removed before it can enter the mine airways. Advantages of postmining methane drainage can be summarized as

1. Generally reduced gas delays in the mine leading to increased safety and higher productivity.
2. Reduced air requirements and corresponding savings in ventilation horsepower.
3. Possible use of mine gas as an additional source of fuel.

16.1 The Gas Emission Space

Fig. 16.1 shows the vertical extents of gas emission space with respect to the mine working according to various authors [1].

Calculations of gas influx are based on the concept that there are finite limits for the gas emission zones above and below the mine workings. The smallest range of gas emission space is given by Gunther [2] in the roof at 300 ft and by Lidin [3] in the floor at 70 ft. The greatest range of gas emission space is given in the roof at 1000 ft and in the floor at 300 ft by Thakur [1]. Winter [4] also observed heavings of 4–7 in. at depths of 400–500 ft below the mined face. This amount of heaving will create sufficient improvement in permeability for a substantial gas influx. In general, larger longwall panels create larger gas emission space. By far, the largest amount of gas encountered in longwall gobs originates from coal seams overlying the mined area.

Fig. 16.1 also shows various estimates of the amount of gas emitted by underlying and overlying gas-bearing strata. Gunther [2] assumes that overlying seams lose all gas to the gob, whereas the loss of gas from underlying seams decreases linearly with distance from the mined seam. Lidin [3] assumes linear relationship between the amount of gas emitted and distance for both overlying and underlying strata. Other authors show exponential and power law relationships between the percentage of total gas emitted and the distance between the gas-bearing zones and mine workings. In general, the percentage of gas lost by the gas-bearing zones reduces with increasing distance from mine workings. The rate of methane flow in the gob is also proportional to the rate of face advance. Hence, methane emission from longwall gob is much higher today than it was in past.

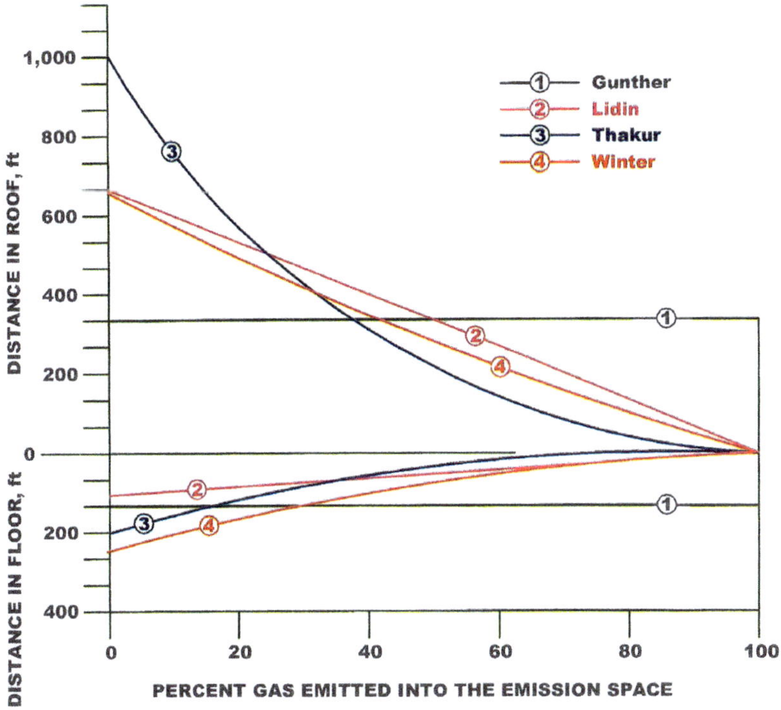

Figure 16.1 Vertical limits of the gas emission space.

Factors which govern the rate of gas emission in the longwall gobs can be summarized as follows:

1. The density and proximity of gas-bearing zones in the gas emission space.
2. The gas content of coal seams and other gas-bearing horizons.
3. The extent to which overlying and underlying coal seams have been mined.
4. The rate of coal face advance. This is the most important factor.
5. The presence of geological disturbances and leftover coal pillars.
6. The subsidence characteristics of the area which determines the permeability of the gas emission space.

Fig. 16.1 showed only the vertical dimension of the gas emission space. The width of the gas emission space is limited to the width of the longwall panel. In deep, gassy mines, chain pillars get crushed, and there is little movement of gases from one longwall panel to the next.

The length of the effective gas emission space is limited by the subsidence of the overlying strata. Fig. 16.2 shows that the gob is fully subsided (up to the full thickness of the coal seam) when the face retreats from the setup entry by a distance equal to 1.3× depth of the coal seam [5].

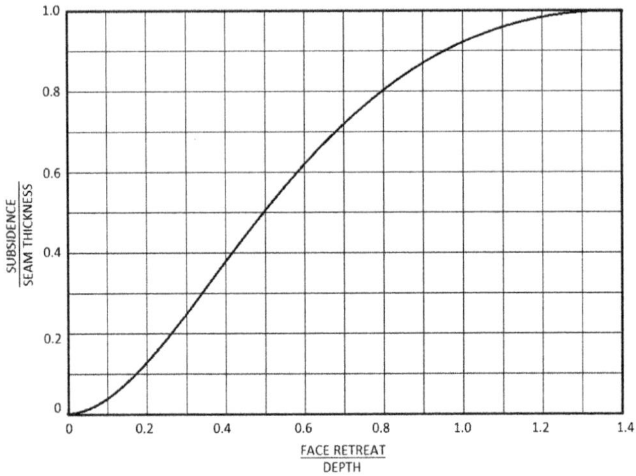

Figure 16.2 Subsidence over a longwall gob as a function of face retreat/depth.

Thus, in a mine with a depth of 2000 ft, the length of the "effective" gas emission space is 2600 ft. This "effective" length is very useful in deciding the spacing of gob wells on longwall panel as discussed later in this chapter.

It is desirable to use this "effective length" for design purposes because it slightly overestimates the number of gob wells per panel. In reality, the gob areas beyond this "effective length" continue to produce methane for a long time, albeit at a lower rate.

Obviously, there can be considerable variations in the gas flows into the longwall gobs because of varying combination of above factors. Experience and data obtained in one geographical area, therefore, can be transferred only qualitatively to other areas. A summary of field data is provided in Table 16.1.

Table 16.1 Typical Longwall Gob Gas Emissions

Coal Seam	Gassiness	Rate of Mining (ft/day)	Total Gob Emissions (MMCFD)[a]
1. Pittsburgh (PA—OH)	Mildly gassy	70—100	1—3
2. Pittsburgh (WV)	Moderately gassy	70—80	3—8
3. Pocahontas #3 (WV)	Moderately gassy	70—80	8—10
4. Pocahontas #3 (VA)	Highly gassy	60—70	25—30
5. Blue Creek/Mary Lee (AL)	Highly gassy	60—70	15—20

[a]Millions of cubic feet per day.

16.2 European Gob Degasification Methods

European coal seams are generally steeply inclined, tectonically disturbed, and seated deeper than US coal seams. Several coal seams are deposited in each basin and, typically, worked simultaneously. East European coal seams are not only gassy but also prone to instantaneous outbursts. Longwall mining system is the most common method of mining coal. Both advancing and retreating longwalls are employed in varying proportions. Most of the methane emission takes place in the gob areas following mining operations and strata movement leading to tremendous improvements in permeability.

European methane control techniques can be broadly classified in the following three groups:

1. The packed cavity method and its variants.
2. The cross-measure borehole method.
3. The superjacent method.

Each of these techniques is discussed below.

16.2.1 Packed Cavity Method and Its Variants

Early methods of methane control consisted of simply isolating the worked out area in the mine using pack walls, partial or complete stowing, plastic sheets, or massive stoppings. A network of pipeline which passed through these isolation barriers was laid in the gob, and methane was drained using vacuum pumps. Fig. 16.3 shows typical layouts for a caving longwall face. Fig. 16.4 shows a similar layout for a partially stowed longwall face.

Lidin [3] has reviewed several variants of this technique. Methane capture ratios quoted by him are shown in Table 16.2.

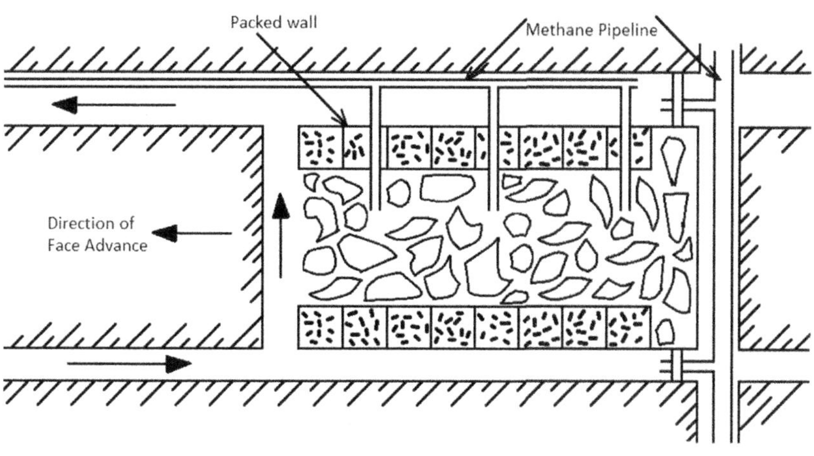

Figure 16.3 Packed cavity method for a caved longwall face.

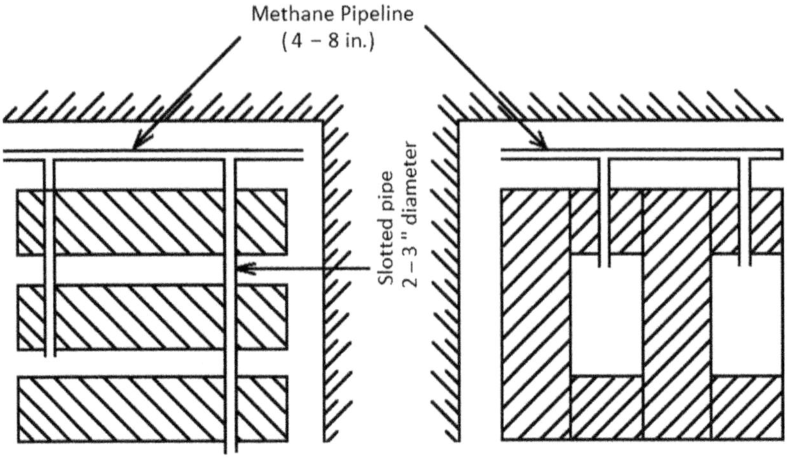

Figure 16.4 Packed cavity method for a partially filled longwall face.

Table 16.2 Methane Capture Ratios for Packed Cavity Methods

Method of Mining	Method of Gob Stowing	Methane Capture Ratio (%)
Longwall advancing	Caving	20–40
Longwall advancing	Partial filling	30–50
Room and pillar	Complete filling	60–80

The ratio generally seems to improve from caving (20%–40%) to fully stowed longwall gobs (60%–80%). The gate roads are protected by a pack wall against the gob. Pipelines are laid through the pack wall to reach nearly the centerline of the gob and are manifolded to a larger diameter pipe in the gate road. Fig. 16.4 shows a partially stowed longwall gob where cavities are purposely left between alternate packs. The overlying strata in that area cracks and provides a channel for gas to flow into these packed cavities. Pipelines are laid to connect the cavity with methane drainage mains. Methane extraction is usually done under suction. The technique is also known as "Roschen" method of methane drainage.

16.2.2 Cross-Measure Borehole Method

This is by far the most popular method of methane control on European longwall faces. Fig. 16.5 shows a plan and elevation of a typical layout for a retreating longwall face.

Boreholes 2–4 in. in diameter are drilled from the top gate to a depth of 60–500 ft. The angle of these boreholes with respect to horizon varies from 20 degrees to 50 degrees, while the axis of the borehole is inclined to the longwall axis at 15–30 degrees towards the gob. At least one hole in the roof is drilled at each site, but several

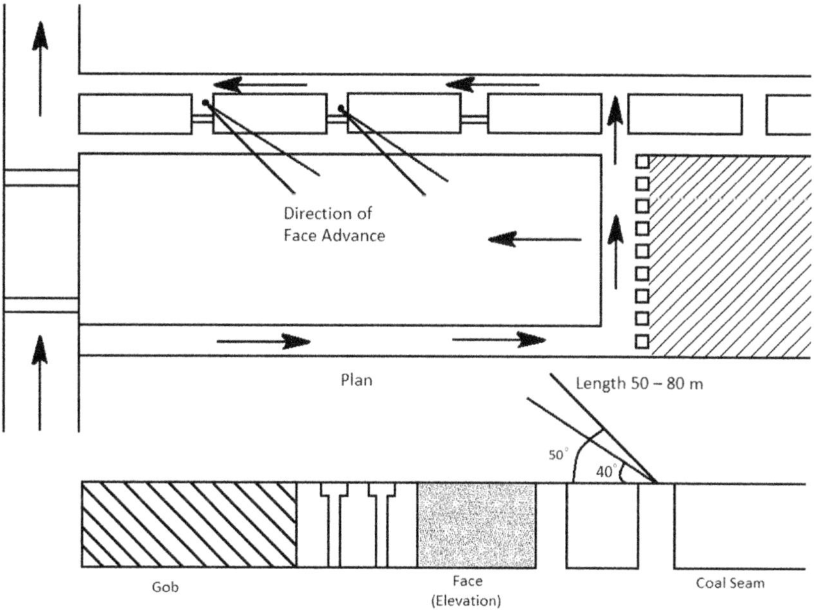

Figure 16.5 Methane drainage with cross-measure boreholes.

boreholes in roof and floor can be drilled at varying inclinations depending on the degree of gassiness. Drill sites are typically 80 ft apart. These holes are then manifolded to a larger pipeline system, and gas is withdrawn using a vacuum pump. Vacuum pressures applied vary from 4 to 120 in. of water gage. The amount of methane captured by the drainage system expressed as a percentage of total methane emission in the section varies from 50 to 90%. Some typical data from British mines are given by Kimmins [6] and shown in Table 16.3.

Table 16.3 Methane Capture Ratios for Cross-Measure Borehole Method

Mine	Specific Methane Emission (ft^3/ton)[a]	Methane Capture Ratio (%)	
		Section	Mine
Ashley Green	3200	60	38
Haig	3000	59	20
Parkside	2800	61	42.6
Point of Ayr	5700	68	47.7
Sutton Manor	3300	70	40.0

[a]This is the total methane emission from a mine divided by the tonnage mined every day. These are highly gassy mines.

The technique is generally more successful for advancing longwall panel than it is for retreat faces. The flow from individual boreholes is typically 20 CFM, but occasionally, it can go up to 100 ft^3/min for deeper holes. Sealing of the surface casing is very important and is usually done with quick-setting cement. Sometimes, a liner (a pipe of smaller diameter than the borehole) is inserted in the borehole and sealed at the mouth to preserve the production from the borehole even if the borehole is sheared by rock movements.

16.2.3 The Superjacent (or Hirschback) Method

This technique is mainly used for retreating longwall faces in very gassy seams. Fig. 16.6 shows a typical layout.

A roadway is driven 60—100 ft above the longwall face, preferably, in an unworkable coal seam. The roadway is sealed and vacuum pressures up to 120 in. of water gage are applied. To improve the flow of gas, inclined boreholes in the roof and floor

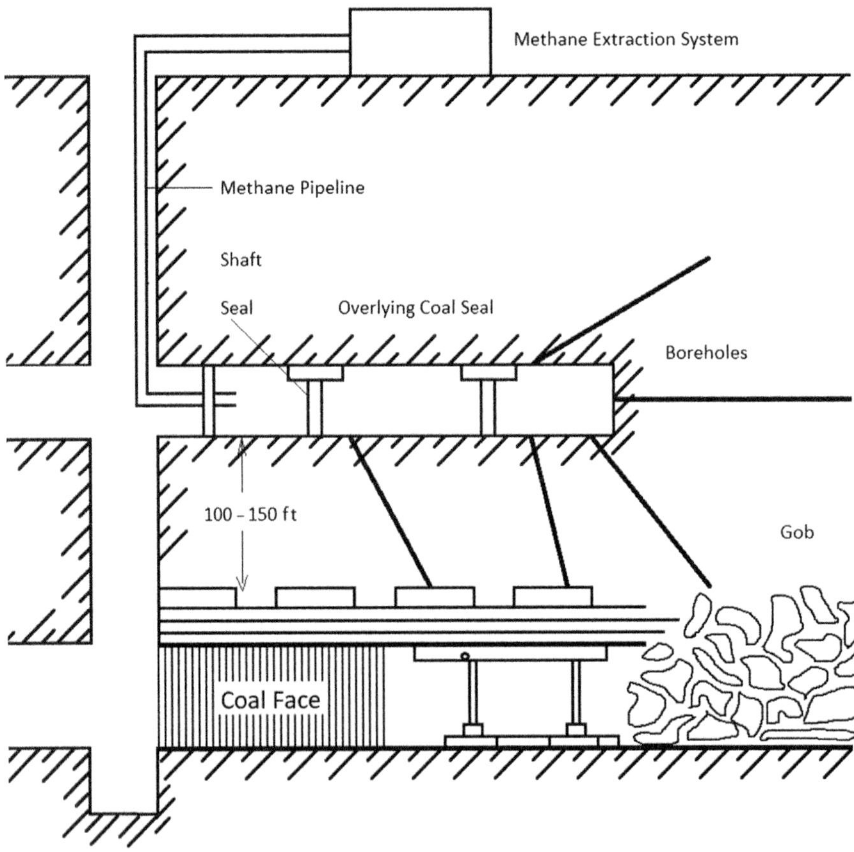

Figure 16.6 Methane drainage by superjacent method.

are drilled to intersect other gassy coal beds. If mining scheme proceeds from the top to the bottom seams in a basin, the entries in a working mine can be used to drain coal seams at lower levels. Methane flow from these entries is high, averaging 700–1000 CFM for highly gassy seams. Nearly 50% of total emission at the longwall face has been captured [7].

16.3 US Gob Degasification Method

US longwall panels are much larger than European longwall panels and are mined at a faster rate. Cross-measure boreholes simply cannot drain enough methane to keep the bleeder entries at less than 2% methane as required by Federal regulations [8]. Thakur [9] estimated the feasible ventilation air quantities in modern longwall mines as shown in Table 16.4.

The bleeder air, therefore, can only capture a small fraction (15%–30%) of total gob emissions. Gob drainage technique must capture 60%–80% of the total gob gas emissions. The only technique that can do this is the use of vertical gob wells. Their size and number will vary with the rate of gob gas emissions and will be discussed later in the chapter.

16.3.1 Construction of a Vertical Gob Well

Fig. 16.7 shows the vertical section of a typical gob well.

A 12–15 in. surface casing is set at the bedrock for a depth of about 100 ft. Next, a $9^{5}/_{8}$–$12^{1}/_{4}$ in. borehole is drilled to a depth below all water aquifers and a $7^{1}/_{2}$–$9^{5}/_{8}$ in. casing is set. Next a $6^{1}/_{4}$–$7^{7}/_{8}$ in. borehole is drilled to the top of the coal seam being mined stopping 30–90 ft above the coal seam. A $4^{1}/_{2}$–$6^{1}/_{2}$ in. diameter casing is lowered in the borehole, and it is anchored in a strong stratum about 300–1000 ft above the coal seam depending on the depth of the coal seam. The casing below this point is not cemented but slotted to allow gob gases to enter the gob well. The slots are typically 1 in. wide × 2 ft high and are cut at different locations on the perimeter of the casing like a spiral. This preserves the integrity of the casing. For shallow coal seams, old casings or pipes can be used but for deeper coal seams, new J55 or K55 casings are recommended. The size of the gob well casing varies from $4^{1}/_{2}$ to $12^{1}/_{4}$ in. outside diameter depending on the volume of gas to be drained.

Table 16.4 Feasible Ventilation Quantities for US Longwall Faces

Category of Coal Seam	Face Intake (CFM)	Tailgate (CFM)	Bleeders (CFM)
Mildly gassy	30,000	25,000	100,000–150,000
Moderately gassy	50,000	40,000	150,000–250,000
Highly gassy	80,000	60,000	250,000–350,000

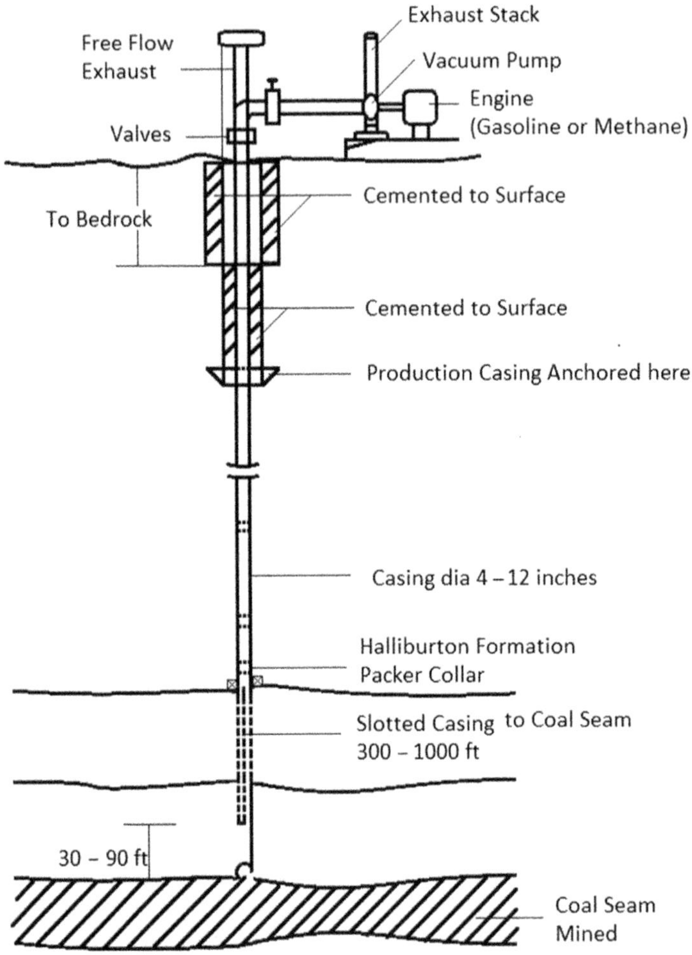

Figure 16.7 Vertical section of a typical gob well.

16.3.2 Location of Gob Wells on the Longwall Panel

The first gob well is usually installed within 50–500 ft from the setup entry. When the longwall face retreats 100–200 ft, the first main roof fall takes place releasing a large volume of methane. Sometimes, it is necessary to have two gob wells in parallel near the setup entry to cope with the onrush of strata gases. Location of other gob wells on the panel must be done in an optimal manner to capture the maximum percentage of total gob gas emissions at the minimum cost.

Optimum gob gas drainage depends on the following:

- The size of the gob well and its production capacity.
- Distance of the gob well from the tailgate.
- Spacing of the gob well on the longwall which is a function of the width of the longwall panel and the rate of mining.

16.3.2.1 The Size of Gob Well and Gas Production Capacity

Coal industry in the United States uses gob wells with diameters ranging from 4 to 12 in. Moderately gassy mines use 4–7 in. diameter casings but highly gassy mines use bigger casings. The most popular size is 12 in. in diameter. The cost of drilling larger diameter gob wells increases exponentially.

All gob wells are assisted in gas production by a well-designed blower. Lampson blowers (Series 600, 800, and 1200) are commonly used blowers and have a good track record. Operating at a suction pressure of approximately 1–3 psi the gas production volumes are as follows:

Casing Diameter	Gas Production
6 in.	1–2 MMCFD
9 in.	2.5–3.2 MMCFD
12 in.	4.5–5.0 MMCFD
15 in.	7.0–8.0 MMCFD

Assuming an average of 70% methane in the gob gas, a single 12 in. diameter gob well can remove 3–4 MMCFD of methane from the gas emission space.

16.3.2.2 Distance of Gob Wells From the Tailgate

For best efficiency, all gob wells must be located between the centerline of the longwall panel and the tailgate. On an average, best results are obtained when the gob wells are located about 100 ft from the centerline toward the tailgate as shown in Fig. 16.8.

Gob wells located on the headgate side of the longwall gob produce significantly lower amount of gas and thus are very inefficient.

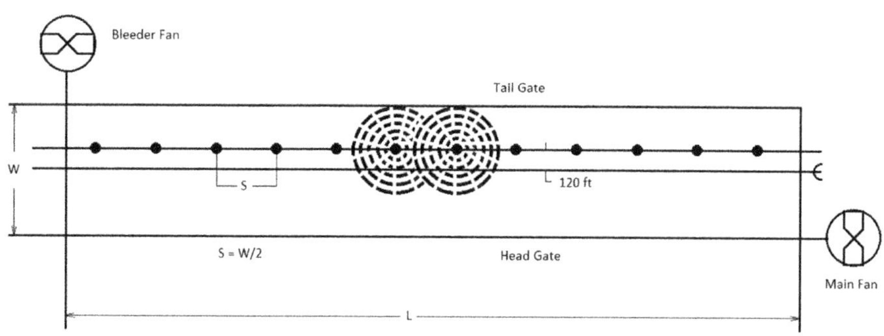

Figure 16.8 Optimum layout of gob wells on a longwall face.

16.3.3 Gob Well Spacing on the Longwall Face

It was earlier shown that the effective length of the gob producing most gas is limited to 1.3× depth immediately behind the face. Gas removed by the bleeders is discounted here to provide a little reserve capacity for the gob wells to handle peak emissions. Knowing the gob well production capacity, the number of producing gob wells in the gas emission space can be calculated by simply dividing the total emissions by the production capacity of the gob well. The total number of gob wells for the panel can be calculated by prorating this number by the ratio, the length of panel/the length of gas emission space.[1] Table 16.5 shows the specific gob emissions as observed for different widths of longwall panels in a highly gassy coal seam in Virginia. Specific gob emissions tend to increase with the width of the panel.

The optimum width of a longwall panel is the width where gob wells are most efficient in draining the gob gases and the total number of gob wells for the panel is the minimum. The most efficient gob drainage is reached when the spacing between the two adjacent gob wells (s) is equal to half the width (w) of the longwall panel. A ratio, s/w that is less than 0.5, indicates inefficient methane drainage. To illustrate this point, spacing of gob wells for longwalls with different widths but the same tonnage of extraction is calculated. The following assumptions are made for this calculation:

1. The rate of extraction is the same for all longwall panels, one acre per day.
2. The longwall face is 10,000 ft long and needs two gob wells near the set up entry.
3. A gob gas capture ratio of 70%–80% will be achieved.
4. Specific gob emission is 30 MMCF/acre.

Table 16.6 shows estimated total number of gob wells for various widths of longwall panels.

Fig. 16.9 shows a plot of s/w against longwall face widths. The most efficient capture of gob gas is obtained when s/w = 0.5 or the longwall width is 700 ft. Because the

Table 16.5 Specific Gob Emissions for Longwalls in Highly Gassy Seams

Width of Longwall Face (ft)	Specific Gas Emission (MMCF/acre)
450	25
600	30
750	33
900	36
1050	40

[1] An example: a longwall panel produces 30 MMCFD at a depth of 2000 ft. Most of the gas is produced from an area 2600 ft immediately behind the face. A 12 in. diameter gob well will produce 4 MMCFD. Hence, we need 6–7 gob wells over a length of 2600 ft. Therefore, the gob well spacing is 370–430 ft or 400 ft on the average. For a 10,000 long panel, total number of gob wells is at least 25.

Table 16.6 Number of Gob Wells Versus Longwall Width

Width of Face (ft)	Number of Gob Wells	Spacing/Width (s/w)
450	24 + 2	0.93
600	26 + 2	0.64
750	28 + 2[a]	0.48
900	32 + 2[a]	0.35
1050	36 + 2[a]	0.26

[a]Total number of gob wells could be slightly higher because of declining capture efficiency.

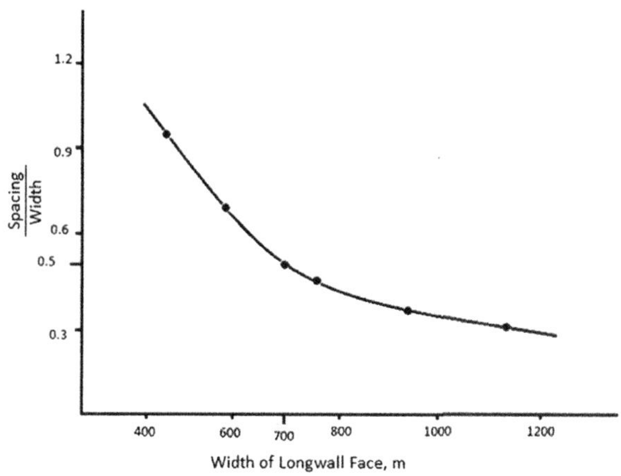

Figure 16.9 Optimal spacing of gob wells on a longwall face.

gob wells are offset from the centerline by 100 ft, the optimum width where gob gas drainage is most efficient is in the range 700–800 ft or 750 ± 50 ft.

16.4 Gas Capture Ratios by Vertical Gob Wells

Vertical gob wells provide a great flexibility in coping with different sized longwall panels mined at low, medium, or high rates. The high cost of longwall equipment requires a high rate of mining (70–80 ft/day) for adequate return on the investment. The capture ratio depends on both the number of gob wells and the rate of gob gas emissions for the longwall gob. Table 16.7 shows typical capture ratios for some US longwall faces.

Number of gob wells per panel and hence the capture ratio goes down, if the longwall is mined at a very slow rate (10–20 ft/day).

Table 16.7 Coal Gas Capture With Vertical Gob Wells

Coal Seam	Gassiness	Total Methane Emissions (MMCFD)	Gob Wells Per Panel	Capture Ratio (%)
Pittsburgh (PA and WV)	Moderately	4–8	5–6	40–50
Lower Kittanning (PA)	Mildly	2–4	3–5	30–50
Pocahontas #3 (WV)	Moderately	5–6	5–8	30–50
Pocahontas #3 (VA)	Highly gassy	25–30	20–35	65–80
Blue Creek/Mary Lee (AL)	Highly gassy	15–20	10–20	60–70

16.5 Gob Well Production Decline

Most gob wells are produced with a vacuum pump. The pump capacity is always matched with the anticipated gas production. Available vacuum pumps (also commonly called blowers) can handle a flow from 100 MCFD to 5 MMCFD and create a vacuum of a few inches of WG to 120 in. of WG. Lampson 650 and 850 are commonly used. The initial gas production may be erratic, but it settles down in 10 days. The decline of production follows a power law given by Eq. (16.1).

$$Q = A\ t^{n} \tag{16.1}$$

where Q is the total production in MCF, A is the initial production in MCFD, t is the time in days, and n is an exponent with a value of 0.8 ± 0.1, a characteristic of the gob emissions space.

Production declines for four gob wells are shown in Fig. 16.10 for longwall panels in the Pittsburgh seam of West Virginia, Lower Kittanning of Pennsylvania, and Pocahontas #3 seam of West Virginia and Virginia. The deeper gob well in Pocahontas #3 seam is about 2000 ft deep.

For design purposes, it was earlier assumed that the gob well will cease production when the longwall face has retreated 1.3× depth (ft), but in reality, they continue to produce some gas until the face has moved beyond 3000 ft regardless of the depth.

A longwall panel in a highly gassy mine becomes a source of low-cost gas. A typical 1000 × 10,000 ft panel with a specific gob production of 30 MMCF/acre will produce 6.9 BCF gas. Even at a low price of $3/MCF, it can create revenue of over $20 million. If this gas is marketed, it can defray the cost of degasification and ventilation of the mines. More of economics will be discussed later in the book in Chapter 20. Methane capture from sealed gob areas will also be discussed in Chapter 20.

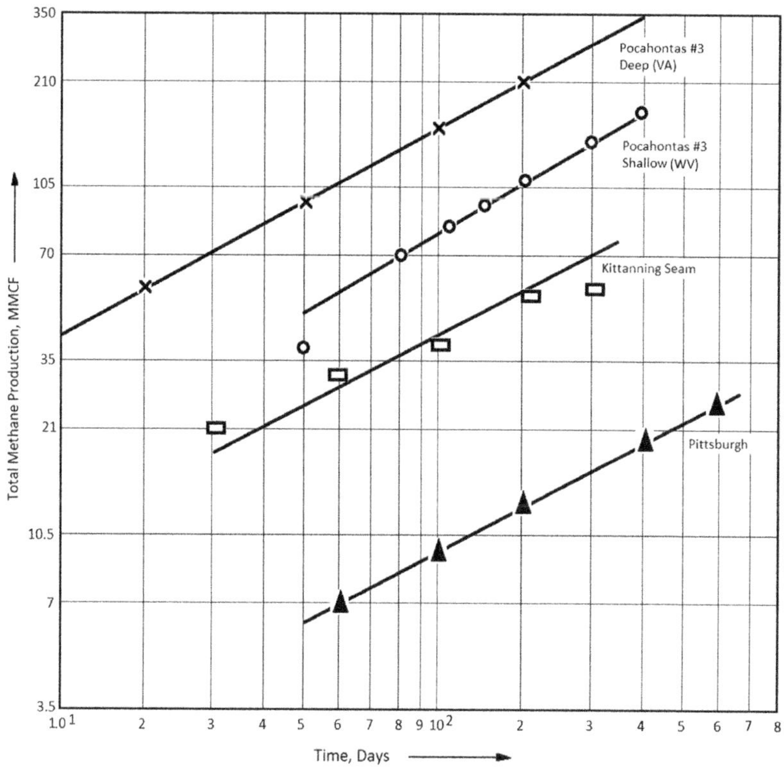

Figure 16.10 Gas production from a single gob well.

Problem

1. Find out the value of "n" and A for the four gob well production curves presented in Fig. 16.10. Calculate the total gas produced by each gob well in the first 100 days.
   Hint: Extend the curves to intercept "y" axis. Divide Q by 10 to get "A" values.

References

[1] Thakur PC. Advanced reservoir and production engineering for coal bed methane. Elsevier; 2017. p. 210.

[2] Gunther J, Belin J. Prevision du degasment du grisou en taille pour le grisements en Plateure. In: 12th international conference on coal mine safety and research. Germany: Dortmund; 1967.

[3] Lidin GD, et al. Control of methane in coal mines. English translation by Israel Program for Scientific Translations, Jerusalem. 1964 [original in Russian, 1961].

[4] Winter K. Extent of gas emission zones influenced by extraction. In: 16th international conference on coal mine safety and research, Washington, DC; 1975.

[5] Holland CT. Mine pillar design, SME mining engineers handbook, vol. 1; 1973. p. 13-06–13-118.
[6] Kimmins EJ. Firedamp drainage in the north western area (of the U.K). Colliery Guardian, Annual review. 1971. p. 39–45.
[7] Thakur PC. Methane control for longwall gobs, longwall-short wall mining. State-of-the-art, Chapter 10. SME, AIME; 1981. p. 81–6.
[8] Thakur PC, Lauer SD, Cervik J. Methane drainage with cross-measure boreholes on a retreat longwall face. SME-AIME Fall Meeting and Exhibit, Pre-print No. 83–398, Salt Lake City, Utah. 1983.
[9] Thakur PC, Zachwieja J. Methane control and ventilation for 1,000 ft-wide longwall faces. In: Proceedings of the longwall USA – international exhibition and conference; 2001. p. 167–80.

Floor Gas Emissions and Gas Outbursts

17

Chapter Outline

In Chapters 15 and 16 of the book, the premining degasification and the postmining degasification of coal mines were discussed in a generic form. Two special cases need a separate and comprehensive discussion. They are as follows:

- Floor gas emissions and
- Gas outbursts.

As discussed in Chapter 16, longwall mining makes the mine floor heave. If there are coal seams in the floor overlain by a strong sandstone or shale, they cannot emit the released gas to the gob emission spaces, until the said overlying strata breaks suddenly releasing a large volume of gas and coal fines into the mine. The volume of gas is so large that it reverses ventilation on the longwall face and explosive mixtures of air and methane flood into the intake airways with nonpermissible equipment. Consequently, a mine explosion occurs costing many human lives. Techniques to avoid this disaster will be discussed here.

Advanced Mine Ventilation. https://doi.org/10.1016/B978-0-08-100457-9.00017-1

Secondly, as mining progresses to deeper levels, the gas content of coal, reservoir pressures, and the diffusivity coefficient appear to increase [1]. Gas outbursts are violent expulsion of gas (and some coal fines) from the working face. These energy release phenomena can have the same catastrophic consequences as the sudden floor emissions. Estimation of dangers and methods of preventing gas outbursts will be also discussed in this chapter.

17.1 Floor Gas Emissions

So far, most of the degasification effort has been confined to the working coal seam and coal seams overlying the working coal seam. Very little attention has been paid to the degasification of coal seams and other gas-bearing strata underlying the working coal seam. Most of the underground mining is at present done by longwall mining. As longwall mining progresses, the roof immediately behind the roof supports caves and the overlying strata subside. Even the underlying strata heave and open up channels that feed gas to the longwall gob. The gas pressure in overlying and underlying coal seams is much higher than the ambient air pressure in the mine. Thus, the longwall face and gobs become pressure sinks into which gas flows from the entire disturbed zone. Chapter 16 summarizes the studies on the dimensions of the gas emission space created by longwall mining and the percentages of gas contents released by various coal seams contained in the gas emission space as a function of its vertical distance from the mined coal seam. The vertical extent of the gas emission space is highly dependent on the width of the longwall panel. In general, the wider the longwall panel is, the larger is the vertical dimensions of the gas emission space and consequently, the higher is the specific gob methane emission (volume of gas produced/acre of gob). In the Appalachian Basin, the problem was not so acute, until the longwall face widths increased to 1000 ft and beyond [2]. The volume of floor gas emissions on the face is seldom large, but it creates local methane concentrations that exceed the statutory limit of 1%. Such areas are also hard to ventilate. When floor emissions occur, the ventilation air to the affected area is increased using curtains. Most emissions are short-lived, but there are cases where the longwall faces are idled for 3–4 weeks. Economic consequences of such production stoppage on modern longwall faces are very serious. Solutions to this problem can be summarized as follows:

1. Premining drainage of the gas from underlying coal seams/other gas-bearing strata.
2. Postmining drainage of the gas emission space underlying mineable coal seams.
3. Drilling short vertical boreholes on the longwall face and pumping grouting material to seal the leakage path.

17.1.1 Premining Methane Drainage

The area to be mined by longwall mining must be drilled from surface to a depth of 270 ft below the mineable coal seam to confirm the presence of other coal seams/gas-bearing strata. Coal cores are carefully collected and gas contents and, sometimes, reservoir pressures are measured. If the gas contents and pressures are high and the coal seams are liable to outburst, they must be adequately degassed prior to mining. Failure to degas the underlying coal seams have resulted in mine disasters in many European and Australian coal mines owing to sudden outburst of gas and coal fines [3]. At present there are two techniques available to predrain these coal seams:

1. Hydrofracturing of underlying coal seams by a vertical well and
2. Horizontal boreholes drilled from surface to intersect all underlying coal seams in the gas emission space, with or without hydrofracturing.

17.1.1.1 *Hydrofracturing of Underlying Coal Seams*

Frac wells are drilled from surface in a grid pattern over the longwall panels liable to floor emissions to intersect the underlying coal seams/gas-bearing noncoal strata. High-pressure water (or other fluids, e.g., nitrogen foam) with sand is pumped into these formations to create a vertical fracture that can run several hundred feet on either side of the frac well (Fig. 17.1). The liquid is pumped out, but sand remains in the

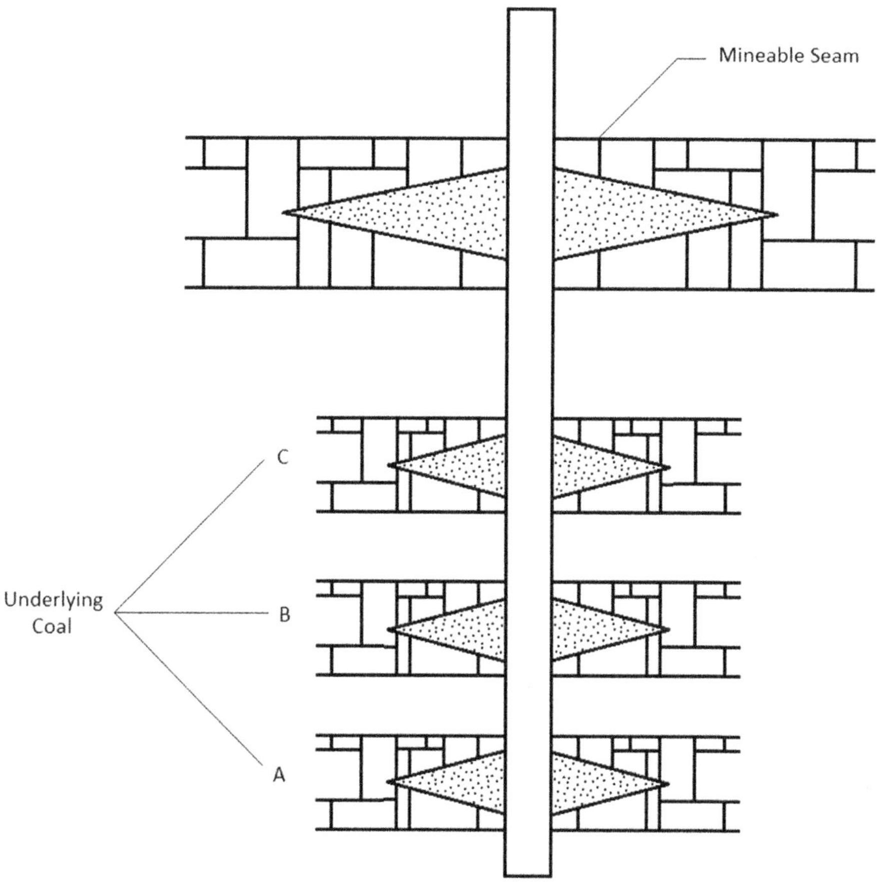

Figure 17.1 Vertical well with multiple seam completion.

fracture and keeps it open for gas to escape to surface. Several coal seams/gas-bearing horizons can be fracked from a single well.

Under ideal conditions, 60%–70% of the gas in the coal seams can be removed by vertical wells if drilled and produced for 5–10 years in advance of mining. Reservoir pressures can be reduced by 80% prior to mining. This technique is ideally suited to deep, very gassy coal seams with low to medium permeability (1–10 md). Its effectiveness may be reduced if the coal seam is deeper than 4000 ft where the coal seam permeability is generally less than one microdarcy.

17.1.1.2 Horizontal Boreholes Drilled from Surface

It is a relatively new technique where a vertical well is drilled from surface, but it is deviated by 90 degrees to make it horizontal and intersect a number of coal seams that needs to be degassed. Fig. 17.2 shows the general layout of a typical drilling scheme.

The horizontal extension can be 3000–10,000 ft depending on the depth of the coal seam and the type of drill rig used. The borehole can be lined with slotted liners to guard against accidental closure due to ground movement. Gas production even prior to mining could be quite brisk if the coal seams are less than 2000 ft deep and have medium to high permeability. Gas production is likely to be much higher when the coal seam overlying these horizontal boreholes is mined out and the underlying coal seam heaves and experiences an order of magnitude improvement in permeability.

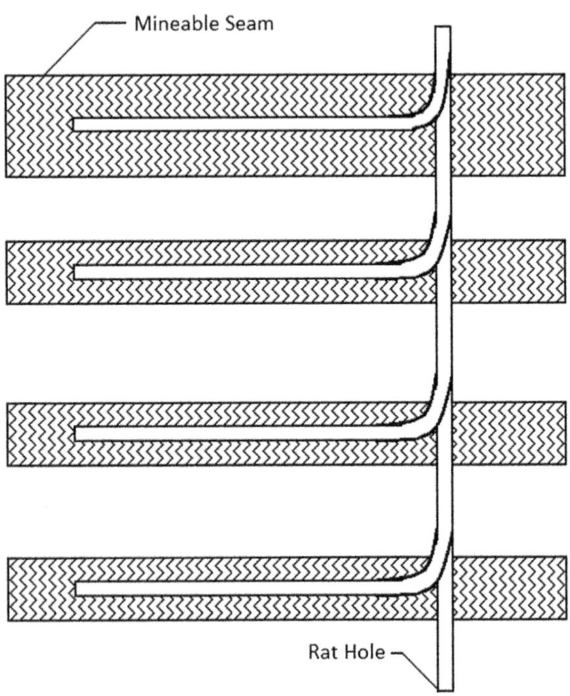

Figure 17.2 Horizontal boreholes in multiple seams drilled from surface.

Gas production can be enhanced by installing an exhausting fan with a negative pressure of 3–5 psi. These boreholes provide a bypass for floor gases, and thus, gas emission on longwall floor is minimized.

17.1.2 Postmining Methane Drainage

Postmining methane drainage from the longwall gobs is perhaps the most effective technique for preventing floor gas emissions on the longwall panels. There are several techniques currently in practice for gob drainage, but the two most suitable for US coal mines are the following:

1. Cross-measure boreholes drilled in the floor and
2. Vertical gob wells completed in the coal seams below the mineable coal seam.

17.1.2.1 Cross-Measure Boreholes in the Floor

This is by far the most popular method of methane control on European longwall faces. Fig. 17.3 shows a typical layout for a retreat longwall face.

Boreholes 2–4 in. in diameter are drilled from the tailgate to a depth of 100–500 ft in the floor. The inclination of these boreholes varies from 20 to 50 degrees, and the axis of the boreholes is inclined to the longwall axis at 15–30 degrees. Sometimes, more than one borehole is drilled from the same drill site. The drill sites are typically 100 ft apart. These boreholes are manifolded to a larger pipeline system, and gas is withdrawn using a vacuum pump. Vacuum pressures vary from 4 to 120 in. of water depending on the local conditions. The amount of methane captured by the

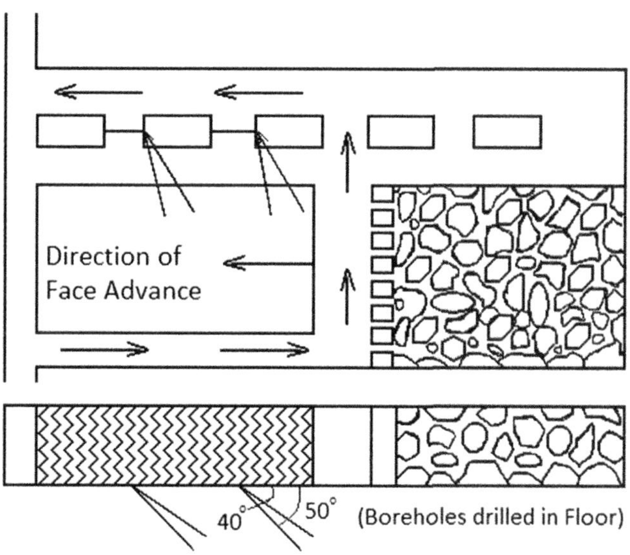

Figure 17.3 Methane drainage with cross-measure boreholes.

cross-measure boreholes varies from 30% to 70% of the total emissions from the longwall panel. Cross-measure boreholes can also be drilled on the headgate side of the longwall panel, but they are generally not as effective as those drilled on the tailgate side of the longwall panel.

17.1.2.2 Vertical Gob Wells Completed in Lower Coal Seams

Vertical gob wells are commonly used to degas longwall gobs in the US coalfields, but they are completed, typically, 50–100 ft above the coal seam being mined. They efficiently capture gas emissions from the overlying coal seams but are not very effective in capturing emissions from the floor. About 20 years ago, they began to drill these gob wells to a depth of about 240 ft below the coal seam being mined, as shown in Fig. 17.4.

The portion of the borehole below the mineable coal seam is filled with coarse sand and small gravel, and a solid cement plug is put across the working coal seam. When longwall face is mined through the borehole and moved 100–150 ft outby, the floor heaves. The partially filled gob well provides a channel for gas to come out in the gob at a point that is 100–150 ft inby of the longwall face. Such gob wells can be placed at close intervals (150–500 ft apart) on the longwall panels that are prone to floor gas emissions. Gas production is assisted by vacuum pumps creating 1–5 psi negative pressure.

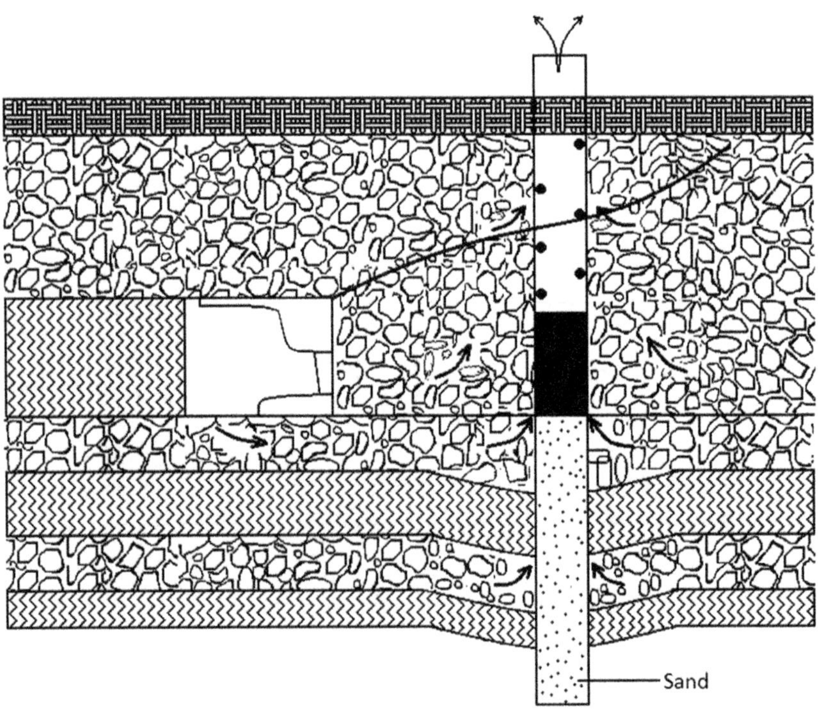

Figure 17.4 Gob well completed 200 ft below working seam.

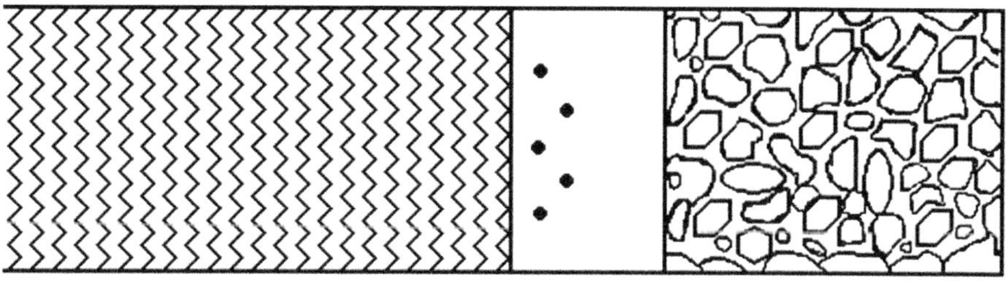

Figure 17.5 Short 10′–15′ deep borehole for grouting on the longwall face.

17.1.3 Injection Boreholes on Longwall Faces

In many mines, the floor gas emission problem is not very serious. There may not be any major coal seam in the gas emission space below the coal seam being mined, but gas may be originating from lenses of coal or other gas-bearing strata, such as carbonaceous shale or tight sandstone with low porosity. These emissions do not last more than a few hours and are best controlled by drilling short (6–10 ft deep) boreholes on the longwall face next to the point of gas emission (Fig. 17.5).

A mixture of micromatrix (very fine) cement and water is pumped into these boreholes to plug the channels of gas entry. Equally good results have been obtained with other mixtures, such as the polyurethane foam used for sealing fractures in the mine roof. Many contractors provide this service to mines on a routine basis. Although not always true, the floor grouting with cement appears to have a compounding effect on floor emissions, i.e., it makes the emissions less severe as the longwall retreats outby.

17.2 Gas Outbursts

Lama [4] provides a number of definitions of "gas outburst," but, in essence, it is a sudden ejection of large volumes of gas (methane, CO_2 or both) accompanied by a large mass of very fine coal from the working face. A classic definition by Hargraves [5] is perhaps the most appropriate and reads as follows:

> *The sudden disintegration of coal, and its projection from the seam, without deliberate initiation and accompanied by, and followed by, enormous gas emissions. The gas has the effect of carrying the broken coal for considerable distances. The projected coal is of fine size. The gas pressures and volumes associated are sometimes sufficient to penetrate the intake airways for considerable distances and blow up stoppings.*

Campoli [6] and Lama and Bodziony [3] list the primary factors contributing to gas outburst as the following:

1. High gas contents and corresponding high gas pressures.
2. Low permeability creating high pressure gradients close to the working face.

Table 17.1 History of Gas Outburst

Coal Field/Country	Depth (ft)	Type of Gas	Volume of Gas (MMCF)	Mass of Coal (ton)	Year
1. Sydney, Canada	2340	CH_4	–	–	1977
2. North Staffordshire, UK	2550	CH_4	–	–	1904
3. Gard, France	836	CO_2	–	1000	1907
4. Cevennes, France	2670	CH_4	14	1300	1947
5. Collinsville, Australia	730	CO_2	–	900	1954
6. Lower Silesian, Poland	2300 (appx)	CO_2	26.5	5000	1958
7. Hokkaido, Japan	2700	CH_4	21.2	5300	1981

3. High ground stress in the rock mass including the coal seam (refer to Chapter 15).
4. Low compressive strength or fractured coal.
5. A very high diffusivity (a sorption time of less than 1 day).
6. A high gas thermal gradient. Temperatures much above the normal coal seam temperature have been noted in the outburst areas [3]. High temperatures cause great increase in diffusivity of the coal and hence large emission of gases in a short time.
7. Geological anomalies, such as faults, igneous intrusions, and clay veins, acting as barriers to normal gas flow and emissions.

Table 17.1 lists some major gas outbursts in the past 100 years. There are only three mining basins in the United States where gas outburst has occurred:

1. Carbondale and Somerset coalfields of Colorado.
2. Harlan coalfields of Kentucky.
3. Wasatch Plateau and Book Cliffs coalfields of Utah.

They are infrequent and not so disastrous. Overseas, it is a major and frequent danger in the deep mines of China, Kazakhstan, Ukraine, Russia, Germany, and Poland.

17.3 Parameters Indicating a Propensity to Gas Outbursts

Imgrund and Thomas [7] provide a very good review of international parameters/criteria for the prediction of gas outbursts. Each country has a slightly different strategy. Some of them are highlighted here.

17.3.1 Australia

Gas outburst risk is assessed by

1. Gas contents (limit 212 ft^3/ton for CO_2; 318 ft^3/ton for CH_4).
2. Gas type (CH_4 or CO_2).
3. Rate of gas emission (200 g of coal sample is crushed in a specified manner for 30 s. The volume of gas released measures the risk of gas outburst. An emission of 900 cc or more indicates that the coal seam is liable to an outburst. Fig. 17.6 shows the details [8].

Drilling ahead of mining (in-mine and from surface) and remote mining have been used as preventive measures.

17.3.2 China

Gas outburst potential is predicted on the basis of four factors:

1. Gas pressure in coal.
2. Tectonic faults.
3. The rate of gas emission.
4. Strength of coal.

Cross-measure boreholes, gas drainage by advance drilling, and undermining a workable coal seam are commonly used protective measures.

17.3.3 Kazakhstan

Kazakhstan mines have had some very large gas outbursts yielding 46 MMCF of gas and large amount of fine coal. They use the following criteria:

1. Depth of coal (which indicates gas content and pressure both).
2. Prior history of gas outburst.

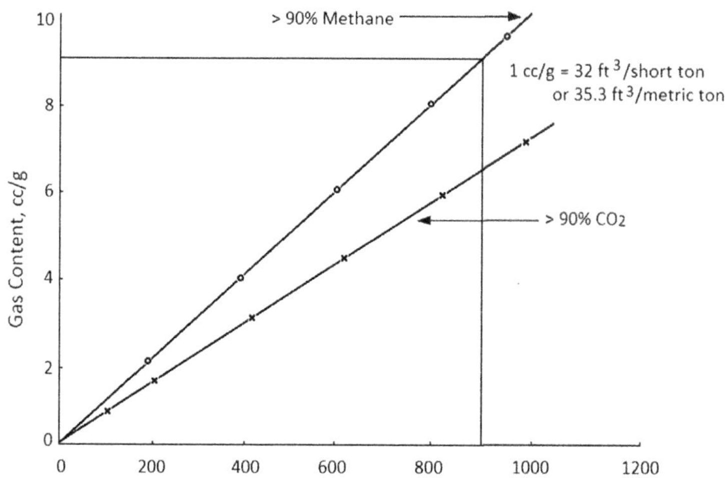

Figure 17.6 Threshold limits for gas contents versus gas emission rates.
Black D et.al. Outburst Threshold limits; Proceedings of the 9th Coal Operators conference, U of Wollongong, Australia 2009.

Advance horizontal drilling is done for pressure relief and gas drainage, but only limited success has been achieved because of low permeability [7].

17.3.4 Ukraine

Coalfields of Ukraine are liable to gas outburst. They use the following criteria:

1. Depth.
2. Coal rank.
3. Gas content.

Similar criteria are also used in Germany (North Rhine Westphalia) and Russia. Pre-mining methane drainage has been done with only limited success because of great depth ($\approx$3000 ft) and consequent reduced permeability.

In summary, we need to concentrate on only three criteria:

1. Depth and gas content all coal seams deeper than 600 ft with a corresponding gas content of 200 ft^3/t or more. Depth and gas content are highly correlated (see Chapter 15).
2. Any coal seam that shows a high rate of gas emission and high pressure gradient. The first is due to coal pulverization and high diffusivity coefficient or sorption time; the second is a function of low permeability and reservoir pressure. A sorption time[1] of less than 1 day (as defined by Thakur [1]) should raise a warning flag. Likewise, a permeability of less than 1 md and a reservoir pressure of 200 psi should signal a warning.
3. The structure and strength/toughness of coal. Chinese research puts great emphasis on it. In general, weaker coal with a compressive strength of less than 3000 psi should raise a concern. Most deep coal seams display this tendency.

17.4 Prevention of Gas Outburst

It is clear from the historical review of global gas outburst events that only moderately gassy and highly gassy coal seams are liable to gas outburst (refer to Table 15.1). There is really one technique for the prevention or, at least, minimizing the risk of gas outburst in all coal mines, and it is a proper degasification of the coal seam(s) prior to mining. Chapter 15 discusses all methods for premining degasification, but it will be further detailed here. A well-degassed coal seam contains less than 100 ft^3/t gas at a pressure of less than 100 psi.

17.4.1 Moderately Gassy Coal Seams (Depth 600–1500 ft)

Such coal seams have good permeability, and in-seam horizontal drilling or horizontal wells from surface will adequately degas the coal seam and minimize the risk of any gas outburst.

Fig. 15.7 shows the way to degas a development heading in advance of mining.

[1] Sorption time is the characteristic time in which a coal sample gives out 63.21% of total gas contained in it.

Fig. 15.8 shows a simultaneous degasification of both development sections and longwall faces.

17.4.2 Highly Gassy Coal Seams (Depth 1500–3000 ft)

These coal seams cannot be properly degassed by in-mine horizontal wells alone. The most economic method to degas these coal seams is to do it in two stages:

Stage 1. Drill the entire property on a grid pattern vertically. A spacing of 40 acres/well may be more economic, but gassy coal seams may need a vertical well at 20 acre spacing. All coal seams (overlying or underlying the mineable coal seam) that are a source of gas are hydrofractured as discussed by Thakur [1]. The hydrofracture of the working seam is so designed that future development headings and the longwall face both are degassed. This is accomplished by keeping the longwall panel width smaller than 750 ft and creating a hydrofracture of at least 1000 ft in length. Typically 50% of the gas in situ can be drained in 5–10 years.

Stage 2. Drill the longwall panels on close intervals (100–150 ft) horizontally from the development sections. This will remove another 20%–30% of in situ gas resulting in 70%–80% of total degasification. Horizontal drilling will also reveal any geological anomalies in the longwall panels, such as faults and washouts. These horizontal boreholes can also be used for dust control by water infusion.

17.4.3 Highly Gassy Coal Seams Deeper than 3000 ft

Mining coal from a depth greater than 3000 ft is expensive and generally not advisable. However, some working mines are 4000–4500 ft deep [7]. The permeability of such coal seams is very low (less than 0.1 md), and a different technique is needed for proper degasification of these coal seams. The entire areas occupied by such deep and highly gassy coal seam needs to be drilled horizontally from surface as shown in Fig. 17.7.

Thakur [1] has proposed this technique for commercial production of gas from deep coal seams. The technique has been extensively used in northeast United States for gas production from Marcellus Shale at a depth of 8000 ft. The drilling process is described here briefly. For details reference should be made to the original work [1]. A drilling pattern as shown in Fig. 17.7 can degas an area of 2300 acres. It is an expensive process, but the revenues from gas sales can defray the cost of degasification. Improved productivity (tons/man-day) and safety of mine workers are added benefits.

17.4.3.1 Horizontal Drilling from the Surface

The technology for horizontal wells drilled from the surface has been developed in the past 15 years. It is an improvement on the in-mine horizontal drilling procedure. It is mainly used for commercial gas production from shallow or deep coal seams. It is much more expensive than in-mine horizontal drilling. Wells in shallow coal do not need hydrofracking because the natural permeability of the coal is high. In deep

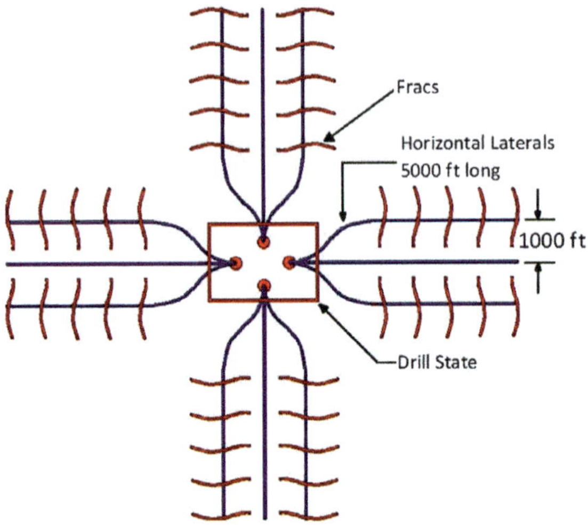

Figure 17.7 Horizontal well completion in deep coal seams.

coal seams or shale, the horizontal laterals are hydrofracked every 250–1000 ft to enhance gas production.

A typical drill site is 4–5 acres in area and has a drill platform for the drill rig, a compressor station for compressed air to get the cuttings out, and a large pond to dump drill cuttings and recover water for reuse. A temporary office is created on-site to provide communications, food, and other facilities for the workers on site. A typical drilling and hydrofracking procedure for a coal seam 8000 ft deep is described here. The coal seam is 60 ft thick and has a gas content of 500 ft^3/ton. It is advisable to complete drilling to the target without any interruptions.

17.4.3.2 Drilling Procedure

A smaller drill rig, such as a Speedstar 185 with top drive and a hook load capacity of 185,000 lbs, is moved to the site and properly anchored.

Fig. 17.8 shows a typical well bore schematic.

First, a 20-in. diameter surface casing is set in place to a depth of 30–40 ft. Then, a $17^1/_2$ -in. diameter well is drilled to a depth of about 600 ft (below all known aquifers), and a $13^3/_8$-in. casing is set with class A cement. Next, a $12^1/_4$ -in. borehole is drilled to a depth of 3000 ft and the borehole is logged for any mineable coal seams. A $9^5/_8$-in. casing is set in the well.

Next, the Speedstar rig is moved away from the site and a heavier rig, such as an IDECO Model H-44 double, capable of handling 318,000 lbs hook load is moved to the site. It also has a top drive. A 5000 lbs, $9^5/_8$-in. casing head is mounted, and a blowout preventer is installed.

Next, drilling starts with an $8^1/_4$ -in. polycrystalline diamond (PCD) bit with $6^1/_2$ -in. drill collars (rods). The well and flow lines are pressure tested and all safety protocols are completed. The production well is drilled to a target depth well below the target

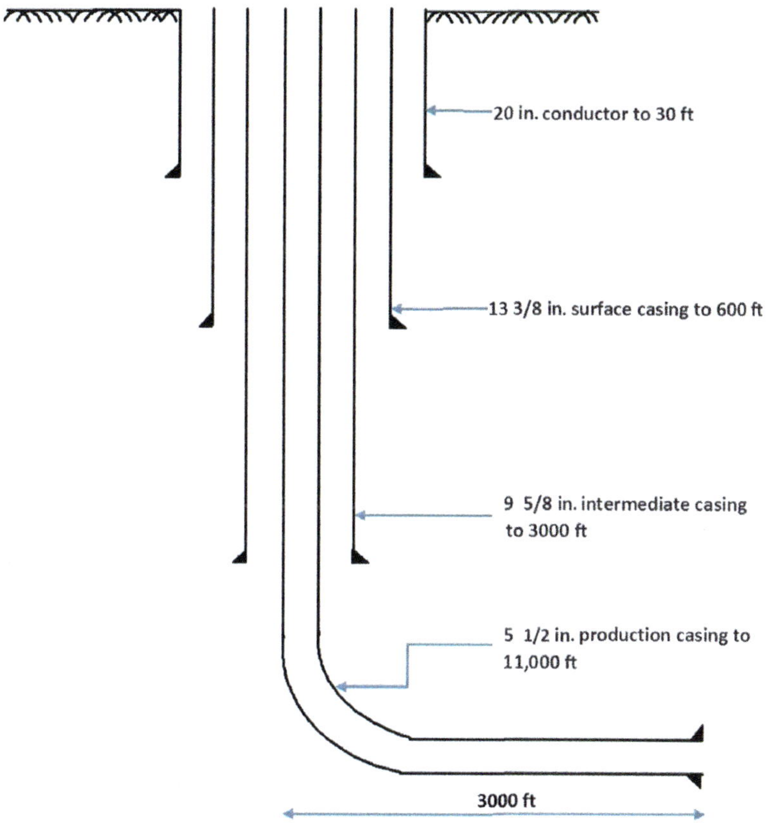

Figure 17.8 A typical horizontal well drilled from the surface for deep coal.

coal seam (usually 100–200 ft below). The well is logged again to choose the location where the deviated borehole will start. Assuming that the kickoff point will be at 7500 ft, the bottom of the vertical well is cemented to a depth of 6000 ft. The directional borehole assembly is lowered in the well and the well is drilled to the target kickoff point of 7500 ft. Next, the curve is drilled with foam mist to make the vertical well a horizontal one. The drill cuttings indicate if the well has entered the coal seam. The rate of angle build is 8–12 degrees per 100 ft. A 2 degrees bent housing is used for this purpose. Directional control of the well is usually provided by professional directional drillers. The horizontal lateral is drilled with a PCD bit of $8^5/_8$-in. diameter and a mud-driven motor (a Moyno pump in reverse).

In coal seams, all drilling is done with foam, but in shale, they use a 12–14 ppg mud (12–14 lbs of mud in a gallon of water). The horizontal drilling continues until the target depth is reached. For a 3000 ft lateral, the target depth would be about 11,000 ft. The drill string is tripped out and a $5^1/_2$ in., 20 lb/ft, P-110 casing is cemented in the entire well.

Table 17.2 Hydrofracking a 3000 ft. Lateral in Five Stages

Stage	Fluid Volume (bbl)*	Sand (lbs)		Rate (bbl/min)
		100 Mesh	40/70 Mesh	
1	20,000	180,000	500,000	102
2	19,000	170,000	510,000	105
3	21,000	190,000	480,000	101
4	18,000	180,000	470,000	106
5	19,000	160,000	510,000	106
Total	97,000	880,000	2,470,000	

*One bbl is equal to 42 US gallons.

17.4.3.3 Hydrofracking of the Lateral

The approximately 3000 ft long horizontal lateral is next hydrofracked through perforations in five sections to enhance the permeability. Typically, slick water (fresh water with a friction reducer, such as polyacrylamide) is used. The hydrofrac should be properly designed using the theories discussed by Thakur [1]. Data for a typical well in Devonian Shale are presented in Table 17.2. No such hydrofracking has been done in a coal seam so far, but the process would be very similar.

This is a massive hydrofracking job using over 4 million gallons of water and 3.35 million pounds of sand.

In a coal seam, the laterals should be drilled parallel to σ_h (the minor horizontal stress) because the fractures will be parallel to σ_H or the face cleat. Assuming two laterals (each 5000 ft long) are drilled from the same location and both laterals are hydrofracked as discussed above, a total of 20 fractures, each of 2000 ft. length, are created. The total length of 40,000 ft can produce 4–6 MMCFD assuming a specific gas production of 10–15 MCFD/100 ft. The specific gas production is a characteristic of the coal seam and the completion procedure (refer to Chapter 15).

References

[1] Thakur PC. Advanced reservoir and production engineering for coal bed methane. Elsevier; 2017. p. 210.

[2] Thakur PC. Floor gas emission mitigation in the Appalachian Basin. In: SME annual meeting, Salt Lake City, Utah; February 2005.

[3] Lama RD, Bodziony J. Outburst of gas, coal and rock in underground coal mines. 1996. p. 497.

[4] Lama RD, editor. International symposium on management and control of high gas emissions and outbursts in underground coal mines, Wollongong, NSW, Australia; 1995.

[5] Hargraves AJ. Instaneous outburst of coal and gas. In: Proceedings of Australian Institute of Mining Metallurgy; 2002. p. 21–72.
[6] Campoli AA, Trevits MA, Mulinda GM. An overview of coal and gas outbursts. In: 2nd US mine ventilation symposium. Reno: Nevada; 1985.
[7] Imgrund T, Thomas R. International experience of gas emission and gas outburst prevention in underground coal mines. In: Proceedings of the 13th coal operators' conference, U. of Wollongong, the Australian Institute of Mining and Metallurgy and Mine Managers Association of Australia; 2013. p. 331–8.
[8] Black D, et al. Outburst threshold limits, - are they appropriate?. In: The 9th coal operators' conference, U. of Wollongong, Australia; 2009.

Gas Transport in Underground Coal Mines

18

Chapter Outline

A growing number of coal mines are degasifying coal seams prior to mining. It is strongly advocated for moderately gassy and highly gassy coal seams. The produced gas, of necessity, has to be transported outside the mine without mixing it with mine air. In successful projects, often, more than 1 MMCFD of gas is carried in these pipelines. It raises many safety concerns about leakage of gas that can create an explosive atmosphere. The subject was thoroughly researched by Thakur [1] in collaboration with Mine Safety and Health Administration of the United States [2]. The subject is divided into three parts to address the safety issues properly:

1. Construction of pipeline and safeguards against potential damage.
2. Gas leakage detection and safeguards.
3. Other preventive measures for safe gas transport.

Advanced Mine Ventilation. https://doi.org/10.1016/B978-0-08-100457-9.00018-3

18.1 Construction of Pipeline

A gas pipeline in underground coal mines is fraught with many dangers as listed below:

1. Roof and rib falls damaging pipeline.
2. Floor heaving and movement of pipeline.
3. Electrochemical corrosion (steel pipes only).
4. Leakage from bad fittings or improper assembly.
5. Water removal from pipes.
6. Dangers from mobile mine equipment that can crush the pipeline.

A good design should protect the pipeline from all above and other unknown dangers.

18.1.1 Pipeline Material

Mild steel pipes and casing (externally flush) were commonly used since 1940s. They suffer from many drawbacks. Some of them are listed below:

1. Very heavy in large sizes (above 5″ in diameter) and liable to hurt workers.
2. Must be on floor and prone to get damaged by both floor heaving and roof fall. The pipes, often, need wooden support to keep them in alignment.
3. In mines, using DC current for haulage, the steel pipes are corroded by ground current. They need cathodic protection in most cases.
4. They are also vulnerable to chemical corrosion inside the pipeline.

During 1970s, the author first replaced steel pipes by high-density polyethylene (HDPE) pipes that had many advantages as listed below:

1. It is lighter than steel and can be supplied in 20–200 ft length depending on the diameter. A crew of four can easily lay more than 1000 ft in a shift of 8 h compared with about 300 ft for heavy steel pipes.
2. It is noncorrosive. It does not need any cathodic protection.
3. It cannot be punctured by roof falls even if it is closed shut by the falling rocks.
4. It is nonconductive and immune to ground current.
5. It can be laid on floor or preferably hung from the roof.

Table 18.1 shows the specifications for some HDPE pipes. Early designs of plastic pipes, such as Aldyl-A by DuPont, were discarded in favor of new HDPE pipes.

18.1.2 Fitting/Couplings on the Pipeline

Couplings on a 20–30 ft long joint are fused onto the pipe by "thermal butt fusion" machines. They melt the pipe and flange and fuse them together. Back up steel flanges are used to bolt the two joints of plastic pipe together. Small diameter pipes, such as 3″ pipes, come in a coiled length of 220 ft and can be laid much faster.

Table 18.1 Specification of High-Density Polyethylene Pipes

	PE80		PE100	
	SDR	**SDR**	**SDR**	**SDR**
	17.6	11.0	17.6	11.0
Nominal Pressure Rating (psi)	28	56	84	140
Diameter (Inch)	Nominal Wall Thickness (Inch)			
2	0.114	0.181	0.114	0.181
4	0.248	0.393	0.248	0.393
6	0.358	0.575	0.356	0.575
8	0.45	0.716	0.45	0.716

Pipes larger than 8 inches (200 mm) in diameter are usually not needed. Most commonly PE100, SDR 11 pipes are used in US coal mines. These HDPE pipes have an excellent safety record for the last 40 years. The in situ coal seam pressure may be as high as 200 psi, but the gas pipeline is seldom pressurized beyond 50 psi. The reason is the short length of stand pipe in the coal rib. At higher pressures the gas begins to leak around the borehole through ribs.

All butt fusion work is done in fresh air because the butt fusion machine is not permissible. All pipelines and fittings are tested to 125 psi pressure to make sure there are no leaks.

Steel pipelines, if used, may also be subject to lateral loading from roof falls. In this regard, the method of supporting the pipe becomes important. Based on calculations of bending strength of 8 inch diameter Schedule 40 steel pipe, a rigid support and a uniform load indicate that supports should be spaced no farther apart than 4 ft to withstand a static load of 18,000 pounds on the pipe midway between the supports. However, to compensate for the effect of joints, and the fact that the load is from impact rather than static, and to provide a suitable safety factor, the pipe should be supported for its entire length on a ballast drainage bed or, where that is not feasible, on wooden blocks on 2-foot centers.

Each joining method has alignment limitations. The allowable deflection angle is specified by the manufacturer for each particular system, and unless the specifications are closely followed, an adverse effect on the strength characteristics of the joint will result, leading to leaks or complete disengagement of the pipes.

Inspection of the pipeline is a necessary safeguard against the development of hazardous conditions along the pipelines. The inspection procedure will involve many worker hours of work, depending, of course, on the length of the pipeline. If experience shows that leaks are occurring frequently, it may be necessary that the frequency of the inspection schedule be increased, and more careful scrutiny be given to the line. If leaks are found to exist during the inspection, the entire pipeline, or a portion of the system, must be shut down to safely repair the leak. The procedure will most likely involve shutting in the wells and purging the gas lines before repair work is begun.

Depending on local conditions and the details of the mining situation, mining may have to cease during this process. It is apparent, therefore, that in determining the type of joining method to be used, reliability of the fitting is extremely important. If it is properly and permanently made initially, later problems are proportionally minimized.

18.2 Gas Leakage Detection and Safeguards

Early warning in case of small leaks or massive ruptures of the pipelines is essential for the safe operation of an underground degasification system. A frequent inspection program is necessary and will possibly detect pipeline problems soon enough to avoid large problems later. However, massive ruptures of the pipeline must be detected immediately—within seconds—and the only practical method of doing so is by properly placed and installed methane detection instruments.

Instruments are available to monitor practically any in-mine environmental condition. The choice of instrument is dependent on reliability, maintenance requirements, and the particulars of the hazard being guarded against. Basically, protection may be required against the following:

1. Gas leakage into the mine from a rupture or leaking pipeline.
2. Oxygen leaking from the mine into the pipeline (especially when a compressor is used).
3. Gas leakage into the mine atmosphere through the coal rib when a well is shut in (it will be discussed later in the chapter).

If leakage from a pipeline is from a massive rupture, a large influx of gas into the airstream will be experienced, and easily measurable changes will occur in the pipeline flow and in the methane level in the entries. This will not necessarily be the case for small leaks. Pinhole leaks caused by corrosion pitting, slight pipe movements, or any of a number of factors, although serious, will not be readily detected. The small leak may go unnoticed until it has progressed to a large leak. Regular inspections of the pipeline will be the best safeguard for such leaks. For this reason, it is important that the pipeline be visible throughout its length. A daily fire boss–type inspection of the pipeline and its components (water traps, valves, etc.) should provide early warning of the development of leak-producing conditions. In this type of inspection, the pipeline can be visually inspected for corrosion and to insure that all water traps are functioning properly—not dry or plugged; that all valves are properly aligned as open or closed as applicable; that the detection equipment is operating; that no roof falls have damaged the pipe; that no other leak-producing conditions are developing; and that ventilation is being maintained in the entries.

A detailed inspection of the pipeline should be made weekly. This inspection should insure that all detection instruments are properly charged and calibrated; that automatic control equipment is functioning properly; and that all components of the pipeline are operating properly and to test them if necessary.

18.2.1 Instrumentation for Detecting Leaks and Ruptures

There are many ways to detect leaks. Most common tools are discussed below.

18.2.1.1 Methane Detectors

In addition to the commonly used hand-held methanometers, recording methanometers with a variety of operating ranges are also available and in use in the mining industry. It may be necessary to use two sensing heads, one downstream and one upstream of a monitored point, to indicate small gas influxes into the airstream. If massive influxes of methane occur, a single detector, or a series of detectors, downstream of the source may be sufficient.

18.2.1.2 Sonic Leak Detectors

An intrinsically safe detector is available to detect pipeline leaks at flow pressures greater than 1 psig by means of the change in sound level caused by gas escaping from a pipeline. Two systems using this principle are available. The first is a hand-held receiver which can be used during pipe inspections. The instrument is now being used by the gas industry and by telephone and electric power companies to detect leaks in pressurized overhead transmission lines. However, its potential for testing mine degasification pipelines is questionable at flow pressures less than 1 psig.

The second type of system is basically the same, except that an audio tone is imparted to the gas stream in the pipeline by an installed speaker. Leaks are then easier to detect with the receiving element. This setup may be suited for use as an automatic detection system whereby the speaker and detector elements would be permanently installed, the detector relaying a signal when a change in the audio tone is observed.

18.2.1.3 Oxygen Detectors

If the pipeline is operated under pressure less than atmospheric, breaks in the line would result in oxygen (mine air) being introduced into the line. Also, there is reason to expect that oxygen may be introduced into the gas stream by migration through the coal rib into the gas well. If oxygen is introduced into the line, an explosive mixture in the line could occur. And if the line is connected to a compressor, serious fire or explosion hazards can manifest. An oxygen analyzer in the line is one method of alerting mine personnel of such leaks. Such analyzers are commercially available and could be used.

18.3 Other Preventive Measures for Safe Gas Transport

18.3.1 Automatic Shut-Off Valves

If a massive rupture of the pipeline occurs, it is imperative that the gas wells feeding gas into the pipeline be shut off. If this is not done, and the gas continues to flow uninterrupted from the ruptured line, the mine entries can be filled with an explosive gas mixture. Automatic shut-off valves are commonly used. Basically, the system involves the installation of an automatic (fail closed) valve as close as possible to the gas well.

The valves are commonly held open by pneumatic pressure supplied from a compressor or compressed air tank via a $^1/_4{}''$ nylon line. As long as pressure is supplied to the valve, the valve remains open, allowing gas to flow. In case of a massive roof fall, the nylon line will be broken first and all boreholes will be shut. This line is fastened firmly on top of the pipeline in such a manner that the pressure line could be broken either by a roof fall or by pipe movement. Loss of pressure would then automatically close the valves.

18.3.2 Sectionalization of the Pipeline

The most important safety feature to be incorporated into the pipeline design is automatic shut-off valves. However, even if the gas wells are shut in at the instant of rupture, a certain volume of gas will be released into the mine if the system is under positive pressure. The volume released will be related to the operating pressures of the pipeline, the diameter of the pipe, and the length of the pipeline. For each atmosphere of positive pressure, a volume of gas equal to the volume of the pipeline will be released. For an 8-inch diameter pipe under 15 psig, the volume released will be 350 cubic feet per 1000 feet of pipe. For all practical purposes, the gas will be released instantly. The amount of gas that can be safely released will depend on the quantity of air flowing in the entry into which the gas is released. It is generally agreed that longer pipelines should be sectionalized by one-way valves in lengths of no more than 1000 ft.

It is unlikely that all gas wells drilled in a panel will operate at the same pressure and flow rate. Also, design of the piping system will probably require that the gas from more than one well will be delivered to the main gas transmission line. The condition may, therefore, develop in which the gas pressure from an inby well may be greater than from the well(s) outby. This may cause a pressure differential from the main transmission line to the gas well instead of vice versa, and in turn may allow gas to flow into the coal seam. To guard against this occurrence, one-way check valves are also installed at the mouth of each degasification borehole.

18.3.3 Water Traps on the Pipeline

Most degasification boreholes produce water with gas. Commercially made water traps are used at each degasification borehole and at the bottom of the vertical venthole. They dump water automatically whenever water builds up in the pipeline. These traps work well when the line pressure is above atmospheric pressure. Water traps are usually installed at a low point in the pipeline. Draining water is easy if the pipeline is hung from the roof bolts.

18.3.4 Flame Arrestors

If the flow of gas is high, it must be discharged on surface. If the flow of gas is small, it can be discharged in the mine airway. A flame arrestor must be used to prevent the flame from entering the mine if the gas is ignited on surface.

Sources for ignition are many. These could be frictional sparks from falling roof or from an incendive spark producing metal; electrostatic discharges; lightning; a carelessly tossed lit cigarette or match; a bullet either from a hunter or someone using the pipe as a target. Compressors have caught fire and have exploded.

The wire gauze in a flame safety lamp is a common example of a flame barrier. The purpose of the gauze is to absorb heat and to restrict the propagation of flame beyond it. There are many other types and designs of barriers; these include crimped metal arrestors, molecular seals, hydraulic traps, inert gas filling, and injection systems and devices that, on sensing flame, discharge an extinguishant into the flame path.

Regardless of the type of barrier used, it is obvious that they are needed. Commercially made flame arrestors do a good job.

18.3.4.1 Some Useful Guidelines

1. Every point of discharge of methane from a pipeline into the atmosphere or into a compressor or pump should be fitted with a flame arrestor.
2. Each device should be securely anchored in an accurate seating so that it cannot be displaced by an explosion and so that flame cannot propagate past the seating.
3. The pressure drop across the flame arrestor should not be high, and it should be monitored. Weekly inspection is necessary, and regular cleaning is recommended.
4. Devices should be constructed of materials having ignition temperatures above 1500°F; in other words, a noncombustible material. The devices should not be constructed of asbestos, wire gauze, or plate perforated with circular holes. Deformation of asbestos by moisture can result in failure of asbestos containing parts to function properly and to provide safety. Compared with other arrestor materials, wire gauzes have a limited effectiveness for quenching flame, are easily damaged, are highly resistant to gas flow, and are readily clogged by dirt and ice. Similarly, perforated plate arrestors are highly resistant to gas flow and thin plates do not protect against violent explosions. Crimped metal arrestors are preferred; they have high resistance to mechanical and thermal shock and low resistance to gas flow.
5. Flame arrestors should be installed within 5 ft but not more than 10 ft from the point of discharge. The most likely source of ignition will be a compressor, lightning, or flame at a pipe outlet. Flame velocity increases with distance and with obstacles such as T's and bends. Up to 10 ft, flame speed is not affected greatly by obstacles.
6. Each device should contain a supporting system that prevents flow of methane through the device after flame has been in the device for 60 or more seconds. Continued flow of methane can result in reignition after extinguishant has been expelled or overheating and damage to an arrestor.
7. Flame arrestors should be approved by the National Fire Protection Association, Factory Mutual Research, or Underwriter's Laboratories. Shop-made devices cannot be relied on because they are not subjected to tests or quality control and at the critical period may not perform as expected.
8. Gas may be discharged underground if the point of discharge is in the return and is surrounded by a fence so located that methane content in the air at each part of the fence is 2% or less. Often, a diffusing-type water spray is incorporated to saturate the air to prevent an ignition.

18.3.4.2 Discharge From the Pipeline

When the gas from the methane drainage boreholes cannot be marketed, it is discharged into the atmosphere. Gas-into-air discharge practices in German coal mines are summarized below.

At German mines, gas may be discharged underground or on the surface. Underground discharge is in a "dilution area" where the methane content is reduced to less than 1% and then mixed into the return air ventilation current. The "dilution area" is a fenced-in zone where the pipelined gas is discharged into metal tubing on the exhaust of an auxiliary blower fan. On the surface, the gas is discharged into a chimney from whence it is released into the air.

18.4 Ventilation

Adequate ventilation in the airways with gas pipeline is the last line of defense against a potential explosive mixture. Accordingly, the amount of ventilation that should be provided is directly related to the amount of gas flowing in the pipeline. Typically, the quantity of air moving over the pipeline is sufficient to dilute all of the gas that could be released by rupture of the pipeline without causing accumulation of over 1% methane in the air current. For example, an air quantity of 70,000 CFM would be needed to dilute a flow of 1 MMCFD in the pipeline. Automatic shut-off valves reduce the gas flow very quickly. If the ventilation air is not adequate, methane monitors and alarms are needed to monitor the situation.

18.5 Corrosion of Steel Pipelines for Methane Drainage

In some mines, steel pipes are still used for gas transport. Damaging corrosion can be expected in metallic pipelines that remain in use for a long period. The definition of "long" depends on the environment, the type of metal, and on the adequacy of the anti-corrosion measures taken. It may range from days to years. The minimum requirements for the protection of metallic pipelines from external, internal, and atmospheric corrosion are available from the natural gas industry.

The gas industry has learned that whenever the surface transmission lines approach coal mining operations, the corrosion problem is accentuated and increased and protective measures have to be taken. The reason for this is that stray electrical currents are common near and in coal mines utilizing trolley haulage systems, waterlines, and communication systems. It is a fact that stray electrical currents will cause corrosion problems in underground degasification systems. In addition to the stray currents, the coal mine water may also be more conducive to corrosion than water in other locations.

Broadly defined, corrosion is an electrochemical reaction between a metal and its environment. The result of that reaction is the deterioration of the metal. Most metals are unstable and tend to revert to a more stable chemical combination such as the original ore, that is, an oxide, sulfide, or carbonate. At the corroding area, direct current electricity flows from the metal into the surrounding electrolyte. The metal from which the current leaves is known as the anode. The metal that receives current from the electrolyte is known as the cathode. A loss of metal results at the anode, while the corroded metal particles deposit on the cathode. The amount of metal loss varies with different metals and current flow; for example, a Ampere current flow for 1 year will cause losses of 20 pounds of steel, 74 pounds of lead, 45 pounds of copper, 23 pounds of zinc, and 6.4 pounds of aluminum [2].

Different metals demonstrate different electropotentials, and a current will flow between two dissimilar metals when immersed in an electrolyte and connected electrically. The galvanic series of metals indicate the electropotential of each metal, and current will always flow from the metal higher in the series to any metal lower in the series when contact is made between the two metals. A similar relationship of anode to cathode exists with iron and steel as shown by Lewicki [3].

1. New steel pipe.
2. Old steel pipe.
3. New wrought iron pipe.
4. Old wrought iron pipe.
5. New cast iron pipe.
6. Old cast iron pipe.

As in the case of the galvanic series, current will flow between a metal connected to another metal lower in the listing. From this listing, it is apparent that new steel will corrode faster than old steel to which it is connected; or old steel pipe will corrode when connected to a new cast iron pipe.

Corrosion currents will be produced on any metal on which conditions create an anode and cathode immersed in an electrolyte. Some conditions which create anodic and cathodic areas are stresses in the pipe, chipped or scratched surfaces, dissimilar metals connected together, dissimilar metal surfaces on the pipe such as caused by accumulations of mill scale, differences in the surrounding electrolyte stray electric currents.

18.5.1 Protection Against Corrosion

Corrosion of the pipeline can be alleviated by preventing the flow of current between anode and cathode. This can be done by using a dielectric coating or by reversing the current by the use of cathodic protection so that the current flows to the pipeline from a sacrificial anode. In the case of stray electric currents in coal mines, corrosion by this source can be prevented by stopping the flow of stray currents into the pipeline or by giving the stray currents a free path from the pipe back to the source (ground) so that the current does not leave the pipe at anodic points.

Before protective measures are undertaken, however, it is important that the conditions to be guarded against are clearly defined. Often it is desirable to obtain measurements of soil resistivity, thus providing an estimate of the strength of electrolyte which will serve as the cathode for the electrochemical cell. If resistivity is high, then not as much current will flow. If the resistivity is higher in one area than in another, more corrosion protection will be indicated in the area of low resistive soil.

Measurements for stray currents can also be made. If the magnitude of the currents are known and the source of the currents are identified, then action can be taken to eliminate the current at the source, reverse the polarity of the pipeline so that the stray currents cannot enter the line, and/or provide the currents with a free path back to the source or to ground.

The problem with dependence on protective coating is the determination of what coating is suitable because it is not always possible to define the long-term corrosive nature of the mine environment.

For example, the corrosivity of the gas transport atmosphere can change rapidly and unexpectedly as will happen should hydrogen sulfide be introduced into the gas. The corrosivity of water in the mine changes, too; it increases with increase in the velocity of flow, with the amount of dissolved oxygen in the water, and with decrease in pH. Soil corrosion is highly localized and is usually caused by electrochemical action resulting from variations in the soil and the moisture therein; thus, the corrosivity of the mine floor material may range considerably from point to point and from time to time.

Other problems affecting dependence on coatings are that

1. The chipped or uncoated area of a coated pipe may become an area of concentrated corrosion. That area becomes an electrochemical cell. Dissimilar materials in electrical contact accelerate corrosion of the one that happens to be anodic (in this case the chipped area). The smaller the anodic area (or chip) in relation to cathodic areas (soil, supports, and other portions of the pipe), the greater is the rate of penetration at the anodic points.
2. In the restricted confines of a mine, it is reasonable to expect that a pipe weighing many hundreds of pounds will be banged and scraped; thus, it is also reasonable to expect that a coating applied before the pipe was brought to its place in the mine will be chipped.
3. It is not reasonable to expect that after a pipeline is installed in a mine adequate anticorrosion treatments and coatings can be applied on all surfaces.
4. Unless pipe is cleaned properly a coating provides little protection against corrosion and may in fact cause accelerated, localized corrosion.
5. Some externally applied coatings, particularly the bitumen, degrade rapidly and unexpectedly. Many of these are combustible and would, if ignited, be the source of copious, toxic smoke. Some are subject to soil stress. That is, soil may cling to the coating and pull it from the pipe as the soil dries and shrinks; or, irregular soil pressures, imposed by both small and large bearing forces, may distort the bitumen and cause thin spots. Some coatings are quickly oxidized and become brittle when protective coatings are under water or are in contact with highly alkaline or acid soils. Most protective coatings have a life expectancy less than 10 years.

18.5.2 Cathodic Protection

The "best" solution to the corrosion problem that is likely to occur in coal mine degasification pipelines is a combination of protective coatings and cathodic protection and means to prevent stray electric currents from affecting the line.

Cathodic protection is a widely used means in the gas industry as a safeguard against corrosion. Basically, it can be applied by imposing a voltage on the pipeline by an independent electrical source or by connecting the pipeline to a buried metal which is higher in the galvanic series than the metal of the pipeline. For steel lines, zinc or magnesium rods are commonly used. The rods are buried alongside the pipeline and bonded to it. Zinc ribbons are often used by laying them alongside the pipeline when it is buried. For coal mine applications, buried rods or zinc ribbons bonded to the line will probably be the most practical application.

18.6 Compressors

If the drained gas is to be sold commercially, a compressor will be necessary to put the gas into the commercial lines. However, transporting the gas to the surface will depend on the size and length of the pipelines and the in situ gas pressures. The gas pressures in the coal seams have been sufficient to allow methane drainage without the need for compressors. However, when the gas is commercially marketed, compressors can be used on surface but the underground pipeline should not be below atmospheric pressure.

Main reasons against installing compressors underground are

1. Any oil-lubricated compressor can catch fire or cause an explosion. Auto ignition is possible because the flash point of the lubricant is 450°F. This temperature can be exceeded if there is (a) loss of coolant water or (b) the discharge pipe becomes blocked. This can be prevented only if all safety devices and switches are properly installed and functioning.
2. Any oil-lubricated compressor can tolerate not more than 130 mg of H_2S in 100 cubic feet of gas.
3. Any oil-lubricated compressor can neither tolerate slugs of water nor can it tolerate pipe scale or other abrasive material.
4. Any oil-lubricated compressor is adversely affected by dust; the dust and oil forms an abrasive emulsion that attacks the vanes and housing which have close tolerances.
5. Cavitation or overheating can destroy a compressor should it continue to operate without gas.
6. At pressure below atmospheric pressure, oxygen is drawn into the line, and explosive condition could develop in the compressor.
7. A negative pressure in the pipeline also makes all water traps inoperative, and the pipeline gets filled up with water stopping all gas production.

18.7 Surface Discharge of Gas

When the gas is not used commercially and discharged on surface, certain safety measures become necessary. Typically a 4″ diameter stack is used. The safety measures are listed below:

1. The vent stack must be about 20 ft high above the ground and the top 10 ft should be nonconductive, such as, PVC.

2. An orifice flange should be incorporated to measure any change in flow.
3. A check valve should be used to avoid air entry in the pipeline.
4. A flame arrestor should be used and be located about 10 ft from the top.
5. A rain top should be used in the stack.
6. Lightning protection should be provided by erecting a pole higher than the gas stack.
7. The area must be fenced and posted for "No Smoking".

18.8 A Typical Application for Mine Safety and Health Administration Approval of a Gas Pipeline System

The following is an example of an application to Mine Safety and Health Administration for the approval of gas pipeline construction.

District Manager
Mine Safety and Health Administration
Coal Mine Safety and Health District
Street address
City, State Zip

Dear Mr. Manager:

The attached material describes part of a methane control project at 123 Section of ABC Mine. Our objective is to degas the mains and the gate roads in advance of mining as shown in the enclosed map. The long boreholes in the longwall panels will be adequately filled up with cement prior to mining through them. This procedure has proved satisfactory at other mines.

The degasification work will be done in accordance with the enclosed guideline that has been established earlier in collaboration with your staff.

Your approval of the project will be appreciated.

Yours very truly,

Mine Executive

\\

PROPOSED GAS PIPELINE SYSTEM.
UNDERGROUND & SURFACE.
ABC MINE.
XYZ COAL COMPANY.

The following material describes the equipment and the installation method proposed for a coal seam degasification system in ABC Mine. Schematic diagrams of the system and ventilation map of the immediate area underground are attached (Figs. 18.1 and 18.2).

1. *Pipeline*: Underground methane pipeline will be made of well-designed plastic or steel.
   a. All underground steel pipelines will be 3 ½ ″ to 4 ½ ″ OD Schedule 40 pipes joined together with threaded couplings. These pipes will be made up tightly using a good grade of thread lubricant. Mill collars will be broken out, doped, and remade. A flange

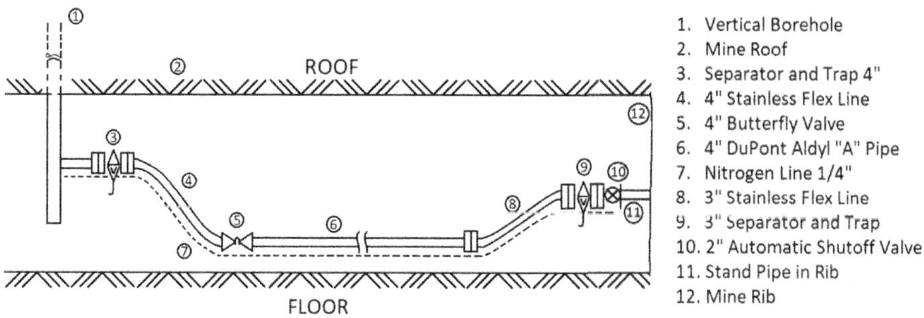

Figure 18.1 Underground methane pipe installation.

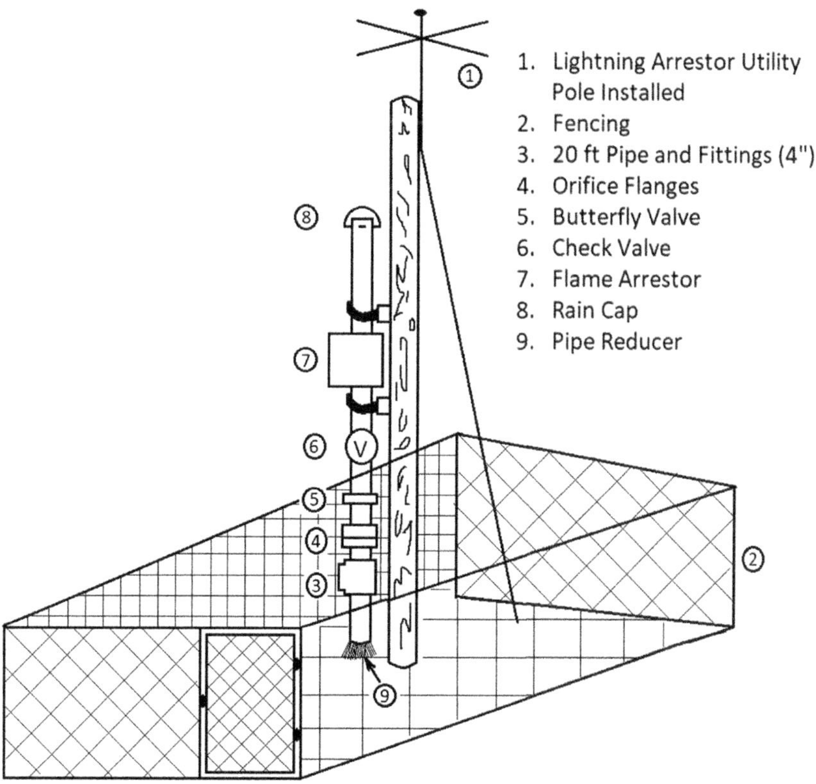

Figure 18.2 Venthole surface installation.

connection will be used every 10 joints (approximately 210 ft apart) so that a section of the pipeline can be removed without cutting the line if one or more joints need to be replaced later.

b. All underground plastic line will be 3″ to 6″ HDPE pipes. Plastic flange adapters will be fusion bonded to the pipe ends in fresh air. Steel flanges back up rings installed prior to fusion bonding will be used to connect plastic to plastic and plastic to steel.

2. *Pressure Testing*: The entire length of pipeline between the bottom of the venthole and well head will be tested to 1.25 times the shut-in pressure of the borehole or 90 psi, whichever is greater.
3. Pipeline will be generally laid in the return and will not be buried. Whenever the pipeline must cross a fresh air entry, it will be conducted through a steel line.
4. No hoses will be used in the system except while a hole is being drilled. Stress-relieving, flexible tubing will be used at critical points such as the well head to pipeline connection. This will be stainless steel tubing with a triple wire braid cover.
5. The steel pipeline will be firmly supported with no unsupported span greater than 2 ft.
6. A gas water separator will be installed at the bottom of the vertical venthole to remove condensation that falls back down the casing. Other separators will be installed on the holes or on the pipeline if water production from coal warrants. All separators will preferably be commercially made. Water drains will be provided on the line wherever necessary.
7. *Safety Devices*: A potential survey will be made after the pipeline is installed, and cathodic protection is provided where needed.
8. Automatic shut-in valves will be installed at each well head. These will be held open by nitrogen or air under pressure contained in a fragile plastic pilot line running parallel to and secured on top of the pipeline. Any roof fall or fires serious enough to damage the pipeline will damage the pilot line first and close the borehole immediately.
9. *Inspection*: The pipeline system will be firebossed daily by a competent person familiar with system operation.
10. *CH4 Monitors*: If the quantity of air flowing over the pipeline is such that a complete rupture of the pipeline and consequent discharge of methane in mine air will raise its concentration above the limits specified by the law, a methane monitor will be used as shown in the enclosed schematic. A commercially made flame arrestor will be installed within 10 ft of the top of the vent stack. A check valve shall be used to guard against reversal of flow. The check valve can be manually defeated if it is desired to purge the pipeline for repairs. An orifice meter may be installed if so needed. All surface installation will be periodically inspected to ensure satisfactory performance.
11. *Surveying:* All boreholes drilled for degasification will be accurately surveyed either during drilling or after drilling is completed using commercially available borehole surveying tools. These boreholes will be accurately plotted on mine maps to ensure the prevention of any inadvertent mining through these boreholes.
12. All boreholes must be plugged prior to mining through them using Class A cement.
13. A compressor will be required at the surface if beneficial use is made of the gas at a future date. Plans for the installation will be discussed with Mine Safety and Health Administration at the appropriate time.

References

[1] Thakur PC, Umphrey RW. Methane drainage and transport in CONSOL underground coal mines. In: Annual meeting of Pittsburgh coal mining Institute of America; November 1979.

[2] Tisdale JE, et al. Piping methane in underground coal mines. MSHA, U.S. Department of Labor; 1978. p. 34. Information Report, 1094.

[3] Lewicki TF. Corrosion protection. In: Proceedings of the 17th annual underground pipe corrosion short course, Bulletin 106, West Virginia University; 1972. p. 34.

Measurement and Monitoring of Mine Gases

19

Chapter Outline

Properties of mine gases were discussed in Chapter 13 of the book and their threshold limit values were listed in Table 1.3. Correct measurement of the concentrations of these gases and, in some cases, continuous monitoring of the concentrations is essential for the health and safety of coal mine workers. In some coal mining countries, it is also a legal requirement.

19.1 Detection Methods

Detection techniques can be classified into the following categories:

1. Catalytic-oxidation detectors.
2. Electrochemical sensors.
3. Optical detectors.
4. Electrical conductivity using semiconductors.
5. Stain tubes: The concentration is usually read on a linear scale on the tube.

Advanced Mine Ventilation. https://doi.org/10.1016/B978-0-08-100457-9.00019-5

Table 19.1 Gas Detection Methods

Gas	Detection Methods
Methane	Flame safety lamp Catalytic oxidation Thermal conductivity Optical (infrared and interferometer)
Oxygen	Liquid absorption Stain tubes Paramagnetic analyzers Electrochemical sensors
Carbon dioxide	Liquid absorption Stain tubes Optical interferometer
Carbon monoxide	Electrochemical sensors Catalytic oxidation Optical and infrared Metal oxide semiconductor Stain tubes
Oxides of nitrogen	Electrochemical sensors Stain tubes
Hydrogen sulfide	Electrochemical sensors Metal oxide semiconductors Stain tubes
Sulfur dioxide	Electrochemical sensors Stain tubes
Hydrogen	Stain tubes
Radon	Radiation detectors

Detection (more correctly spot readings) methods for different gases are listed in Table 19.1 [1].

Many instruments have been developed over the past 100 years to measure the instantaneous concentrations of all important gases listed in Table 19.1. The oldest is the flame safety lamp—a symbol of safety in mines. It detected both the lack of oxygen (above 13%) and small concentrations of methane (below 5%) and provided light as well for miners to work safely. It is almost obsolete by now. Instruments specially suited to measure the concentration of important gases are discussed below.

19.1.1 Methane Measurement

Two handheld instruments are most commonly used. The first one, which is cheaper, uses the catalytic-oxidation technique. It works on the Wheatstone bridge principle:

one leg of the bridge is used to burn methane catalytically raising its temperature. The imbalance in the current flow is a measure of methane concentration. Such detectors are also sensitive to the presence of higher hydrocarbons, hydrogen, and carbon monoxide, but fortunately, these interfering gases are present in mine air only in parts per million (ppm) and do not introduce serious error in methane measurements. Catalytic-oxidation type instruments are liable to get damaged if methane concentration exceeds 5% because of excess heat.

Optical detectors for methane are called "interferometer." These detectors compare the speed of light through pure air with that in air contaminated by methane. Light traveling through both media is combined again producing "interference fringes." The position of these fringes indicates methane concentrations. These instruments are also sensitive to other hydrocarbons and hydrogen. Ethane and propane present in air indicate a higher concentration of methane, but CO and H_2 have an effect of lowering the actual methane concentrations. It is also sensitive to water vapor and carbon dioxide, but these gases are scrubbed out by passing the inlet gas through a column that absorbs both of them. One percent of CO_2 in air will indicate 1% methane even if there is no methane there. Lack of oxygen also impacts an interferometer. Each 1% decrease below 20.95% (normal O_2 in air) results in 2% methane reading even if there is no methane in air. As such, it is not a very reliable instrument for 0%–5% methane. It is much more useful to measure higher concentrations, 5%–100% of methane.

19.1.2 Oxygen Detectors

A flame safety lamp was the most commonly used device to measure oxygen deficiency in past, but it is now replaced by new instruments that work on various principles, such as liquid absorption, paramagnetic, or electrochemical cells. Many instruments measure both methane and oxygen concentrations over the range 0%–5% and 0%–21%, respectively. Stain tubes are also available to measure O_2 concentration in mine air.

19.1.3 Carbon Monoxide Detectors

As listed in Table 19.1, five different techniques are available to measure CO in mine air. Ambient CO concentration is normally checked by a handheld instrument that uses catalytic oxidation, but the readings can be seriously impacted by other gases present in air, especially higher hydrocarbons. For detecting spontaneous combustion, reliable measurements are given only by gas chromatographs (GCs). Stain tubes are good indicators, but it must be followed by at least two samples of the mine air for GC analysis. In many cases, when the handheld CO monitor read 100–200 ppm, the actual CO concentration as measured by a GC was only 5–10 ppm.

19.1.4 Oxides of Nitrogen

With the introduction of diesel engines in coal mines some 50 years ago, the need to measure both NO and NO_2 has become urgent. Most commonly stain tubes with a handheld pump were used for this purpose, but now electrochemical cells have become available for this purpose. When NO, NO_2, or even CO enter a special cell; they react with the electrolyte and create a light signal, "a photon." A bank of photomultiplier tubes picks up the signal and translates it into specific gas concentrations. The precision of such instruments is quite good at $\pm$ 1 ppm.

Earlier instruments were bulky, such as Ecolysers [2], and were difficult to carry, but new instruments are light and can be handheld. An instrument named "Passport" can measure five gases in a single unit as listed below.

CO (1–50 ppm)
O_2 (1%–20%)
CH_4 (1%–5%)
NO (1–25 ppm) and
NO_2 (1–5 ppm).

It is strongly advised that for conclusive results, all gas samples must be analyzed in a laboratory by trained technicians using a GC.

19.1.5 Radon

Radon is a gaseous, chemically inert, radioactive product of the disintegration of radium. Found primarily in uranium mines, although present in trace amounts in several coal mines, radon diffuses from the rock strata into the mine environment, where the decay process continues. Table 19.2 gives the disintegration process for uranium-238 to become lead-206. Shown in this table is the type of radiation given off by each decay process and the half-life of each element in the series. The half-life of a radioactive substance is the time required for a given amount of that substance to lose one half of its radioactivity. The half-life of uranium-238 is approximately 4.5 billion years; radium, 1622 years; and radon, 3.8 days.

Once radon is released into the mine environment, the decay process continues with the formation of radium A, which decays to radium B, which produces radium C, and so forth. The products formed by the decay of radon are referred to as *radon daughters*. The radon daughter products are atoms of solid matter having relatively short half-lives. During the decay process, either alpha or beta particles are emitted. These emissions may also be accompanied by gamma ray activity. It is the short-lived alpha particles and potential alpha emitters such as radon and its daughters that are of prime concern to the ventilation engineer. Because it is a gas and has a relatively long half-life, inhaled radon is exhaled before large amounts of alpha particles are emitted. The daughter products, however, attach themselves to the dust that is present in the environment and when inhaled, tend to be deposited in the respiratory system. It has been estimated that when both radon and radon daughters are inhaled, only about 5% of the alpha radiation received is contributed by the radon [3].

During radioactive decay, the individual members in the series are decaying and being formed at the same time. At some point in time, equilibrium is reached, and the quantity of each member in the series remains constant. At this time, each member

Table 19.2 Uranium Disintegration Process

Common Name or Symbol	Isotope		Type of Radiation	Half-Life
Uranium	238 92	Uranium	Alpha	4.49×10^9 year
UX_1	234 90	Thorium	Beta	24.1 days
UX_2	234 91	Protactinium	Beta	1.17 min
Uranium-234	234 92	Uranium	Alpha	2.48×10^5 year
Ionium	230 90	Thorium	Alpha	8×10^4 year
Radium	226 88	Radium	Alpha	1622 year
Radon	222 86	Radon	Alpha	3.825 days
Radium A	218 84	Polonium	Alpha	3.05 min
Radium B	214 82	Lead	Beta, gamma	26.8 min
Radium C	214 83	Bismuth	Beta, gamma	19.7 min
Radium C′	214 84	Polonium	Alpha	2.73×10^{-6} min
Radium D	210 82	Lead	Beta, gamma	22 year
Radium E	210 83	Bismuth	Beta	5.02 days
Radium F	210 84	Polonium	Alpha	138.3 days
Radium G	206 82	Lead	–	Stable

in the series is being generated at the same rate which it is decaying. The time required for the radon daughters through radium C′ to reach equilibrium from a given quantity of radon is approximately 3 h. In approximately 40 min, the alpha energy reaches approximately 50% of maximum.

Exposure to excessive concentrations of radon and radon daughters has been linked with a high incidence of lung cancer. The maximum exposure limit for radon daughters has been set at 1.0 working level (WL), with a yearly cumulative exposure of 4 working level months (WLM). A working level is defined as that concentration of short-lived radon daughter products in a liter of air that will yield 1.3×10^5 million electron volts of alpha energy in decaying through radium C'. WLM is a cumulative measure of exposure that is calculated by multiplying the average WL of exposure during a given timed period by the time of exposure and dividing by 173 (the number of working level hours per WLM) [4].

An example calculation is as follows:

Given the following exposures during a shift:

4 h 0.4 WL
2 h 0.2 WL
2 h 0.1 WL

Find the WLMs of exposure.

Solution:

$$\text{Average WL} = \frac{4 \times 0.4 + 2 \times 0.2 + 2 \times 0.1}{8}$$

$$= 0.275 \text{ WL}$$

$$\text{WLM} = (0.275 \text{ WL})(8 \text{ h})/(173)$$

$$= 0.013 \text{ WLM}$$

19.2 Monitoring of Mine Gas

In the United States, CO monitoring in the belt entry is required by law for fire detection. This was the beginning of mine monitoring. It slowly developed into atmospheric monitoring system (AMS). While the CO-monitoring system monitors only CO concentrations in strategic locations, the AMS monitors many other parameters and locations. Parameters that are monitored comprise CO, CH_4, O_2, NO, H_2 concentrations, smoke, air velocity, and air temperature. The locations that are monitored comprise mine airway, battery charging station, fans, fan houses, electric equipment, pumps, and coal storage.

A monitoring system, as defined by the Mine Safety and Health Administration, is a network of hardware and software meeting the requirements of 30CFR 75.301 and capable of performing the following functions:

1. Measure the required atmospheric parameters.
2. Transmit the data to a surface location.
3. Provide alert and alarm signals.
4. Process and store measured data.
5. Create reports by analyzing the data.

A recent report [5] surveyed 235 US mines and major findings are presented below.

19.2.1 US Mine Survey Results

These mines had 204 CO systems and 33 AMS systems. The parameters monitored and number of sensors are listed in Table 19.3 [5].

It is apparent that the CO systems make up the majority. It may be because it is cheaper and has fewer regulatory requirements for inspection, operation, and maintenance.

19.2.2 System Manufacturers

There are several manufacturers of the monitoring systems in the United States, but the four major manufacturers are (1) Pyott-Boone, (2) AMR, (3) Conspec, and (4) Matrix. Table 19.4 shows their share of the market.

Table 19.3 Distribution of Monitors in US Mines

Parameter Monitored	Percent of Mines	Number of Sensors
CO	100	2–300
CH_4	17	1–20
O_2	6	1–20
NO	1	1–3
H_2	2	1–3
Smoke	2	1–14
Air velocity	9	1–20
Heat	2	4–191

Table 19.4 Manufacturers of the Monitoring Systems

Manufacturer	Number of Systems Installed	Average Number of CO Sensors/System
Pyott-Boone	161	27
AMR	32	54
Conspec	25	84
Matrix	15	45
Others	9	NA

19.3 Wireless Communication and Monitoring System

After the Sago Mine disaster in 2006 where 12 miners died, MINER Act was passed by the US Congress. It required underground coal mines to install two-way communication and tracking systems. The need to communicate is universal but essential in an emergency in underground coal mines.

Several manufacturers have developed communication and tracking systems but one of them, the Innovative Wireless Technologies (IWT), shows the promise to transmit gas concentration data also [6]. IWT earlier developed the SENTINEL system which is comprises line-/battery-powered network mesh modes. It supports voice, text, tracking, data, personnel and vehicle tracking tags, two voice handset models, dispatch and tracking stations. In 2017, they developed an HDRMesh system that is wireless extension or booster for a fiber optic network. It can provide long-range, reliable communications in hard-to-reach working areas with no fiber interconnect between them. The ability of this system is claimed to be better than radio-frequency identification tracking systems.

In early 2017, IWT expanded the use of their system to carry CO concentration data from various locations without any cable; that is wirelessly. The CO monitor has a 6-month battery and a 1000 ft range. These units can be placed every 1000 ft along the belt, and they can transfer the data to the next sensor, 1000 ft away until the surface is reached. It is also claimed to be cheaper than the cabled system. Extensive field testing is necessary to confirm the claims of the manufacturers.

19.3.1 Intrinsically Safe CO Detectors

One such detector, called Sentro 1, is manufactured by Trolex Ltd. (UK) and marketed in the United States by Strata Worldwide. The latter has a wireless communication system that works on batteries. A combination of the intrinsically safe CO sensors with a wireless communication system provides a very good alternative to the present cabled system.

A wireless system is defined as one that needs no external power, cables, repeaters, or splitters. Battery life is 40–60 days. Monthly calibration is needed.

19.4 Special Arrangements for Monitoring in Mines Liable to Spontaneous Combustion

Mine AMS becomes especially important when the coal seam being mined is liable to spontaneous combustion. Brady [7] recommends a combination of three kinds of systems, namely (1) real-time monitoring, (2) tube bundles, and (3) on-site ultrafast GCs. Even an aggressive approach like this cannot prevent a fire, but it does offer means to identify the problem early and a chance to contain the fire before it becomes too big to control and the mine has to be shut down.

19.4.1 Real-Time Monitoring

Real-time sensor systems are ideal for telling us what is happening now. The sensors must be located where the gas needs to be measured, and the measurement signal is sent to the surface. This means having multiple sensors underground. These sensors are exposed to the harsh underground environment which is not ideal for precise analytical measurements. This is not really a major problem as these systems are used to detect step changes, such as the onset of a fire, a sudden increase in a mine gas in the general body, or reduction in oxygen. They offer real-time warning and are the best system for identifying a sudden event such as a belt fire. The situation is reported when it happens. Generally, sensors included are for methane, carbon monoxide, carbon dioxide, and oxygen.

These types of sensors employed underground tend to have limited measuring ranges: carbon monoxide is often only capable of being measured up to 50 ppm, methane to 5%, and carbon dioxide to several percent. This range is fine while no problems exist and indeed to alert the onset of a problem. But if a fire or other major incident involving generated gases occurs, these sensors may quickly reach full scale and become unable to return a true indication of the concentrations.

Most of these sensors require the presence of oxygen to work and are therefore unsuitable for monitoring areas of low oxygen concentration such as sealed or nonventilated gobs.

As each individual sensor needs to be calibrated regularly (at least monthly), they are not suited to being located for long-term monitoring in inaccessible areas such as the gob.

Some of these sensors also suffer from cross sensitivities, as the reactions they rely on to give a response that can be common to other gases found underground, such as carbon monoxide sensors being cross sensitive to hydrogen sulphide and hydrogen.

In the case of an explosion, it is likely that the real-time monitoring system will be rendered inoperable, requiring other techniques for the determination of the status of the underground environment.

19.4.2 Tube Bundle

Tube bundle systems draw gas samples from designated sampling locations underground to the surface through plastic tubes using vacuum pumps and analyzed sequentially using infrared and paramagnetic techniques. Gases measured are carbon monoxide, carbon dioxide, methane, and oxygen.

Because the analyzers are on the surface, tubes can be located in the gob as once positioned there is no requirement to access the end sampling point.

Tube bundle systems are suited to long-term trend analysis. Very good analytical equipment is available and can be housed in dedicated air-conditioned rooms on the surface with the samples dried and passed through particulate filters prior to entering the analyzer.

Generally tube bundle systems are set up to measure oxygen, carbon monoxide, carbon dioxide, and methane. Given their ability to measure carbon monoxide down to

1 ppm, the long-term stability of these analyzers, and the frequent sampling, this technique is best for long-term trending of carbon monoxide to identify a spontaneous combustion event. With respect to measuring range, it is normally only carbon monoxide that presents problems, with most systems capable of measuring to only 1000 ppm. Because methane and oxygen concentrations can be measured over all expected concentrations ranges, this technique is the best for automated monitoring of explosibility of an area so long as a fire or heating does not exist.

To get this improved stability and analytical capability, the immediate availability of the results is sacrificed. The samples need to be drawn to the surface prior to being analyzed, meaning the data being generated can be from samples collected from over an hour before. There is only one bank of analyzers, so only one sample is analyzed at a time. Depending on the number of tubes in the system and the programmed sampling sequence, each point may only be sampled once every 30–60 minutes. Add this to the time taken to draw the sample from underground, which may be as long as an hour, and it is obvious that this technique is not suitable for the instantaneous detection of an incident such as a fire.

Because the analyzers in these systems rely on infrared absorbance and paramagnetic attraction, the gas matrix is not important, making this technique suitable for the analysis of gases from oxygen-depleted areas such as the gob. The measurement of oxygen using paramagnetic analyzers is flow rate dependent, and the flow from each tube must be balanced to be the same, including any calibration gases used. Otherwise, it is possible that two locations could in fact have the same oxygen concentration, but because of more resistance in one of the tubes, the flow through the analyzer is at a lower flow rate and as such results in a lower reading than a location with the same concentration but flowing through the instrument at a faster rate.

In the event of a mine explosion, the tube bundle monitoring system may still appear to be functional, but the location from which tubes are sampling may not be the same, due to damage to the tubes. A good tube bundle system will include monitoring of the vacuum pressure in each of the tubes, so following an explosion this data can be used to determine whether a tube has been compromised or not. It is also useful during routine operation for identifying increased restriction or sudden leakage in a tube, both of which can compromise the operation of the system.

If the tubes are damaged and not providing any valuable information, it may be possible to make use of boreholes and connect new tubes to locations of interest as the surface equipment will still be operational. This is the preferred technique in the United States, although quite expensive.

19.4.3 Gas Chromatography

Gas chromatography, with regard to gas analysis, involves the separation of all sample components followed by their measurement on relatively nonspecific detectors. Specificity is obtained by virtue of the separation process rather than detection.

The use of a GC expands analytical capabilities to include gases crucial in the interpretation of spontaneous combustion events, particularly ethylene and hydrogen. The GC provides a complete analysis of the gases expected underground and is the only one of the three techniques capable of measuring hydrogen, nitrogen, ethylene, and ethane. Determination of nitrogen is particularly important for determining oxygen deficiency in some spontaneous combustion indicating ratios (refer to Chapter 21).

Similar to the tube bundle, problems exist with bringing the samples to the GC. The significance of time delays in getting results is dependent on what the results are being used for. GC is not going to be suitable for detection of a belt fire because of the time delay between collection of the sample and analysis, but the delay is acceptable for confirmation of other results or for evidence and trending of spontaneous combustion indicators.

Like the tube bundle system, the gas matrix of the sample does not affect GC analysis. So long as appropriate calibration gases are available, this technique is capable of measuring gases at any concentration above their detection limit. This eliminates the problems seen with the other techniques, particularly for carbon monoxide concentrations greater than 1000 ppm.

The ultrafast GCs in use in Australian mines allow the analysis of most of the components expected underground in approximately 2 minutes.

This increased speed of analysis is invaluable during emergency situations, particularly when assessing the safety of the underground atmosphere for reentry or during reentry by mine rescue teams. In these cases, what makes this assessment more effective is that GC is on-site and can be operated by mine personnel. There is no delay in determining the status underground, while waiting for external providers to arrive or transporting samples away from site for laboratory analysis.

References

[1] Hartman HL, et al. Mine ventilation and air conditioning. 2nd ed. John Wiley and Sons; 1982. p. 59–60.

[2] Thakur PC. Computer-aided analysis of diesel exhaust contamination of mine ventilation systems. Ph.D. thesis. The Pennsylvania State University; 1974. p. 234.

[3] Holaday DA. Control of radon and daughters in uranium mines and calculations of biologic effects. Washington, DC: Publication No. 494, U.S. Department of Health, Education and Welfare; 1957.

[4] Thakur PC. In: Darling P, editor. Gas and dust control in SME mining engineers handbook; 2011. p. 1595–609.

[5] Rowland JH, Harteis SP, Yuan L. A survey of atmospheric monitoring systems in US underground coal mines. Mining Engineering 2018;70(No. 2):37–40.

[6] Morton J. New wireless tech for underground mines could save lives, costs. Coal Age April 2018:20–3.

[7] Brady D. The role of gas monitoring in the prevention and treatment of mine fires, Coal Operator's Conference, the AusIMM. February 2008. p. 202–8.

Economics of Coal Mine Degasification

20

Chapter Outline

The economics of coal mine degasification can be viewed in two different ways: (1) overall cost and benefit of degasification and (2) gas drainage/production as a stand-alone business. Main benefits of coal mine degasification are three-fold: (1) safety of the mine workers, (2) reduced cost of mining through improved productivity, and (3) revenues from the marketing of produced gases. The chapter will discuss the three benefits and the related costs of degasification in detail.

20.1 Safety in Mines

A good mine degasification plan can minimize, if not eliminate, chances of a mine explosion. It is not possible to place a monetary value on human life. It is truly priceless. The economic gravity of a mine disaster can be realized by the fact that the companies, that owned Sago Mine (2006 disaster) and Upper Big Branch (2010 disaster) mines,

Advanced Mine Ventilation. https://doi.org/10.1016/B978-0-08-100457-9.00020-1

went bankrupt. This truly exemplifies the benefits of coal mine degasification. The most valuable commodity that comes out of a coal mine is a healthy coal miner.

20.2 Reduced Cost of Mining by Improved Productivity

Next to the safety of the mine workers the biggest financial benefit of degasification is the improved productivity. In a highly gassy mine in Virginia, USA, the productivity improved from 14 tons/man-day to 40 tons/man-day when nearly 70% of the in situ gas was drained from the coal seam prior to mining and nearly 80% of gas was captured from the gob areas. The cost of mining was reduced by at least 25%. For example,

Let us suppose the mine produces 6 million tons of clean coal per year.
Sale price of coal: $150/ton. (It is a metallurgical coal.)
Annual gross: $900 million/year.
Estimated cost of mining: $300 million/year.
Avoided cost by degasification: $75 million/year.

The savings owing to degasification is thus a significant part of the net income. Sometimes, the revenue from gas alone makes the mines profitable.

20.3 Revenues From Drained Methane

A typical mine produces 5–30 MMCFD gas depending on the thickness of coal seams and their gas contents. At an assumed price of $5/MCF, the annual gross revenue can range from $9 million to $54 million. The actual net profit will be less when the cost of degasification and gas processing are included.

The actual cost of degasification will depend mainly on the gassiness of the coal seam (see Chapter 13) and the degree of degasification needed for an uninterrupted coal production. Table 4.1 defines the three categories of coal seams based on their gas contents. Premining and postmining schemes were discussed in Chapters 15 and 16, respectively.

For all underground longwall mining, a generalized scheme of degasification depending on the gassiness of the coal seam has been proposed by Thakur and Zachwieja [1]. The following assumptions were made

1. The longwall panel is 1000 ft wide and 10,000 ft long.
2. The coal seam has an average thickness of 6 ft.
3. The coal block to be degassed is 1300 ft by 10,000 ft, assuming that the width of chain pillars is 300 ft.
4. The cost of contract drilling for the in-mine horizontal drilling is $50/ft.
5. The cost of a gob well is $50,000 to $200,000, depending on the depth of the mine and the size of the borehole.
6. The cost of a well with hydrofracturing is $250,000.

If the total cost of in-mine drilling, including all of the underground pipeline costs, all vertical frac wells, and all other gob wells, is added and then divided by the tons of coal in the longwall block, the result is the cost of coal seam degasification per ton of coal. An annual clean coal production of 3.1 million tons is assumed.

20.3.1 Estimated Cost for Mildly Gassy Coal Seams (Gas Content Less Than 100 ft^3/ton)

Premining degasification: For coal seams with gas contents less than 100 ft^3/ton, there is generally no need for premining degasification.

Postmining degasification: Two gob wells are recommended for the longwall panel. The first gob well should be installed within 1000 ft of the setup entry and the second one installed in the middle of the panel.

The total cost is $100,000 or $0.03/ton.

20.3.2 Estimated Cost for Moderately Gassy Coal Seams (Gas Content 100–300 ft^3/ton)

Premining degasification: The longwall panel should be drilled horizontally at 1000 ft intervals, and development boreholes should be drilled to degas development sections. Total in-mine drilling footage for a typical panel may total 25,000 ft.

Postmining degasification: In moderately gassy coal seams, a proposed longwall panel may need 5–6 gob wells. The diameter and size of exhaust fans will depend on local conditions.

The total cost is approximately $1.55 million or $0.50/ton.

20.3.3 Estimated Cost for Highly Gassy Coal Seams (Gas Content Over 300 ft^3/ton)

Premining degasification: Highly gassy coal seams must be drained several years ahead of mining with vertical frac wells (wells that have been hydraulically fractured). These frac wells can be placed at about a 20-acre spacing. Frac wells drilled about 5 years ahead of mining can drain nearly 50% of the in situ gas prior to mining, but this may not be sufficient. Additional degasification with in-mine horizontal drilling can raise the gas drained to nearly 70%. Horizontal boreholes are drilled 200–300 ft apart to a depth of 900 ft. Assuming a 200 ft interval, nearly 45,000 ft of horizontal drilling and about 15 vertical frac wells may be needed to properly degas the panel.

Postmining degasification: Because of very high gas emissions from the gob, the first gob well must be installed within 50–100 ft from the setup entry. Subsequent gob wells may be drilled at a 6- to 15-acre spacing, depending on the rate of mining and the gas emission per acre of gob. In the US states of Virginia and Alabama, two states with some highly gassy coal seams, gob wells are generally 9–12 inches in diameter. Powerful exhaust fans capable of a suction of 5–10 inches of mercury are needed to capture up to 80% of gob gas emissions.

The total cost of degasifying a longwall panel in a highly gassy coal seam is approximately $11 million or $3.52/ton. Coal seam degasification is needed for mine safety and high productivity, but in highly gassy mines, it becomes quite expensive. In these mines, the processing and marketing of coal mine methane becomes necessary to defray the cost [2].

20.4 Gas Production From Coal Seams—A Stand-Alone Business

Modern coal mines cover a large lease of 100,000–300,000 acres. A mineable coal seam of 6 ft thickness can be overlain and underlain by several thinner (3–5 ft) coal seams giving a total coal thickness of 30 ft to a mineable depth of 3000 ft. A simple calculation shows that the coal lease has a gas reserve of 2–6 TCF assuming a gas content of 500 ft^3/ton. Assuming a low recovery factor of 40%, the recoverable gas volume is 0.8–2.4 TCF. The gross revenue from this gas is $2.4 to $7.2 B even at a low cost of $3/MCF for natural gas. In author's opinion, "Every large coal mine lease is a small to medium size gas field." A typical example to produce 100 MMCFD of gas with an economic analysis is presented to show that marketing coalbed methane (CBM) can not only defray the cost of degasification but also substantially increase the net profit from the combined mining and gas production venture. A highly gassy mine can generate a net profit of $50–60 million/year at $5/MCF.

20.4.1 Reserve Estimate

The reserve classification for natural gas wells is divided into three categories:

1. Proved reserves: They are the estimated volumes that geological and engineering data demonstrate with reasonable certainty to be recoverable in the future with existing operating and economic conditions. These reserves are further divided into (1) producing reserve, (2) nonproducing reserve, (3) shut-in reserve, and (4) undeveloped reserve [3].
2. Probable reserves: It is less certain than the proved reserves and can be estimated with a degree of certainty sufficient to indicate they are more likely to be recovered than not.
3. Possible reserves: They are less certain than probable reserves and can be estimated with a low degree of certainty, insufficient to indicate whether they are more likely to be recovered than not.

The gas contained in a coal lease will fall in the category of proved reserve because the entire property is drilled on a 500-acre spacing to withdraw a complete core to a depth that contains all coal seams. Proximate analysis, gas content measurements, and other reservoir property measurements are carried out to prove the coal and gas reserves.

The gas in place, GIP, is given by Eq. (20.1)

$$GIP = A \times H \times \rho \times Gc \tag{20.1}$$

where A is the area of the lease in acres, H is the thickness of the coal seams that will be produced, ρ is the coal density in tons/acre-ft, and Gc is the gas content of coal seams (ft^3/ton).

Each coal seam in the lease is evaluated separately, and a composite gas isopach map is prepared that shows the gas content contours of the lease in MMCF/acre or BCF/section. A section is equal to one square mile (640 acres). Areas with higher values of BCF/acre are generally very profitable to produce.

20.4.2 Production Techniques and Gas Production Forecast

At present, there are only two techniques for commercial CBM production:

1. Vertical wells with multiple completions.
2. Horizontal wells drilled from surface into multiple coal seams.

In active coal mines, there is a third source of gas that is gas recovered from longwall gobs. It can range from 5 to 10 MMCFD in moderately gassy coal fields to 20 to 30 MMCFD in highly gassy coal fields. This is almost "free" gas because the cost of production is very small compared with the cost of production from virgin coal.

Thakur [4] has discussed these techniques in great detail in a recent publication. Reference should be made to the original work for greater details.

20.4.2.1 Vertical Well Production Forecast

The steady-state production from a vertical well is estimated from Eq. (20.2) [4].

$$q = \frac{707.8\ kh\left(P_e^2 - P_w^2\right)}{\mu zT \ln\left(r_e/r_w\right)} \tag{20.2}$$

where q = cubic feet/day at 60°F and 14.67 psia, k = permeability in darcy, h = thickness of each coal seam in feet, P_e = pressure at external radius (r_e), P_w = pressure at the well radius (r_w), μ = gas viscosity, z = compressibility factor (can be assumed 1.0 to a mineable depth of 2500 ft), and T = temperature of coal seam in degrees Rankine (Fahrenheit + 460).

For example, let us assume the following:

k = 0.015 darcy (15 md)
h = 40 ft
μ = 0.2 cp
z = 1.0
T = 520 R
r_e = 1000 ft (length of the fracture created)
r_w = 0.25 ft
P_e = 500 psi
P_w = 50 psi

Using Eq. (20.2)

$$q = \frac{707.8 \times (0.015) \times 40\left(500^2 - 50^2\right)}{0.02 \times 520 \times \ln\left(\frac{1,000}{0.25}\right)}$$

$$= 1.22 \text{ MMCFD}$$

Let us drill 84 wells at 80-acre spacing (2.78 BCF per well) to have an initial production of 102.48 MMCFD. CBM wells continue to produce for a long time. A life of 30 plus years is not uncommon in a reserve that will not be mined. Hence, let us assume an annual decline of 3%.

Table 20.1 shows the production rate and cumulative production for the first 5 years only. This is a typical life of a well in a working mine.

20.4.2.2 Horizontal Well Production Forecast

Steady-state gas production from horizontal wells will be discussed by the author in a book to be published in 2019 in great detail [5]. For the current project, let us assume the coal seams of the project have a specific gas production of 15 MCFD/100 ft [4].

Hence, we need to drill 17 wells with two laterals of 5000 ft in four coal seams with an average thickness of 10 ft each.

Each lateral of 5000 ft will produce 750 MCFD and hence each well will produce 6 MMCFD. With 17 wells, the total production will be 102 MMCFD. The gas production rate and cumulative gas production per year will be much the same as shown in Table 20.1.

It is deliberately kept close for a fair comparison of the two techniques.

20.4.3 Gas Gathering and Processing

The CBM as produced cannot be sold without processing owing to the presence of impurities. Table 20.2 shows the typical composition of CBM. Gas pipeline (buyers)

Table 20.1 Gas Production in Five Years of the Project

Year of Production	Gas Rate (MMCF/Year)	Cumulative Production (MMCF)
1	36,893	36,893
2	35,786	72,679
3	34,712	107,391
4	33,671	141,062
5	32,661	173,723

Table 20.2 Coalbed Gas Composition

Component	Typical Analysis (%)	Range (%)
Methane	95	85–96
Ethane	0.5	0.2–2.5
Propane	Trace	Trace
Isobutane	Trace	Trace
Normal butane	Trace	Trace
Isopentane	Trace	Trace
Nitrogen	1.5	1–6
Carbon dioxide	3	0.5–10

specifications vary, but in general, they limit the percentage of noncombustibles to less than 4% and a minimum BTU of 960/ft^3.

The produced gas has to be gathered and processed before it can be sold.

20.4.3.1 Gathering Gases

Several CBM-producing wells are connected to a small compressor that feeds into a bigger compressor or the main compressor at the processing plant. These compressors can be run by gas engines, but it is preferable to have electrical drives for reliability. Most often HDPE (high-density polyethylene) lines are used for gathering gases. The line pressure is typically less than 50 psi.

20.4.3.2 Gas Processing

The raw gas is seldom of pipeline quality. Most often, it is processed to meet the marketing pipeline specifications.

The general scheme is to pass the raw gas through "intermediate" compressors raising the pressure to 200–300 psi.

The gas processing line diagram is shown in Fig. 20.1.

The compressed gas first goes through a cyclone where free water and solids are removed. Next, it goes in a chamber where oxygen is removed by burning some methane catalytically. The gas goes through a "diamine" unit next where all CO_2 is removed. Next, the gas goes through a dehydration unit that contains ethylene glycol. The gas is now ready for commercial pipelines unless there is more than 4% nitrogen. In this case, a nitrogen rejection unit (NRU) is used to remove most of nitrogen in the gas. There are two kinds of NRU: (1) the molecular gate process and (2) the cryogenic process. In the molecular gate system, unique molecular sieves have pore-size openings that can be controlled with a precision of 0.1 A°. The pore size is precisely

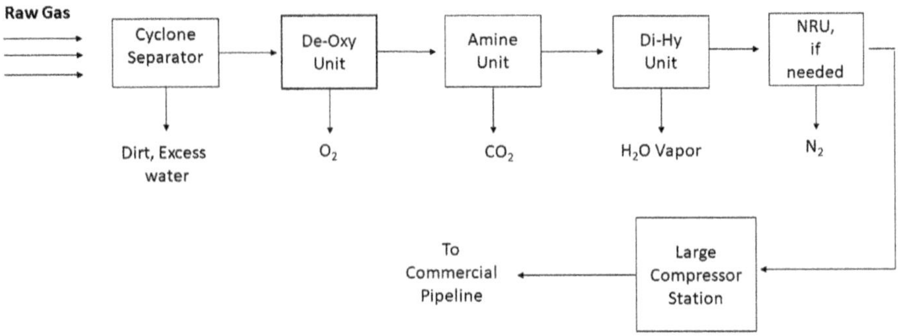

Figure 20.1 Line diagram for gas processing.

adjusted to allow smaller molecules of CO_2, N_2, and O_2 to pass through, but not the CH4 molecules of 3.7 A°. CO_2 molecules are the smallest at 3.3 A°. N_2 and O_2 molecules are 3.6 A°. It is economic to use molecular gate system up to a gas flow of 5 MMCFD.

For higher flows that may have a significantly large amount of nitrogen (20% or higher), a cryogenic system is preferred. The gas is raised to a high pressure (typically above 600 psi) and cooled to liquefy all gases. The liquid is then distilled, and gases separate into individual components. The nitrogen leaves the system as a vapor at mid-range pressure, while methane separates at a lower pressure.

Reference can be made to a book, *CBM from Prospect to Pipeline*, written and edited by the author [2] for additional information.

20.4.3.3 Water Management

Almost all coal seams are water saturated. Water production is the highest immediately after the well is opened. Water is pumped out of the coal seam, by the reservoir pressure. Water handling strategies include trucking and pipeline systems. The flow can vary from a few barrels per day to 100s of bpd. The quality of water is rarely good when it can be discharged into a local stream. Most often, water is treated to remove impurities.

Two common techniques are to use (1) an evaporation pond or (2) deep hole injection. In the evaporation pond, the produced water is aerated often with some chemical to precipitate all dissolved solids. The water mostly evaporates. When the flow is large, the produced water is often cleaned using a "reverse osmosis" process to get the dissolved solids out. Next, it is injected into deep wells. Such wells may need hydrofracturing to increase their capacity to hold water [2].

20.5 Economic Analysis

Three measures of profitability are commonly used to decide if a CBM project is economically viable. They are

1. Cash flow method.
2. The net present value (NPV).
3. Discounted cash flow rate of return (DCFROR).

One or more methods are used to finally decide if the project is profitable [6].

20.5.1 Cash Flow Method

In this method, the capital investment is the most important parameter. It indicates when the investment will be returned. It is important for small, independent operators especially if they are operating in politically unstable areas.

Undiscounted cash flow (payout) is described by Eq. (20.3).

$$\sum_{x=1}^{n} NCF_x = I \tag{20.3}$$

where n = number of years when $I = \sum_{x=1}^{n} NCF_x$, x = years, NCF_x = net cash flow for year x, and I = total investment.

Table 20.3 shows the capital investment for a gas production of 100 MMCFD for 5 years. Total gas production in 5 years is 180 BCF. Assumed price of gas = $5/MCF.

The cash flow over 5 years is shown in Table 20.4.

Table 20.5 shows the cash flow for drilling the same area horizontally. The total investment in this case is $381.6 million.

Table 20.3 Capital Investment for Vertical Wells

Item	Investment (in Millions)
G & G acquisition	7.2
Drilling and completion	39.6
Lease equipment and gas gathering	21.6
Water disposal	14.4
Well stimulation	21.6
Engineering and G & A	10.8
Compressors	36.0
Gas processing	180.0
Total	331.2

Table 20.4 Cash Flow Method for Vertical Wells

Year	Production (MMCF)	Revenue ($)	Operating Costs ($)	Investment ($)	Tax ($)	Profit Net Cash After Tax ($)
0	0		0	331,200,000		−331,200,000
1	36,892.8	184,460,000	2,025,000	0	17,708,160	+164,726,456
2	35,786.8	178,934,000	2,025,000	0	17,177,664	+159,731,336
3	34,712.45	173,562,250	2,025,000	0	16,661,976	+154,875,274
4	33,671.06	168,355,300	2,025,000	0	16,162,108	+150,168,191
5	32,660.93	163,304,650	2,025,000	0	15,677,246	+145,602,404
Total						+443,903,661

A positive value from cash flow method shows that the project is profitable.

Table 20.5 Cash Flow for Multilateral Horizontal Wells

Year	Production (MMCF)	Revenue ($)	Operating Costs ($)	Investment ($)	Tax ($)	Profit Net Cash After Tax ($)
0	0		0	381,600,000		−381,600,000
1	37,094	185,470,000	1,600,000	0	17,805,120	+166,064,880
2	35,982	179,910,000	1,600,000	0	17,271,360	+161,038,640
3	34,902	174,510,000	1,600,000	0	16,752,960	+156,157,040
4	33,855	169,275,000	1,600,000	0	16,250,400	+151,424,600
5	32,839	164,195,000	1,600,000	0	15,762,720	+146,832,280
Total						+399,917,440

Again, the project is financially viable, but the total cash flow in 5 years is a bit lower. The calculation can be easily extended to the anticipated life of 30 years for these wells if they are in a nonmineable area.

20.5.2 Net Present Value Method

NPV is another measure of profitability that is based on the present value of cash flows discounted on an average rate of i in excess of the present value of investment. It is defined by Eq. (20.4).

$$\mathrm{NPV} = \sum_{x=0}^{n} \frac{\mathrm{NCF_x}}{(1+i)^x} \tag{20.4}$$

where i is the average rate of discount; the rest of the variables have the same meaning as defined earlier in Eq. (20.3). This method introduces the true value of money into the analysis based on an interest rate, i, representative of the company's reinvestment opportunities. If the NPV is a positive value, a viable investment is indicated.

Tables 20.6 and 20.7 show the NPVs for the two methods of drilling and completion, respectively. The interest rate is assumed to be 15%.

Table 20.7 shows similar data for the horizontal wells.

20.5.3 Discounted Cash Flow Rate of Return Method

The DCFROR gives profitability in terms of a compound discount rate which can be compared with interest rates of borrowing money or to internal rates generated by concurrent projects. It is the interest rate necessary to make the sum of the present value of investment equal to the sum of the present values of each year's net cash flow.

Mathematically, Eq. (20.5) defines it.

$$\sum_{x=0}^{n} \frac{\mathrm{NCF_X}}{(1+i)^X} = 0 \tag{20.5}$$

The value of i that makes Eq. (20.5) valid is the DCFROR for the project.

Tables 20.8 and 20.9 show the DCFROR for the vertical well and horizontal well options, respectively.

Most projects require a minimum DCFROR of 15%. As such, both production techniques offer a good rate of return on investment. Based on the above analysis, there is only a minor difference in the total profits using vertical or horizontal wells for gas production.

Problem

Extend the life of the CBM wells to 30 years and calculate the revenues for both methods of drilling using the three measures of profitability. Assume 3% production decline per year and a price of gas at $5/MCF.

Table 20.6 Net Present Value (NPV) for Vertical Wells

Year	Total Gas Production (MMCF)	Revenue ($)	Operating Costs ($)	Financial Costs ($)	Net Revenue ($)	NPV ($)
1	36,892.8	184,460,000	2,025,000	17,708,160	94,866,200	82,492,347.83
2	35,786.8	178,934,000	2,025,000	17,177,664	91,992,680	69,559,682.42
3	34,712.45	173,562,250	2,025,000	16,661,976	89,199,370	58,650,033.70
4	33,671.06	168,355,300	2,025,000	16,162,108	86,491,756	49,451,942.21
5	32,660.93	163,304,650	2,025,000	15,677,246	83,865,418	41,695,934.73

Total NPV for 5 years is $302 million which is a bit less than the initial investment of $331 million, but these wells will produce for 30 years or so indicating a higher positive NPV.

Table 20.7 Net Present Value (NPV) for Multilateral Horizontal Wells

Year	Total Gas Production (MMCF)	Revenue ($)	Operating Costs ($)	Financial Costs ($)	Net Revenue ($)	NPV ($)
1	37,094	185,470,000	1,600,000	17,805,120	95,612,400	83,141,217.39
2	35,982	179,910,000	1,600,000	17,271,360	92,721,200	70,110,548.20
3	34,902	174,510,000	1,600,000	16,752,960	89,913,200	59,119,388.51
4	33,855	169,275,000	1,600,000	16,250,400	87,191,000	49,851,737.24
5	32,839	164,195,000	1,600,000	15,762,720	84,549,400	42,035,994.66

Total NVP for 5 years is $304 million. A positive value again indicates that the project is viable.

Table 20.8 DCFROR Method for Vertical Well Scenario

Year	Net Cash Flow ($)	i = 0.1	i = 0.2	i = 0.3	i = 0.4
0	−331,200,000	−331,200,000	−331,200,000	−331,200,000	−331,200,000
1	+164,726,456	149,751,323.6	137,272,046.6	126,712,658.46	117,661,754.29
2	+159,731,336	132,009,368.6	110,924,538.8	94,515,583.4	81,495,579.59
3	+154,875,274	116,360,085.6	89,626,894.68	70,493,979.97	56,441,426.38
4	+150,168,191	102,566,895.1	72,419,073.6	52,578,057.91	39,090,012.29
5	+145,602,404	90,407,637.08	58,514,340.44	39,214,960.58	27,072,500.22
Total		259,895,310	137,556,894	52,315,240	−9,438,727

DCFROR appears to be 0.3 and 0.4. By approximation, it appears to be 38.5%.

Table 20.9 Discounted Cash Flow Rate of Return (DCFROR) Method for Multilateral Horizontal Wells

Year	Net Cash Flow ($)	i = 0.1	i = 0.2	i = 0.3	i = 0.4
0	−381,600,000	−381,600,000	−381,600,000	−381,600,000	−381,600,000
1	+166,064,880	150,968,072.73	138,387,400.00	127,742,215.38	118,617,771.43
2	+161,038,640	133,089,785.12	111,832,388.89	95,289,136.09	82,162,571.43
3	+156,157,040	117,323,095.42	90,368,657.41	71,077,396.45	56,908,542.27
4	+151,424,600	103,425,039.27	73,024,980.71	53,017,961.56	39,417,065.81
5	+146,832,280	91,171,293.57	59,008,600.18	39,546,202.06	27,301,176.59
Total		214,377,286	91,022,027	5,072,912	−57,192,872

DCFROR appears to be between 0.3 and 0.4. By approximation, the DCFROR value is about 30.8%.

References

[1] Thakur PC, Zachwieja J. Methane control and ventilation for 1,000 ft-wide longwall faces. In: Proceedings of longwall USA, International Exhibition and Conference; 2001. p. 167−80.
[2] Thakur PC, et al., editors. Coalbed methane from Prospect to pipeline. Elsevier; 2014. p. 420.
[3] Miller MJ, Watson DA. Economic analysis of coalbed methane projects. in Coalbed Methane from Prospect to Pipeline, Ed. P. Thakur, Elsevier, 2014:261−272.
[4] Thakur PC. Advanced reservoir and production engineering for coalbed methane. Elsevier; 2016. p. 210.
[5] Thakur PC, editor. CBM theory and application. Elsevier; 2019. to be published.
[6] Stermole FJ, Stermole JM. Economic evaluation and investment decision method. Golden, Colorado: Investment Evaluation Corporation; 1987. p. 479.

Section Four

Mine Fire Control

Spontaneous Combustion of Coal 21

Chapter Outline

Because coal is not only a combustible material but also liable to spontaneous combustion, mine fire is a clear and present danger in all coal mines. As coal is exposed to ventilation air, it begins a slow oxidation process that produces heat. Mostly such produced heat is dissipated by large volumes of circulating air but there are cases where air supply is limited and the oxidation process continues to raise the coal temperature until it catches fire. There are many other sources of fire in the mines. Table 21.1 lists the major sources recorded over a 10-year period (1990—99) in US coal mines.

Mine explosions (not listed) are another cause of mine fire. They are rare but have the greatest fatalities. Mobile electrical equipment is being replaced by diesel-powered

Advanced Mine Ventilation. https://doi.org/10.1016/B978-0-08-100457-9.00021-3

Table 21.1 Number of Fires in Underground Coal Mines [1]

Source	Number of Fires	Percent (%)
Electrical Short/Arcing	28	32
Flame Cutting/Welding	17	20
Spontaneous Combustion	15	17
Conveyor Belt Friction	15	17
Frictional Ignitions and Other Sources	12	14
Total	87	100

equipment minimizing the risk of fire. Flame cutting and welding is mostly a housekeeping issue that can be made safe by training and enforcement of laws. Belt fires are mainly due to friction and will be discussed under frictional ignitions. Major remaining issues are the following:

1. Spontaneous combustion.
2. Frictional ignitions.
3. Gas and dust explosions.

These subjects will be discussed in detail in this and the following chapters.

21.1 Spontaneous Combustion of Coal

It is generally accepted that the spontaneous combustion propensity is related to the rank of coal: the lower the rank, the more the coal is liable to spontaneous heating. Other factors that may impact this phenomenon include heat of wetting, temperature of coal, ventilation pressure differential, oxygen content of coal, particle sizes, geological anomalies, and mining methods. It is always advisable to determine the liability to spontaneous heating in the laboratory prior to planning the mining method for the coal seam.

At present, there are three different techniques to measure a coal seam's liability to spontaneous combustion:

1. Adiabatic heating oven.
2. Temperature differential methods (also called crossing-point temperature method).
3. Oxygen sorption methods.

21.1.1 Adiabatic Heating Oven

Smith and Lazzara [2] ran an experiment on 24 coal samples, ranging from anthracite to lignite, to determine the minimum temperature necessary to start the process of spontaneous combustion and called it, "minimum self-heating temperature" or SHT. Fig. 21.1 shows a schematic of the experimental setup.

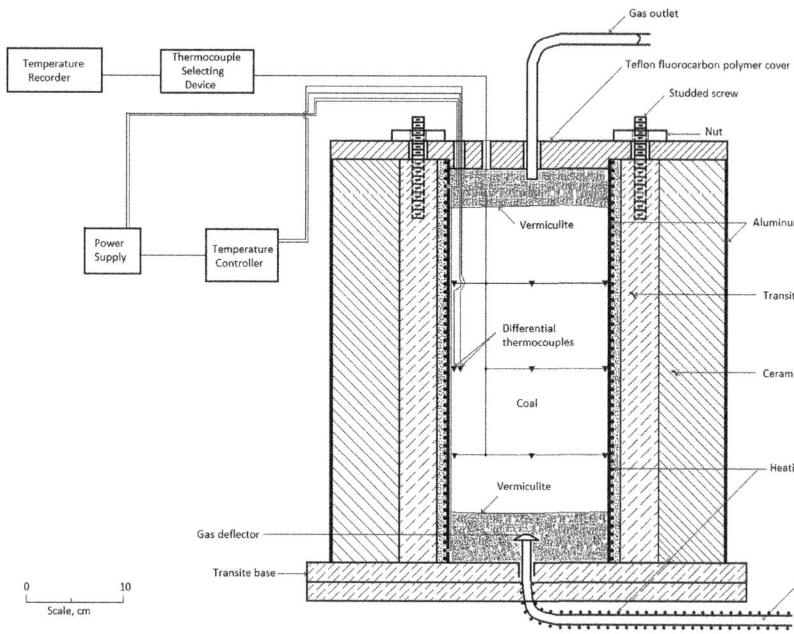

Figure 21.1 A schematic of the spontaneous combustion apparatus.

The coal sample, about 6.0 pounds (stacked 12″ high), consisted of 61% minus ¼ inch plus 10 mesh, 33% minus 10 mesh plus 50 mesh, and 6% minus 50 mesh. It was insulated at both ends by a 1.5- to 2-inch thick vermiculite layer of insulation. The coal sample was predried using dry nitrogen at 67°C flowing at 900 cc/min. Dry nitrogen was circulated through the apparatus, as shown, to stabilize the coal temperature at 67°C. All temperatures are measured with thermocouples.

Humid air was next introduced in the test chamber and temperature rise in coal was recorded. At an initial temperature of 70°C, there was a rise in coal temperature to 80°C in about 6 h (Fig. 21.2) but then it leveled off.

Next, the initial temperature was raised to 75°C. This time the coal temperature increased steadily, reaching a thermal runaway in 30 h (Fig. 21.2). The crucial temperature was next raised to 80°C and 90°C and the coal began to self-heat in 11 and 4 h, respectively. Thus the SHT of the coal was established at 75°C.

This procedure was repeated for 22 coal samples. The results of the experiment on 22 samples are shown in Table 21.2 with proximate and ultimate analysis of coal.

A regression analysis of SHT as a dependent variable and dry ash-free oxygen as the independent variable yielded a linear relationship.

$$\text{SHT (minimum)} = 139.74 - 6.57\,(O_2) \tag{21.1}$$

where: SHT is in degree centigrade and O_2 is the dry ash-free oxygen content of coal.

Fig. 21.3 shows the plot of actual SHT against the values predicted by Eq. (21.1).

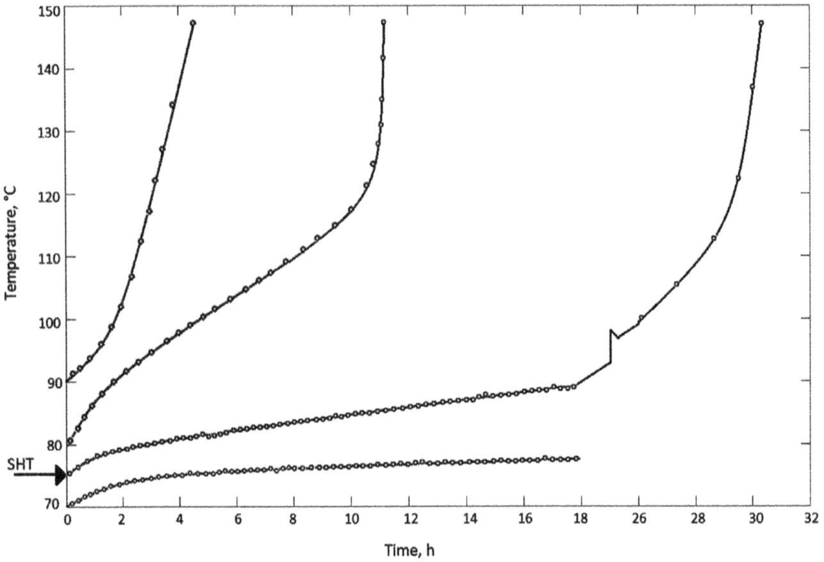

Figure 21.2 Temperature rise with time for a typical coal. *SHT*, minimum self-heating temperature.

In a slight modification of the above technique, Beamish [3] used oxygen to initiate self-heating. Fig. 21.4 shows their results for seven coal samples in Australia.

It was noted that initially the coal temperature rose linearly but after 80°C the temperature rise became exponential. Beamish [3] proposed that the gradient of the linear portion of the curve was a good index of spontaneous combustion propensity and called an R70 index of coal, expressed as degrees/h. The values for Australian coals range from 0.06 to 3.73°C/h [4].

The rate of temperature rise (R70 scale) can be analyzed mathematically by Arrhenius equation, where:

$$\frac{dT}{dt} = A\left(e^{-E/RT}\right) \tag{21.2}$$

where: $\frac{dT}{dt}$ is the rate of temperature rise; A is a characteristic for coal; E is the activation energy; R is the gas molar constant; and T is temperature on absolute scale (Kelvin).

Taking a logarithm of Eq. (21.2) on both sides, gives:

$$\ell n\left(\frac{dT}{dt}\right) = \frac{-E}{RT} + \ell n\ A \tag{21.3}$$

So a plot of $\ell n\left(\frac{dT}{dt}\right)$ against 1/T will yield a straight line with an intercept on y-axis equal to $\ell_n A$ and the slope of the line is $\frac{-E}{R}$ from which E for each coal can be calculated.

Table 21.2 Coal Seam Characteristics

Rank and Seam	State	SHT °C	Proximate Analysis (wt pct)				Ultimate Analysis (wt pct)					
			Moisture	Volatile Matter	Fixed Carbon	Ash	Hydrogen (H)	Carbon (C)	Nitrogen (N)	Sulfur (S)	Oxygen (O)	Dry Ash-Free Oxygen
Lignite (Lig):												
Beulah-Zap	ND	60	27.3	29.9	31.8	11.0	6.0	43.4	0.5	3.9	35.2	17.7
Lehigh bed	ND	35	42.6	32.8	19.5	5.2	7.1	37.9	0.6	0.5	48.6	20.7
hvCb Coal:												
F	CO	45	11.4	40.9	45.3	2.4	6.0	66.4	1.4	0.3	23.4	15.5
No. 6	IL	70	2.2	41.8	42.7	13.2	5.3	65.5	1.2	4.2	10.6	10.4
No. 80-1 hvc	WY	35	7.6	38.3	44.1	10.0	5.4	64.3	1.5	0.4	18.4	14.2
No. 80-2 hvc	WY	40	11.0	39.5	43.9	5.6	5.9	65.2	1.6	0.7	21.0	13.5
hvAb Coal:												
B-1	CO	70	2.8	38.3	49.6	9.3	5.5	72.0	1.4	0.5	11.2	9.9
B-2	CO	75	3.9	40.1	53.8	2.2	5.7	77.4	1.6	0.5	12.5	9.6
Clarion	OH	75	4.8	43.0	44.4	7.7	5.6	70.0	1.2	3.5	12.0	8.8
E–1	CO	65	3.9	39.7	54.6	1.8	5.8	76.5	1.3	0.7	13.9	11.0
E–2	CO	65	3.2	40.5	54.4	1.9	5.7	76.9	1.5	0.7	13.2	10.9
Lower Kittanning	PA	80	1.0	31.7	44.9	22.4	4.6	62.4	1.0	3.9	5.9	6.5
Lower Sunnyside—1	UT	85	3.2	36.2	54.2	6.4	5.4	74.3	1.5	0.5	11.8	9.9
Lower Sunnyside—2	UT	80	2.3	38.6	56.2	2.9	5.6	78.6	1.5	0.4	11.0	9.4
Pittsburgh	PA	90	1.7	38.8	53.9	5.6	5.4	78.0	1.6	1.3	8.2	7.2

Continued

Table 21.2 Coal Seam Characteristics—cont'd

Rank and Seam	State	SHT °C	Proximate Analysis (wt pct)				Ultimate Analysis (wt pct)					
			Moisture	Volatile Matter	Fixed Carbon	Ash	Hydrogen (H)	Carbon (C)	Nitrogen (N)	Sulfur (S)	Oxygen (O)	Dry Ash-Free Oxygen
mvb Coal:												
Coal Basin—1	CO	120	0.7	22.0	70.4	6.9	4.8	82.6	1.8	0.7	3.3	2.9
Coal Basin—2	CO	120	0.9	21.6	69.5	8.0	4.8	81.0	1.5	0.5	4.2	3.7
lvb Coal:												
Blue Creek	AL	135	1.1	19.2	72.7	7.1	4.5	82.0	1.7	0.5	4.3	3.6
Mary Lee	AL	135	1.3	20.2	71.3	7.3	4.7	82.2	1.7	0.6	3.5	2.6
Pocahontas 3-1	VA	135	0.9	18.7	73.2	7.3	4.2	82.6	1.3	1.7	2.9	2.3
Pocahontas 3-2	VA	130	0.6	17.6	67.5	14.3	3.9	77.0	1.1	1.1	2.6	2.5
Anthracite (an):												
Anthracite	PA	140	1.7	4.7	84.7	8.9	2.1	83.9	1.1	0.6	3.5	2.3

SHT, minimum self-heating temperature.
Adapted from Smith A, Lazzara CP. Spontaneous combustion studies of US coal, USBM R.I. 9079 1987:28.

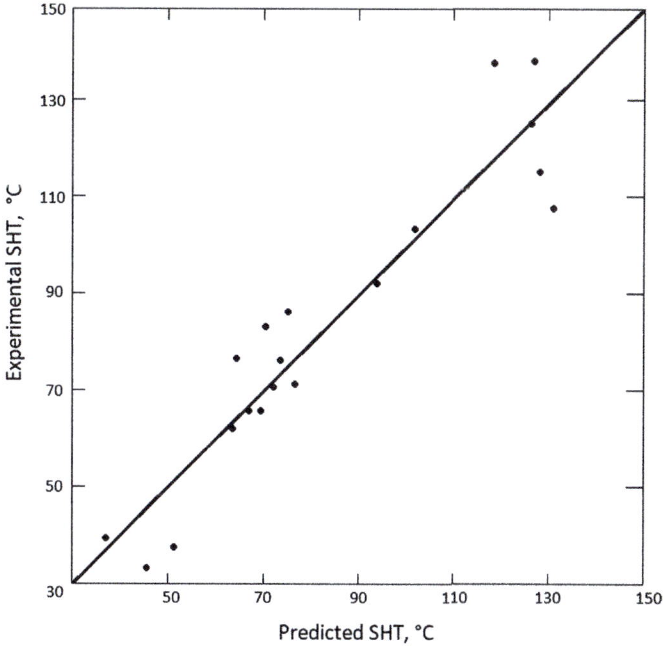

Figure 21.3 Predicted versus experimental minimum self-heating temperatures (SHTs) of bituminous coals.

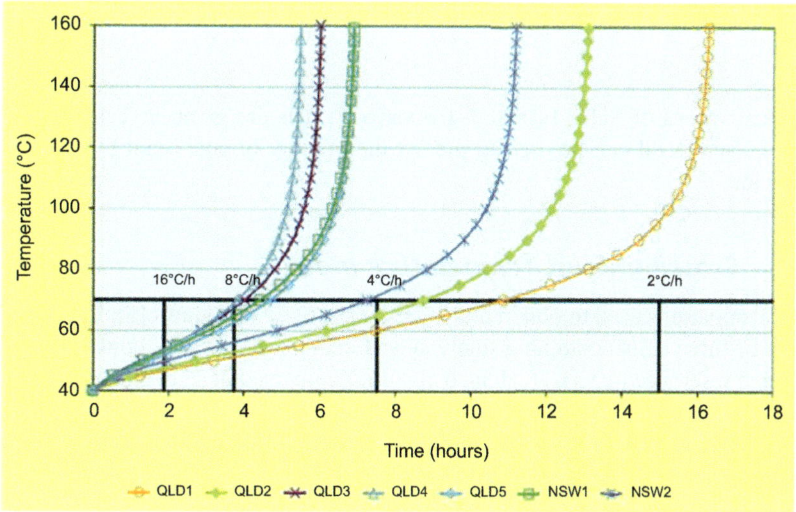

Figure 21.4 Temperature rise Vs oxidation time for Australian coals. [3]

In general, $\frac{dT}{dt}$ decreases with the rank of coal and has a corresponding higher value of SHT. E (K cal/mol) and A $\left(\frac{\text{Kelvin}}{\text{s}}\right)$ also appear to increase with the rank of coal.

Table 21.3 Coal in Eq. (21.2)

Rank of Coal	SHT (°C)	E (K Cal/mol)	A (K/S)
Lignite:			
Beulah-Zap	60	16.8	5.1×10^6
Hvc			
No. 6 Illinois	70	14.5	9.2×10^5
F Seam	45	12.6	3.1×10^5
hvA			
Clarion, Ohio	75	20.4	3.4×10^8
Lower Sunnyside	85	21.1	5.9×10^8
Pittsburgh	90	21.1	4.4×10^8
mvb			
Coal Basin—1	120	–	–
lvb			
Pocahontas#3	110	20.8	1.1×10^7
Mary Lee	135	20.2	1.4×10^7
Blue Creek	135	22.6	4.6×10^8
Anthracite	>140	–	–

SHT, self-heating temperatures.
Modified from Smith A, Lazzara CP. Spontaneous combustion studies of US coal, USBM R.I. 9079 1987:28.

Typical values of SHT, E, and A for various coals are given in Table 21.3. R70 values, E, and A all can be used to predict the liability of coal seam to spontaneous combustion.

21.1.2 Crossing-Point Temperature Index

A typical apparatus used to obtain this index is described by Gouws [5]. The apparatus consists of three cells containing finely sieved coal and three cells containing a thermally inert reference material (calcined alumina) immersed in an oil bath that is heated at a constant rate of 1°C/minute. Oxygen is supplied to the coal cells by means of an air compressor at the rate of 400 cc/min of air. The temperatures of the coal and the inert reference material are measured at 15 s intervals.

Initially, the reference material tends to heat up faster than the coal, which loses heat due to the evaporation of moisture and desorption of gases. At higher temperatures the coal heats up faster than the reference material and a "cross-over temperature" is reached. A typical crossing-point temperature plot for a coal sample is shown in Fig. 21.5. In general, higher-ranked coals have a higher crossing-point temperature.

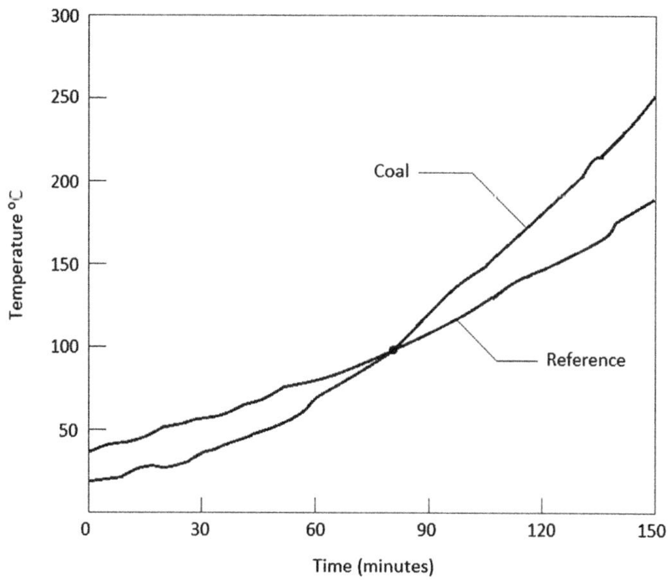

Figure 21.5 Typical crossing-point temperature graph.

21.1.3 Oxygen Absorption Index

This method of obtaining an index of spontaneous combustion propensity is used by United States, Chinese, Indian, and Russian researchers. The basic principle involved is to measure the amount of oxygen absorbed by a known quantity of powdered coal in a closed vessel. The technique is summarized in the following description from Karmakar [6].

About 40 g of fine coal powder (1−3 mm) was taken in a glass vessel of about 60 cc volume and spread in a thin layer over a glass chip bed. The bottom end of the vessel is dipped into sodium sulphite (Na_2SO_3) solution to facilitate equalization of air pressure inside and outside the vessel. Allowing a time period of 24 h for absorption of oxygen by the coal powder from the air within the vessel, the gas over the coal is drawn in a gas sample collector for analyzing the concentration of CO_2, CH_4, and O_2. The oxygen absorbed by the coal is expressed in cc of oxygen absorbed per hour per gram of coal (cc/h/g). It is the oxygen absorption index (OAI) index which is a measure of susceptibility to spontaneous combustion. The OAI index is given by the expression:

$$\mathrm{OAI} = \frac{-\mathrm{V}(\mathrm{B}-P)}{\mathrm{Wt}\;760}\mathrm{Ln}\left[\frac{(1-\mathrm{C_o})\,\mathrm{C_R}}{\mathrm{C_o}\,(1-C_R)}\right] \tag{21.4}$$

where: V = Volume of air enclosed in the vessel (cc); B = Barometric pressure (mm Hg), (760 mm of Hg converts to 1 standard atmosphere); P = Saturation vapor pressure at room temperature (mm Hg); W = Weight of coal sample (g); t = Time of absorption (hr); C_o = Concentration of oxygen in fresh air, it is taken to be 0.2093; and

C_R = Residual concentration of oxygen in the air inside the glass vessel after time, t; it is given by:

$$C_R = \frac{O_2}{1 - CO_2 - CH_4} \text{ as a fraction.}$$

O_2, CO_2, and CH_4 represent the concentration of the respective gases after 24 h absorption. The higher the value of OAI for a coal, the more it is liable to spontaneous combustion.

21.2 Detection of Spontaneous Combustion

Monitoring of mine atmosphere by either continuous monitors or by handheld instruments can help detect spontaneous heating in its early stages and allow preventative actions to be taken. Monitoring systems and instruments are already discussed in Chapter 19. Besides monitoring air quality, there are some physical indicators of spontaneous combustion.

21.2.1 Physical Indicators of Spontaneous Combustion

They can be classified as follows:

Sweating
Haze
Smoke
Heat, and
Smell.

Sweating or condensation of water on cooler surfaces is the earliest indication of a hot spot in the mines. The next stage may be the appearance of haze, which is condensation of volatile gas given out by coal in air. As the temperature rises, smoke may become visible. Smoke is mostly ultrafine dust particles (less than 1 mm) or soot. The coal temperature has to be above 300°C before smoke can be seen.

As heating produces smoke, it also creates an acrid smell. Expert nose can detect this characteristic smell at parts in billion level. But it is a very individualistic criterion. The smell is very similar to the smell given out by brick kilns.

In many cases, the smell is absent and spontaneous combustion is detected by heat. A pillar with glowing fire can testify that there is a fire. Some infrared sensors can detect incipient heating before the flame appears. Their use is strongly advocated.

21.2.2 Gas Analysis for Detection of Spontaneous Combustion

As coal heats many gases are given out in sequence. The following is an approximate chronological list:

1. Carbon dioxide: may be difficult to draw a conclusion because mine air has CO_2.
2. Carbon monoxide: this is accompanied by CH_4, H_2, C_2H_4, and C_2H_6.
3. Pyrolysis products of coal: higher hydrocarbons as propane and butane.

Many indicators based on the mine air analysis have been developed over a long period from 1910 to 1960. Most used indices are presented here [7].

21.2.2.1 CO/CO$_2$ Ratio

It was developed in the United Kingdom around 1910. It is defined as follows:

$$\frac{CO}{CO_2}\text{ratio} = \left[\frac{CO_f - CO_i}{CO_2f - CO_2i}\right] \tag{21.5}$$

where: i and f refer to initial and final stages, respectively.

The ratio directly indicates the temperature of coal. A value of 0.02 is considered normal but a value between 0.15 and 0.35 may indicate a coal temperature of 150°C, which exceeds the SHT for most coal as discussed earlier.

21.2.2.2 Graham Ratio

It was developed in the United Kingdom around 1921. It is defined as follows:

$$\text{G. Index (CO Index)} = \frac{100(CO)}{\Delta O_2} \tag{21.6}$$

where: CO is the concentration of CO in percent and ΔO_2 is the oxygen deficiency in percent.

An example illustrates the calculation of Graham Index:

Assume: The gas analysis is as follows:

CO:	100 ppm (assume the base level of CO in mine is zero) (0.01%) (1% = 10,000 ppm)
CO_2:	15.4%
N_2:	70.4%
O_2:	15.2%

Calculate Graham Index.

First calculate the oxygen equivalent of N_2 present. In fresh air $O_2/N_2 = 0.265$. Hence:

Equivalent $O_2 = 0.265 \times 70.4 = 18.66\%$
O_2 deficiency = 18.66−15.20 = 3.46%

Hence:

$$\text{Graham Index} = \frac{100\ (0.01)}{3.46} = 0.289$$

Based on experience, the following conclusions are made:

Less than 0.4	Normal.
0.4–1.00	Look for spontaneous combustion.
>1.00	Heating in progress.
>2.00	Serious heating.
>3.00	Fire; coal is burning.

It is by far the most reliable and most used index for the detection of spontaneous combustion in mines.

21.2.2.3 Young's Ratio

Again the Young's Ratio was developed in the United Kingdom around 1924. It is defined as follows:

$$\text{Young's Ratio} = \frac{CO_2}{\Delta O_2} \tag{21.7}$$

It is not a very reliable index because the mine always naturally produces some CO_2 and it can be dissolved in mine water.

21.2.2.4 Jones–Trickett Ratio

This index was also developed in the United Kingdom around 1955. It predicts exactly what is burning in the mine. It is defined as follows:

$$\text{Jones–Trickett Ratio} = \frac{CO_2 + 0.75\ CO - 0.25\ H_2}{\Delta O_2} \tag{21.8}$$

Normally accepted interpretation of Jones–Trickett ratio is as follows:

< 0.4	Normal.
<0.5	Methane is on fire.
<1.0	Coal is on fire.
<1.5	Timber on fire.
>1.6	Not possible (check air analysis).

It is generally agreed that all these indices work well when measurements are made in a flowing air current and not in sealed areas. There is lower limit for oxygen deficiency (about 0.3%) for useful interpretation. A trend analysis of these indices often gives a better understanding of the state of a fire in the mine.

21.3 Mine Design for Coal Seams Liable to Spontaneous Combustion

Underground low-rank coal mines of Northern Illinois, Colorado, and Utah were initially designed in a manner similar to existing successful mine designs of eastern United States (Pennsylvania, West Virginia, Ohio, and Virginia) that mined higher-rank coal seams not liable to spontaneous combustion. This created many mine fires. Several design changes are necessary to mine these coals safely and minimizing the risks of mine fire. The subject will be discussed under the following headings:

Mine Development
Secondary Extraction
Mine Ventilation and Methane Control
Inerting the Gob With Progressive Sealing

21.3.1 Mine Development

In all underground mines, the property is first developed by driving a set of headings as "mains" and later submains are driven to create longwall panels. Fig. 21.6 shows a typical mine layout.

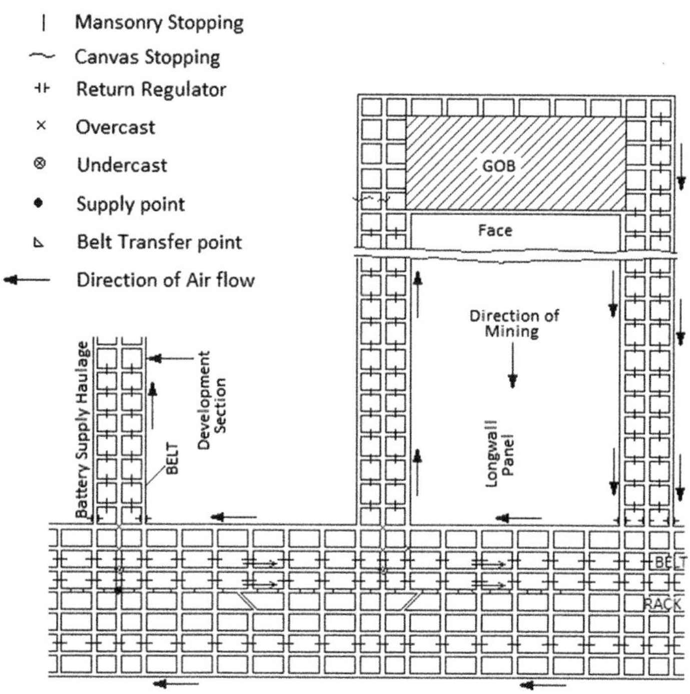

Figure 21.6 Typical mine layout.

The intake airways are separated from return airways by "stoppings." It is preferable to mine the full thickness of the coal seam leaving no coal in the roof or floor to minimize the risk of spontaneous combustion.

21.3.1.1 Roadways and Pillar Size

As far as possible (consistent with ground support requirements) the roadways should be reasonably wide and pillars should be large with few crosscuts. Most heating in pillars is because it is too small and air can leak through it. Small airways require higher pressure to supply the mine ventilation air and exacerbate air leakage through the pillar. Crosscuts at regular intervals are needed but they should be preferably 300 feet apart. The stoppings in the crosscut used to separate intake air from return should be substantial and leakproof.

21.3.1.2 Direction of Mains

These main roadways are usually closest to the mine fan and are exposed to higher pressure differential. Besides making the pillars large, it is desirable that they are driven at 45° to the "face" cleat to minimize spalling and fracturing. Loose coals in the roadways are potential sources of spontaneous combustion.

21.3.1.3 Roof Supports in the Airway

All development entry roofs are reenforced to make them strong. Timber or concrete blocks for roof support should be avoided because they increase the pressure differential. The preferred roof support is a long roofbolt.

21.3.1.4 Premining Infusion of Coal With Silicate Gel to Reduce Permeability

Considering the fact that most fires owing to spontaneous combustion start in coal pillars, it would be desirable to infuse the coal prior to mining with a silicate gel that makes it impermeable. In a typical three-heading development, the central entry is horizontally drilled to 500 feet on weekends. Two other horizontal boreholes are dilled at an angle to cover the headings on either side. A mixture of sodium silicate gel and sodium bicarbonate (as discussed in methane control section earlier) is pumped into the coal to be mined. Typically the amount of gel is 1%−1.5% of the volume of coal 500 feet ahead of the injection point. The technique was very helpful in preventing the reoccurrence of spontaneous heatings in mines overseas.

21.3.2 Secondary Extraction

There are many ways for secondary extraction of coal but two most commonly used techniques are as follows:

1. Pillar extraction by continuous miners.
2. Retreat longwall mining.

In United States, slightly more than 50% of underground coal mining is done by the retreat longwall method. Main benefits of longwall mining are as follows:

- Improved safety and reduced injury rates because of improved longwall-to-development coal ratios and fewer longwall moves.
- Improved recovery of in situ coal.
- Improved productivity and reduced cost per ton.

21.3.2.1 Pillar Extraction by Continuous Miners

This technique is a poor choice to control heatings. There are many variations of this technique but, generally speaking, the coal pillars are further subdivided by the continuous miners, using "split and lift" technique. A lot of coal is left in the gob and small remnant pillars are exposed to increased abutment pressures resulting in pillar failures. If heating occurs, it is difficult to seal the area. Most often, the entire panel has to be sealed.

21.3.2.2 Longwall Mining

Longwall blocks of coal are commonly developed to have a width of 700–1500 feet and a length of 10,000–15,000 feet. To get enough air to ventilate, usually a three-entry development is used in United States. Chain pillars are as large as possible with minimum crosscuts. This is by far the safest method of mining coal especially when it is prone to spontaneous combustion. Longwall mining of low-rank coal, liable to spontaneous heating require that no bleeders/bleeder shafts are used; the newly created gob is progressively sealed and in most cases the gob is inerted with nitrogen. These subjects will be discussed next. In summary, general guidelines for safe secondary recovery are as follows:

1. Minimize coal lost in the gob.
2. Maximize consolidation of gob areas.
3. Maintain a steady mining rate.
4. Do not use bleeder ventilation fan.
5. Design the extraction panels for quick sealing.
6. Make sure the gob areas are monitored and kept inert, that is, oxygen at less than 4%, preferably 2%.

21.3.3 Mine Ventilation and Methane Control on Longwall Faces

Earlier discussed mine ventilation and methane control techniques need modifications if the coal seam is liable to spontaneous combustion. Main changes are listed below:

1. The district (number of panels to be mined contiguously) must be limited to 3–4 longwall panels so it can be completely sealed in 3–4 years.
2. The longwall panels should be modest in size (1000 ft × 10,000 ft) so it can be mined in less than a year.

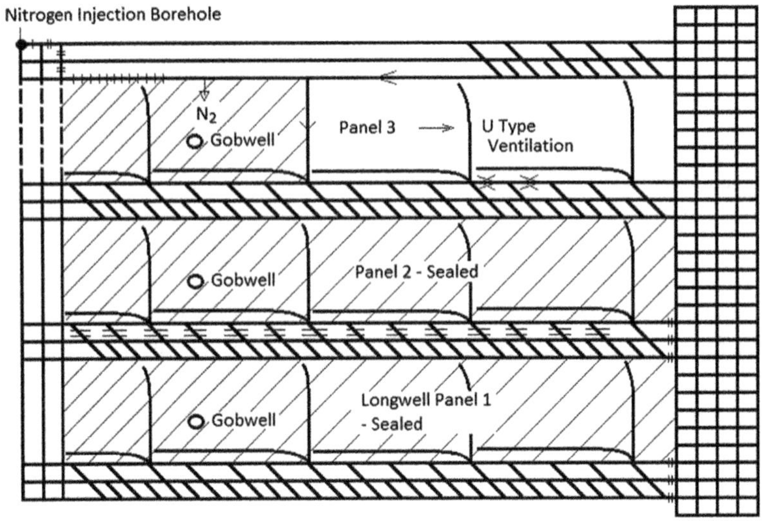

Figure 21.7 Typical series of longwall panels.

3. Bleeder ventilation should not be used.
4. The active gob must be progressively sealed.
5. Inerting gases should be injected into the active gob.
6. Gob wells, if needed, must operate naturally or with very low negative pressure (no more than 10 inches of water gauge).

Fig. 21.7 shows a typical series of longwall panels that are progressively sealed and isolated from each other. When all the panels are mined out, the district is sealed with barrier seals and isolated from active mine workings.

21.3.4 Inertization

The surest way to prevent spontaneous combustion of fallen coal in longwall gobs is to seal the area and maintain a low oxygen content of 4% or lower. The inertization can be as follows:

- Self-inertization or
- Induced inertization.

21.3.4.1 Self-Inertization

In room and pillar mining, pillar extraction can be so designed that mined out area is submerged in water. This provides excellent inertization but is possible only in highly inclined coal seams.

Most commonly, self-inertization is provided by the methane gas released from coal seams in the gas emission space. Again, this is possible only in rare cases (such as mines in Buchanan County, Virginia) where the gob produced 15 to 30 million cubic feet per day of methane and gob air contains 80% or higher methane-reducing oxygen content to less than 4%. With such high emission of methane, gob methane drainage is almost essential. The gob wells must be monitored to make sure that excessive air is not produced.

21.3.4.2 Induced Inertization

Most mildly gassy and moderately gassy coal seams do not produce enough methane to inert the gob area. Induced inertization becomes necessary in these mines. Nitrogen is most commonly used for this purpose. Carbon dioxide can also be used but it is expensive and requires expert handling when in liquid form. Nitrogen is cheaper and much safer in underground mines.

Compressed or liquid nitrogen was purchased and service companies would inject them in mines from surface. Lately, a much cheaper source of nitrogen has been developed. It is called "pressure swing adsorption" or PSA technique. Atmospheric air is compressed to about 100 psi and made to pass through especially designed molecular sieves that can separate oxygen from nitrogen because of a slight difference in the size of their molecules. Successive filtration can yield 98% nitrogen at practically the same pressure as the inlet pressure. The low pressure nitrogen is conducted in PVC pipelines and delivered 1000 to 1500 feet behind the active longwall faces. Vertical gob wells on older panels can also be used to monitor the gob gas composition and inject nitrogen, if necessary.

Problems

1. Calculate the SHT for three different ranking coals using the data in Table 21.1 and comment on the effect of rank on SHT.
2. Calculated the Graham Index and comment on the status of spontaneous combustion using the air analysis data given below:

   $CO = 500$ ppm
   $CO_2 = 3\%$
   $O_2 = 17.5\%$
   $N_2(+AR) = 79.5\%$
   $H_2 = 200$ ppm

   Also calculate Jones–Trickett ratio and determine what is burning.

References

[1] NIOSH. Analysis of mine fires for all US underground and surface coal mining categories (1990-1999). IC 9470. 2004. p. 36.
[2] Smith A, Lazzara CP. Spontaneous combustion studies of US coal, USBM R.I. 9079. 1987. p. 28.

[3] Beamish BB and Theiler J. Contrast in Self-Heating Rate Behavior for Coals of Similar Rank, 15th Coal Operators' Conference, University of Wollongong, The Australasian Institute of Mining and Metallurgy and Mine Managers Association of Australia, 2015, 300–304.

[4] Beamish BB, Theiler J. Improved understanding of the role of pyrite in coal spontaneous combustion. In: Proceedings of the 16th US Mine Ventilation Symposium; 2017. p. 18. 19–26.

[5] Gouws, Wade. The self-heating liability of coal: predictions based on simple indices. Mining Science and Technology 1989;9(1):75–9.

[6] Karmakar NC. Methods of estimation of spontaneous combustion of coal. Journal of Mines, Metals and Fuels 1989;37(1):21–3.

[7] Cliff D, Rowlands D, Sleeman J. Spontaneous combustion in Australian underground coal mines. SIMTARS; 1986. p. 165.

Prevention of Frictional Ignitions

22

Chapter Outline

Frictional ignition at the coal face is defined as rapid oxidation of either (1) a mixture of methane and air or (2) a mixture of methane, coal dust, and air with a visible flame that can last from several seconds to several minutes. Field studies done by the Mine Safety and Health Administration indicate that in recent years, on the average, 71 ignitions occur in US coal mines per year as shown in Table 22.1. Mine fires caused by conveyor belt friction will be discussed separately and later in the chapter.

It is interesting to note that most of the ignitions occurred in gassy mines of Alabama, Virginia, Pennsylvania, Utah, and West Virginia. Since the early 1970s, there has been no fatality in US mines because of frictional ignitions but injuries have occurred. However, there are known cases overseas when a coal mine explosion was caused by a frictional ignition when multiple fatalities were caused [1]. Besides, production losses for at least a shift for investigation of causes of ignition can be very expensive. Thus, frictional ignitions have a potential to create a mine disaster as well as production losses, and every effort should be made to minimize them.

Necessary conditions for a frictional ignition to occur require (1) an ignitable medium, usually a mixture of methane and air, (2) a minimum source of energy, generally created by coal cutting bit striking coal, sandstone, or pyrite, and (3) minimum temperature to ignite the mixture.

Two common denominators in most frictional ignitions are the presence of excessive methane and the presence of sandstone in the roof, floor, or as a middleman.

Advanced Mine Ventilation. https://doi.org/10.1016/B978-0-08-100457-9.00022-5

Table 22.1 Frictional Ignitions in US Coal Mines (Anthracite Mines Not Included)

	States													
Year	**AL**	**CO**	**IL**	**IN**	**KY**	**NM**	**OH**	**OK**	**PA**	**TN**	**UT**	**VA**	**WV**	**Total**
1983	23	1			2				11		13	5	4	59
1984	30	3			5				8	2	5	2	9	64
1985	56				3				10		1	5	3	78
1986	25				2				13		4	15	9	69
1987	15	2	3		5		1		8		24	10	16	83
1988	20	2	1		3				13		12	19	5	75
1989	23				6				18		16	13	4	80
1990	33		2	2	16				10		4	17	6	90
1991	62		2	3	2		2		8		3	29	5	116
1992	49				3				8		2	36	5	103
1993	52	2							7		3	9	2	75
1994	43			6	4	1	1		4		3	14	1	77
1995	29	1				1	1		5		1	17	2	57
1996	48								8			12		68
1997	38	2	6				1		11		1	17	3	79
1998	38				1		1		5		5	2	3	55
1999	45	2	1		1		6		13		3	4	1	76
2000	10	1					5		15		8	1	4	44
2001	16				2		2		13			5	10	50
2002	23				2	2		2	13			3	17	60
2003	16	2			5	2	2		11		4	7	9	58
2004	7	3	4			1			22		1	3	9	50
Total	701	21	19	11	62	7	22	2	234	2	113	245	127	1566

If the gas content of coal seams exceeds 100 ft^3/t and mining conditions encounter sandstone, special precautions should be taken to prevent frictional ignitions. Preventative techniques have been discussed in the literature extensively, but the authors have mainly emphasized bit design and cutting machine parameters. In view of the progress made in coal seam degasification, current preventative techniques can be classified, in order of importance, as follows:

1. Coal seam degasification to reduce the seam gas content to less than 100 ft^3/t.
2. Ventilation to prevent layering of methane.
3. Wet cutting/water-jet-assisted cutting.
4. Machine design to minimize ignitions.

Machine design parameters that influence frictional ignitions include (1) optimization of cutting speed, (2) optimum depth of cut, and (3) bit designs incorporating polycrystalline diamonds for both drilling roof bolt holes and coal cutting.

22.1 Coal Seam Degasification

This subject has been discussed in detail in Chapters 15, 16, and 17. In summary, coal seams containing more than 100 ft^3/t of gas can be broadly classified as (1) moderately gassy (gas contents, 100–300 ft^3/t) or (2) highly gassy (gas contents 300–700 ft^3/t) coal seams. Moderately gassy coal seams should be degassed before mining to reduce their gas content to less than 100 ft^3/t. Such seams are generally less than 1500 ft in depth and display high permeability. Fig. 22.1 shows a degasification scheme with in-mine horizontal boreholes.

Boreholes, 3–4 inches in diameter, are drilled parallel to and in advance of development headings to degas them. These boreholes are 1300 to 2000 ft in length. Similarly, cross-panel boreholes are drilled into the longwall panels to degas them. The length of these boreholes is typically 50–100 ft less than the width of the longwall panel. If these horizontal boreholes are drilled promptly and produced efficiently, nearly 50% of in situ gas in coal can be drained before mining. The outbye boreholes are drilled at 1000 ft intervals, but inbye boreholes should be spaced closer to expedite degasification. Detailed descriptions of the technique are available in the literature [2,3].

Highly gassy coal seams cannot be properly degassed with in-mine drilling alone for lack of adequate time to degas the coal. These coal seams must be drained several years ahead of mining with vertical, hydrofracked wells as shown in Fig. 22.2.

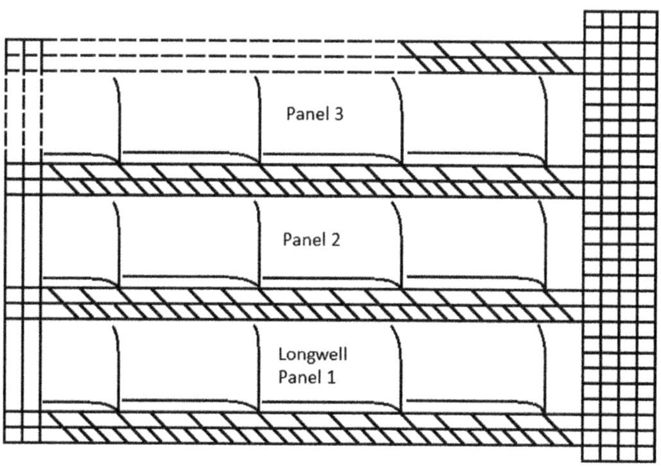

Figure 22.1 Degasification scheme for a moderately gassy coal seam.

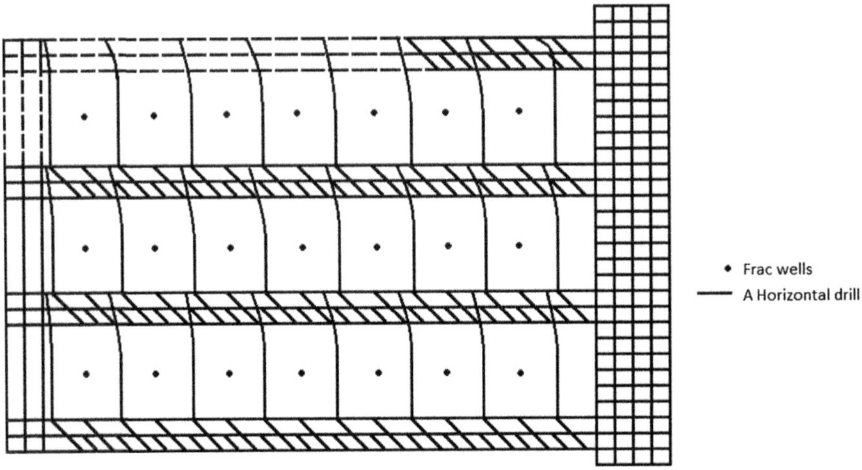

Figure 22.2 Degasification scheme for a highly gassy coal seam.

Frac wells are typically drilled at a 20-acre spacing, 5 to 10 years ahead of mining, and can drain nearly 50% of the gas contained in coal. Additional degasification is generally necessary to remove another 20 to 30% of the in situ gas contents. The secondary degasification is done with horizontal drilling across the longwall panel at 100–200 ft intervals. These boreholes can be extended beyond the longwall panel to intersect and degas the next set of development headings. Additional details of these techniques can be found in the literature [4] as well as Chapters 15 to 17 in this book.

22.2 Ventilation

Even after the coal seam is properly degassed, development headings and longwall faces must be properly ventilated to prevent dangerous accumulations of methane. Methane concentrations in mine air are restricted to 1% by law everywhere except in the bleeders where the methane concentration can go up to 2%. A minimum velocity of 60 ft/min and a minimum ventilation air quantity of 3000 CFM at the face in the development heading is also required by the Federal laws. In coal seams containing less than 100 ft^3/t of methane, the abovementioned requirements may be sufficient, but for gassier coal seams a different criterion must be used to calculate ventilation air quantities and prevent the layering of methane in roof areas. Methane layering in mine airways, especially near the working faces, can occur if there are high methane emissions and ventilation air velocity may not be high enough to create good turbulent mixing. These gas layers can be easily ignited by hot cutting tools resulting in frictional ignitions.

Gas layering in a mine airway is governed by methane emission rate, air velocity, and the width of the mine entry. The gradient of the airway has a significant impact on gas layering, but it can be ignored for most US coal mines because the coal seams are relatively flat. The gas layering number (GLN) is mathematically expressed as follows [5,6]:

$$\mathrm{GLN} = \frac{6\,\mathrm{V}}{(\mathrm{Q}/\mathrm{D})^{1/3}} \tag{22.1}$$

where Q is the emission rate of methane (L/s); V is the velocity of air (m/s); D is the width of mine airway (m).

In the FPS system, GLN $= \frac{0.0258\ \mathrm{V}}{(\mathrm{Q}/\mathrm{D})^{1/3}}$

Q is the emission rate of methane (CFM); V is the velocity of air (ft/s); D is the width of mine airway (feet).

A minimum value of five for GLN is considered necessary to prevent layering. The higher the value of GLN, the less likely it is that gas layering will occur. However, the formula provides only an empirical guideline, and it must be verified by physical measurements of methane in mine entries, and ventilation air quantities should be increased to avoid gas layering.

To illustrate this point further, two cases are discussed here.

22.2.1 Development Heading in a Highly Gassy Mine

Assuming a methane emission rate in a 20-ft-wide heading to be 50 CFM and using the abovementioned formula, a minimum velocity of 263 ft/min is needed. Assuming a seam height of 6 ft, the required ventilation air to prevent gas layering is 31,560 CFM. This quantity is almost 10 times the minimum quantity required by law at the face and illustrates the need for premining degasification and adequate ventilation when the gas contents of coal seams are high.

22.2.2 Longwall Face in a Highly Gassy Mine

The typical value of methane emissions from a longwall face in degassed but highly gassy coal seams is 500 CFM. Assuming a width of 10 ft and a height of 6 ft, the minimum air quantity needed to prevent layering is 42,782 CFM. Ventilation air quantities on longwall faces should be further increased to provide a safety factor of at least 1.5.

An innovation by the erstwhile US Bureau of Mines known as "shearer-clearer spray system" has also proved very successful in providing increased ventilation to minimize methane concentrations in areas difficult to ventilate and thereby minimize frictional ignitions [7]. In this system, a number of water sprays are so mounted on and around a longwall shearer as to create additional flow of air at the face and dilute methane concentrations (as well as respirable dust) to safe levels (Fig. 22.3).

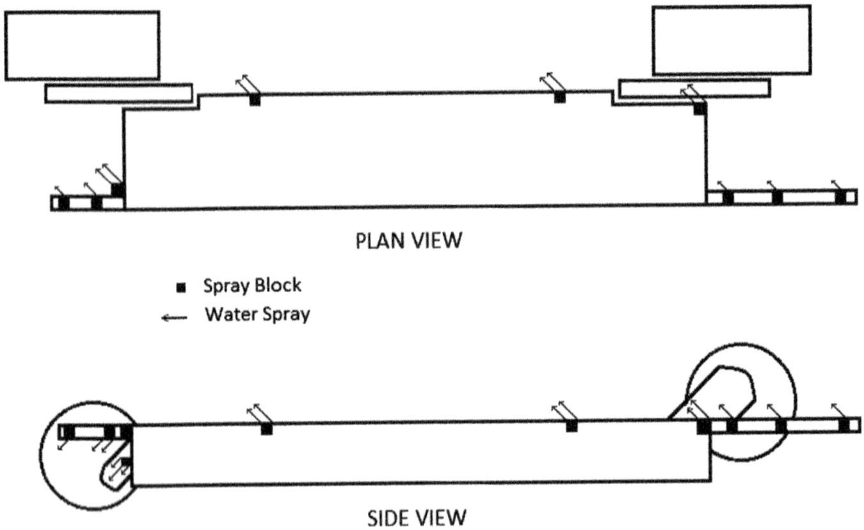

Figure 22.3 An arrangement of air-moving sprays.

22.3 Wet Cutting or Water-Jet-Assisted Cutting

Coal seam degasification and provision of adequate ventilation removes the medium that can be ignited. Drilling for degasification is usually done with water to remove the cuttings and drive the drill motor. This process indirectly infuses the coal seam with water to a degree, but these horizontal boreholes can be used for active water infusion to reduce the generation of fine coal dust and prevent its ignition.

Although water sprays have been used for a long time in conjunction with coal cutting to suppress respirable dust and cool the cutting bits, only wet cutting has proven very beneficial in reducing frictional ignitions [8,9]. For wet cutting (or a continuous miner/shearer to be called a wet-head machine), back flushing sprays are installed behind each cutting bit to reduce frictional ignitions. Most frictional ignitions are caused by hot material (sandstone or steel from bits) ejected from an area behind the cutting tool. A solid cone spray bit delivering 0.5 to 1 gpm of water at 50 to 120 psi at the rear of cutting bits proved very effective in preventing frictional ignitions [7]. Fig. 22.4 shows a typical arrangement for such a spray. Bit flushing rear sprays should create a spray zone of 30° as shown in Fig. 22.4.

Another innovation in wet cutting is water-jet-assisted cutting. Minnovation of the United Kingdom modified a longwall shearer to include a water pump that delivered high-pressure water (up to 10,000 psi). High-pressure water was fed to sprays located in front of the cutting bits. High-pressure water jets reduced the specific energy of cutting needed to cut coal (kwh/t) and thus reduce the chances of frictional ignition. Besides, it can make mining of a very hard coal seam a real possibility and reduce machine vibrations.

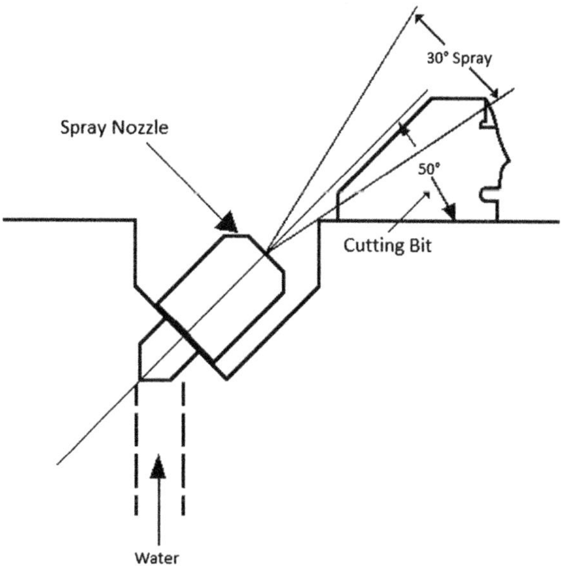

Figure 22.4 A typical antiignition back spray.

In the United States, an attempt was made by the author to modify a Joy continuous miner to include a high-pressure pump and assist coal cutting with high-pressure (up to 5000 psi) water. The design failed for the following two reasons: (1) considerable leakage through and frequent failure of seals and (2) blockage of pump and nozzles because of impurities in water supply. Considering the fact that most frictional ignitions take place on continuous miner faces, efforts to develop water-jet-assisted coal cutting for continuous miners should be encouraged.

Benefits of water-jet-assisted cutting in conjunction with wet cutting can be summarized as follows:

1. Minimized frictional ignitions,
2. Reduced respirable dust concentration,
3. Increased bit life,
4. Reduced machine vibration leading to increased equipment life, and
5. Improved product (raw coal) size ($-^1/_4$ inch size of coal is significantly reduced).

Fig. 22.5 shows a bit with a combination of a high-pressure jet in front of the bit to assist cutting and a rear, bit-flushing spray to suppress ignitions. The system combines the benefits of both wet cutting and water-jet-assisted cutting.

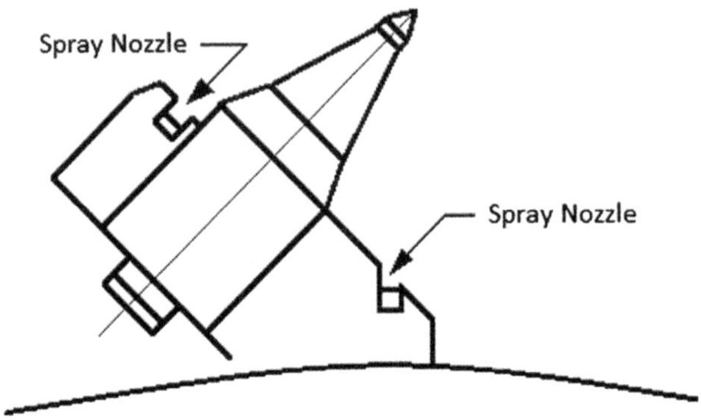

Figure 22.5 A typical arrangement for antiignition back spray and a water-jet cutting spray in front of the bit.

22.4 Machine Design Parameters

Machine design parameters that have an influence on the probability of a frictional ignition are as follows:

1. Speed of the bit,
2. Size of the bit or depth of cut,
3. Angle of attack, and
4. Material of construction for the bit tip and bit block.

22.4.1 Speed of the Bit

It is generally agreed that when the bit speed is reduced below 40 ft/s, there is a significant reduction in the probability of ignition [10]. However, such low speeds may lower the coal production rate and may require deeper cuts and larger motors.

22.4.2 Size of the Bit

A large bit with a depth of cut of at least 1.5 inch is desirable. Deeper cuts at lower speeds also tend to reduce respirable dust production. It would appear from above that there is an optimum for the combined parameters of speed of cutting and depth of cut to yield the maximum coal output per unit of energy consumed.

22.4.3 Angle of Attack

Fig. 22.6 shows the definition of angle of attack, θ_A, bit angle, θ_T, and angle of clearance, θ_C [7].

Field experiments confirmed laboratory findings that when θ_A is greater than 57°, the probability of ignition was substantially reduced [9]. Frictional ignition probability

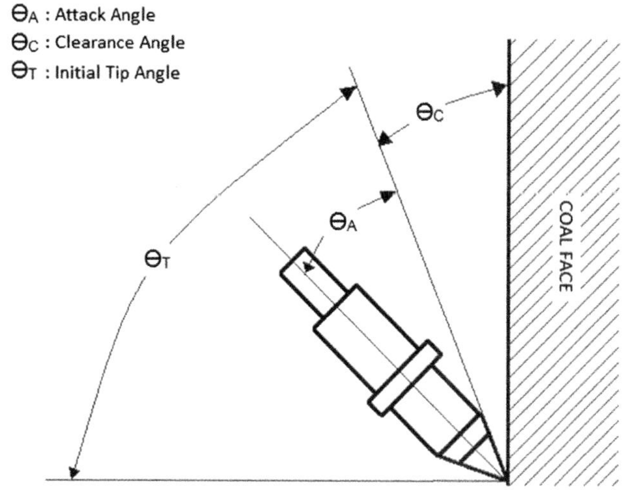

Figure 22.6 Geometry of a conical bit for cutting coal.

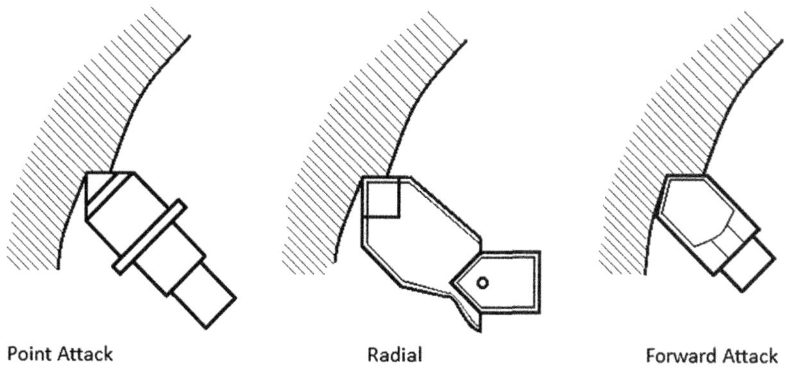

Figure 22.7 Typical coal cutting bits.

was also reduced by increasing the initial clearance angle, θ_C, and decreasing the internal bit angle, θ_T, but these two parameters keep changing as the bit wears. They merely postpone the time when a worn out bit becomes very liable to cause a frictional ignition.

Fig. 22.7 shows three main types of bits used by the coal industry [10]. Radial bits, with an angle of attack of almost 90°, are the best even if they require increased horsepower to cut coal. The larger size of the radial bits and bit blocks also reduce maintenance costs for repair of broken bits and bit blocks.

22.4.4 Material of Construction

As far as the body of the bit is concerned, 4140 steel is the most commonly used material. It can be made stronger by switching to stainless steel of proper grades to

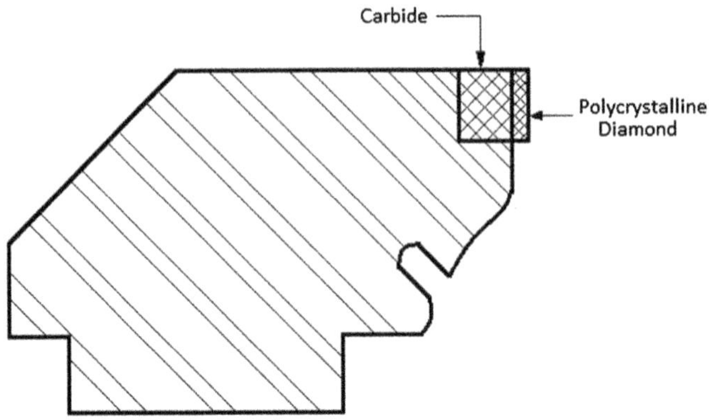

Figure 22.8 A suggested design for a polycrystalline carbide bit.

minimize broken bits, which are very liable to cause frictional ignitions. Bit tip is generally made of tungsten carbide. Because experiments show that the hot steel body ignites methane air mixture more easily than the carbide itself, larger carbide tip bits became popular. Even the grade of carbide was changed to optimize the design. Tungsten carbide containing 15% cobalt was strong but too soft and wore out quickly, but carbide containing 8% cobalt was wear resistant and brittle. Sandvik made some dual property bits that had an outer body of hard carbide and an inner body of tougher carbides, but it did not reduce the frictional ignitions. Other attempts to cover normal carbide with an outer layer of ceramic (e.g., Syalon) or cubic boron nitride were also not successful. Some ceramic material, in fact, increased the ignition frequency when cutting sandstone. The future improvements in bit design will probably result by

1. incorporating polycrystalline diamond tips or
2. directly depositing polycrystalline diamond on tungsten carbide tips

Polycrystalline diamonds have been used on rotary drill bits with excellent results. The life of a rotary carbide bit improved 100–200-fold when all carbide tips were covered with a very thin (1 mm) wafer of polycrystalline diamond. A suitable design for a continuous miner bit or a longwall shearer bit is yet to come, but a suggested design is shown in Fig. 22.8. Deposition of polycrystalline diamond directly on carbide tips is feasible, but the present cost is prohibitively high.

22.5 Summary and Conclusions

Frictional ignition frequency in US coal mines is on the decline, but it still is a cause of concern because it has a potential to cause a mine explosion leading to a large-scale mine fire. Prevention of frictional ignition is based on the following:

1. Coal seam degasification to reduce original gas content of the coal seam by 50–80%. Horizontal boreholes drilled for degasification can also be used for water infusions.
2. Using ventilation quantities that will not only dilute methane to statutory limits but also prevent gas layering near the roof or floor.

3. Incorporating wet cutting on all mining machines and eventually design machines that assist coal cutting with high pressure water.
4. Using larger cutting tools made of high-grade stainless steel to provide deep cutting without frequent breakdowns.
5. Using high attack angle bits, such as radial or forward attack bits. The machine should be originally designed to incorporate these bits. A retrofit is generally unsuccessful.
6. Bit tip material made of polycrystalline diamonds, which can be directly deposited on tungsten carbide or bonded to it as a thin wafer.

The suggested steps will not only make mining safer but also more economic by enhancing coal productivity and creating a source of extra revenue from coalbed methane. Loss of production due to mandatory suspension of mining subsequent to an ignition will also be minimized.

22.6 Frictional Ignitions Caused by Belt Conveyors

Belt conveyors are quickly replacing all old methods of coal transport. As mentioned in Chapter 21, they create nearly 17% of mine fires but they are usually contained quickly. An exception is the Wilberg Mine Fire [11], where 27 lives were lost due to a belt fire.

Francart [12] reports an average of 5.7 incidents of belt fire per 1000 mines over the period 1980 to 2005. A breakdown of belt conveyor ignition sources is shown in Table 22.2.

Belt friction accounts for nearly 50 percent of all belt fires.

The belt drives and conveyor transfer points are the first primary ignition source for many reasons. Electrical sources are likely to exist near the drives. Spillage of coal at such location is also very likely. Belt slippage can cause the belt fabric to heat up and ignite itself or the spilled coal underneath.

Table 22.2 Ignition Sources of Belt Fires

Ignition Source	Number of Fires	Percent
Friction at belt drive	11	18
Friction along belt	11	18
(Stuck) hot rollers	6	10
Electrical	8	13
Cutting and welding	5	8
Hydraulic fluid	2	3
Undetermined	18	30

22.6.1 Detection of Belt Fires

The old technique for detecting belt fire is known as point-type heat sensors (PTHSs). They are simply a bimetallic sensor that measures a rise in temperature above a base of 57°C (134.6° F). Because of the low ambient temperature in mines, the fire volume has to be large to activate the alarm. By then it may be difficult to contain it.

Current technology to detect a belt fire depends on carbon monoxide concentration in the air. It works in conjunction with PTHS but can detect a fire as far as 1500 ft away. Usually, an audible alarm sounds and it can activate a water spray system that extinguishes the fire.

22.6.2 Preventing Belt Fires

Preventing belt fires requires the following measures:

1. Use belts that can withstand heat without catching fire.
2. Clean all spilled coal promptly to avoid belt rubbing on coal.
3. Prevent slippage of belt by maintaining adequate tension.
4. Maintain the idler/rollers. Damaged rollers should be removed and replaced promptly.
5. Check for misalignment and tension periodically.

Current regulations require all the above precautionary measures. Slippage switches are installed at belt drives that can deenergize the drive motor if the roller is turning without belt movement. Most conveyor slip switches are designed to work on one of the three principles:

Magnetism,
Centrifugal force, or
Photoelectricity.

The most common device is a slip switch proximity sensor. They detect any interruption in the magnetic field by targets installed on a rotating shaft or a roller. Another device monitors the voltage and current for the main drive motor. Such data are communicated to the mine office and corrective actions are immediately taken.

Electrical sources of fires near a belt conveyor account for 13% of ignitions and fires. Cutting and welding on belt structure contributed 10% of reported belt fires. These fires can be minimized by proper training and maintenance of equipment. Safe work practices must be developed and workers should be annually refreshed.

References

[1] Elfstrom RH. Explosion in No. 26 colliery, Glacier Bay, Nova Scotia, on February 29, 1979. Canadian Department of Labour; 1980. p. 122.

[2] Thakur PC, Davis JG. How to plan for methane control in underground mines. Mining Engineering 1977:41–6.

[3] Thakur PC, Poundstone WN. Horizontal drilling technology for advance degasification. Mining Engineering 1980:676–80.
[4] Thakur PC. Methane drainage from gassy mines – a global review. In: Proceedings of the 6th International Mine Ventilation Congress; 1997. p. 415–22.
[5] National Coal Board. Ventilation in coal mines: a handbook for colliery ventilation engineers. 1979. p. 90–2.
[6] Leeming JR, Yates CP. Current British practice for methane ignition prevention in coal mine headings. In: DeSouza E, editor. Mine ventilation. Exton (PA): A. A. Balkema Publishers; 2002. p. 487–90.
[7] Courtney WG. Frictional ignitions with coal mining bits. US Bureau of Mines IC 9251; 1990. p. 25.
[8] Belle BK. An improved wet-head system: prevention of incendive ignitions and dust control. In: DeSouza E, editor. Mine ventilation. Exton (PA): A. A. Balkema Publishers; 2002. p. 479–86.
[9] NcNider TE, Grygiel TE, Hanes J. Reducing frictional ignitions and improving bit life through novel pick and drum design. In: Proceedings of the 3rd U.S. Mine Ventilation Symposium, SME; 1987. p. 119–25.
[10] Powell F. Design guidelines for picks. Colliery Guardian July 1991:139–41.
[11] Huntley D, et al. Report of investigation. USA: Wilberg mine fire, Mine Safety and Health Administration; 1986.
[12] Francart R. Reducing belt entry fires in underground coal mines. In: Mutmansky J, editor. The 11th U.S. Mine Ventilation Symposium. The Pennsylvania State University; 2006. p. 303–8.

Gas and Dust Explosions

23

Chapter Outline

The history of coal mining is replete with mine explosions resulting in great loss of lives. In the United States alone, at least 8000 lives have been lost to mine explosions [1]. Table 23.1 provides a list of some major disasters in the past (1936–93).

As discussed earlier, coal seams and methane in coal are syngenetic in origin. Methane is released when coal is mined. Methane air mixture becomes explosive at 5% at the lower level. If this mixture of gases meets an ignition source, such as an electric spark, or an open flame, it explodes creating a gas explosion. If not controlled, it can build on intensity and become a "detonation" that travels at a speed faster than sound. This creates a pressure front (shock wave) that kicks coal dust into air. If the

Advanced Mine Ventilation. https://doi.org/10.1016/B978-0-08-100457-9.00023-7

Table 23.1 Major Coal Mine Explosions in the World (Arranged Chronologically 1936–93)[a]

Country	Mine	Type of Explosion	Number of Deaths	Year
India	Poidih	Mixed	209	1936
India	Chinakuri 1 and 2	Mixed	176	1958
South Africa	Coalbrook	Gas	684	1960
China	Laobadong	Gas	680	1960
Germany	Luisenthal	Gas	299	1962
Japan	Mikawa	Gas	457	1963
Japan	Yamona	Gas	431	1965
India	Dhori	Coal dust	375	1965
Yugoslavia	Orasi	Gas	144	1965
USA	No. 9	Gas	78	1968
Zimbabwe (Rhodesia)	Wankie	Gas	426	1972
Yugoslavia	Kreka	Gas	169	1990
Ukraine	Ukraine	Gas	65	1992
South Africa	Middle Belt	Gas	58	1993

[a]There were explosions with even greater fatalities, but good reports are unavailable (such as an explosion in China in 1942 with 1549 fatalities).

coal dust concentration is in the explosive range, the coal dust cloud explodes creating a coal dust explosion. The latter is often more powerful and causes serious damage to mine structures as well as causing large number of fatalities. Hence, the subject will be discussed separately as gas explosions and coal dust explosions.

23.1 Gas Explosions

A few terms need a clear definition for better understanding of the subject.

23.1.1 Definitions [2]

Gas: It means a substance that can exist only in the gaseous state at standard pressure (1 atm) and temperature (32°F).

Vapor: It emanates from a substance that is liquid at standard temperature and pressure.

Flammability Limits: It is variously called flammable, explosive, or explosion limits. In some books, such limits refer to flammability limits of gas and vapors only. Dust is explosive, and it also has an upper and a lower limit. However, many books use flammability and explosive to mean the same thing.

Explosion: An explosion is the result, not the cause of rapid expansion of gases caused by a physical change or a chemical reaction.

Deflagration: It is a reaction that propagates in the unreacted material at a speed that is less than the speed of sound.

Detonation: It is an exothermic reaction that propagates in the unreacted material at a speed greater than the speed of sound.

23.1.2 Flammability Limits of Gas–Air Mixtures

Coward and Jones [3] produced a flammability curve for methane and air as shown in Fig. 23.1. The lower and upper limits of methane in air are 5% and 14.5%, respectively. It is famously known as "Coward's Diagram." It clearly shows that if methane concentration is less than 5% or oxygen concentration is less than 12%, the mixture cannot be flammable. Mixtures containing more than 14.5% methane can become explosive if mixed with air. The flammable/explosive mixture is shown as a triangle.

In a typical mine on fire, the mine atmosphere may contain methane, hydrogen, and carbon monoxide as combustible gases and nitrogen and carbon dioxide as inert gases. Zabetakis [4] developed a flammability diagram for such gas mixtures. All combustible gases were combined and called

$$\text{Effective Combustibles} = [CH_4\% + 1.25\ H_2\% + 0.4\ CO\%] \tag{23.1}$$

The factors 1.25 and 0.4 are the ratios of lower flammable limit for CH_4 to that of H_2 and CO, respectively. Likewise, all inerts were combined and called

$$\text{Effective Inerts} = [\text{Excess}\ N_2\% + 1.5\%CO_2] \tag{23.2}$$

where excess N_2 is defined as

$$\text{Excess}\ N_2\% = [100 - CH_4\% - \text{Air}\%] \tag{23.3}$$

Fig. 23.2 shows a flammability diagram for all these five gases.

A new term, composition point, R is defined as

$$R = \frac{CH_4\%}{CH_4\% + H_2\% + CO\%} \tag{23.4}$$

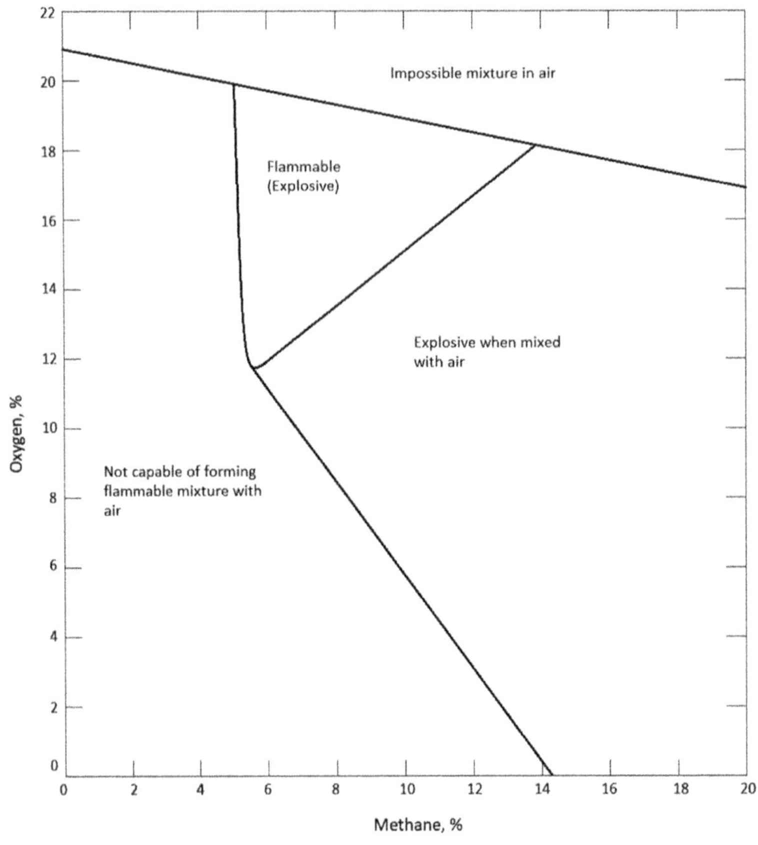

Figure 23.1 Flammability curve for methane–air mixture [3].

It defines the various regions in Fig. 23.2. Thus area A_1 has a value of R equal to 1.00 and has only methane in it. Area A_2 has an R value of 0.8 and so on. The gas analysis data can be used to find the composition point on the diagram and determine if the atmosphere is explosive.

An example:

The following gas analysis was received from a mine on fire.

Gas	Concentration (Volume %)
CO_2	12
O_2	8
CH_4	3 27%
CO	3
H_2	1
N_2	(100–27) 73

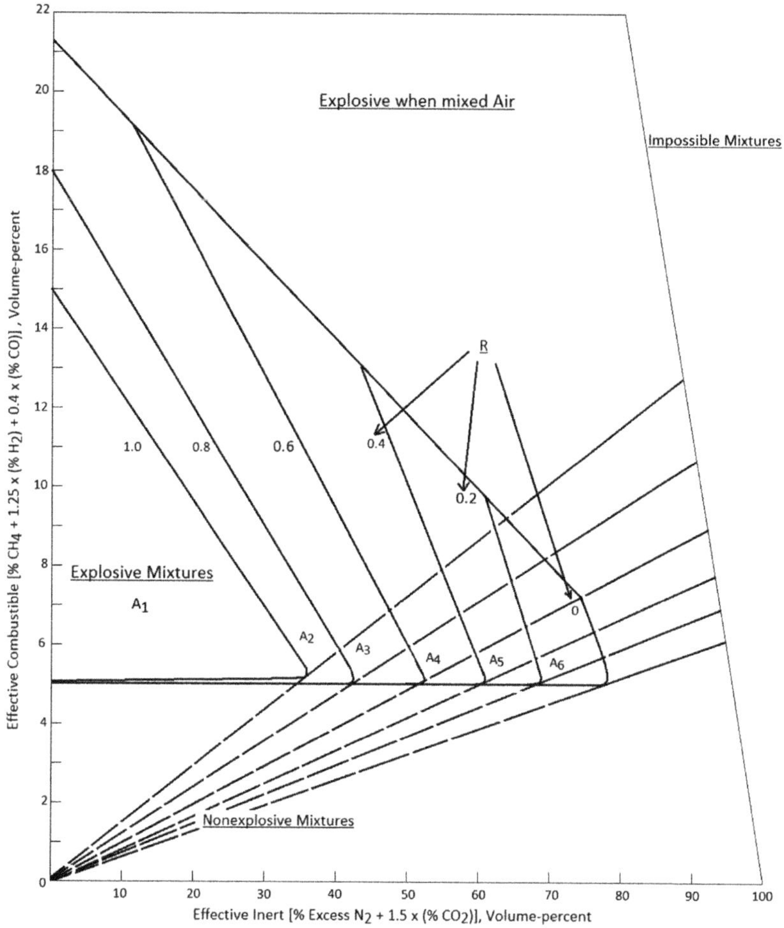

Figure 23.2 Flammability diagram for methane–hydrogen–carbon monoxide–nitrogen–carbon dioxide–air mixtures [4].

Find out if the mixture is flammable:

Excess $N_2\% = 73 - (8 \times 3.8) = 42.6\%$. (Normal air has a ratio of 3.8 for N_2/O_2.)

Effective inert $= (42.6 = 12 \times 1.5) = 60.6\%$.

Effective Combustible $= (3 + 1.25 \times 1 + 0.4 \times 3) = 5.45\%$.

$R = 3/5.45 = 0.55$.

Fig. 23.3 shows the composition point for $R = 0.55$. The point corresponding to 60.6% effective inert and 5.45% effective combustible lies outside the explosive mixture area, hence the mixture is nonexplosive.

Maximum allowable oxygen: The gas composition as discussed above can also be used to determine the maximum allowable oxygen percentage to prevent an explosion. The oxygen content must be higher than the maximum allowable oxygen for an explosion to occur. This is a shortcut to predict if the mixture is flammable.

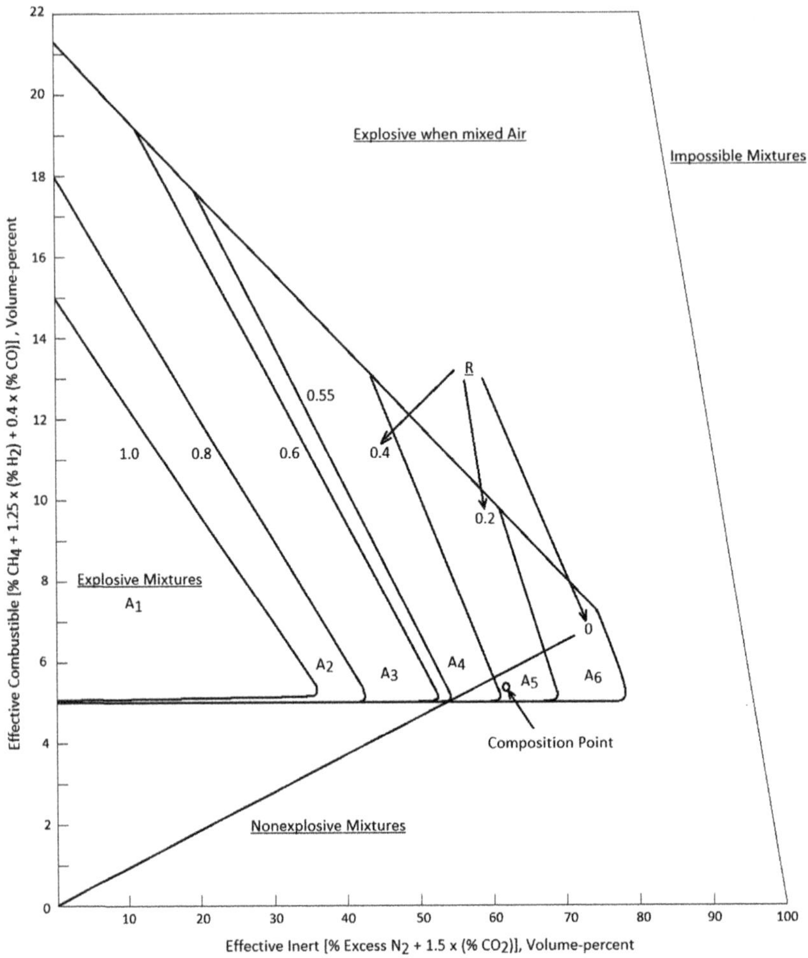

Figure 23.3 Flammability for methane–hydrogen–carbon monoxide–nitrogen–carbon dioxide–air mixtures (example).

$$\text{Max Oxygen Concentration} = 5 + 7\text{R} \tag{23.5}$$

In the above examples, R = 0.55; this allows the maximum O_2 concentration to be 8.85% but the actual oxygen is only 8%. Hence the mixture is nonexplosive/nonflammable.

Table 23.2 shows the upper and lower limits of flammability for gases generally found in coal mines. Minimum oxygen concentrations for combustion are also shown. All numbers are for standard pressure (1 atm) and temperature (60°F).

If the flammability limits of the components of a mixture are known, the limit of flammability for the mixture can be calculated using Le Chatelier's law [6]. The lower limit of flammability for the mixture, L, is given by Eq. (23.6).

Table 23.2 Lower and Upper Limits of Flammability for Mine Gases

	Limits on Flammability in Air		
Gas	**Lower Limit %**	**Upper Limit %**	**Minimum Oxygen for Combustion**
Acetylene	2.5	100	0
Butane	1.8	8.4	12
Carbon monoxide	12.5	74	5.5
Ethane	3.0	12.4	11
Ethylene	2.7	36	10
Hydrogen	4.0	75	5
Methane	5.0	15.0	12
Propane	2.1	9.5	11.5
Propylene	2.4	11	11.5

Adapted from Kuchta JM. Investigation of fire and explosion accidents in chemical, mining and fuel related industries – a manual, USBM Bulletin #686 1985:84.

$$L = \frac{100}{\frac{C_1}{L_1} + \frac{C_2}{L_2} + \frac{C_3}{L_3} + \cdots\cdots \frac{C_N}{L_N}} \tag{23.6}$$

where C_1, C_2 .... are the concentrations of each gas in the mixture, and L_1, L_2 .... are their corresponding lower limits of flammability.

The Le Chatelier law may be used to calculate only an approximate upper limit of flammability of a mixture of gases because of more complex reactions at higher concentrations.

The lower limit of flammability for most hydrocarbons decreases linearly with rising temperature, reaching a limit at about 1300°C. The upper limit of flammability tends to increase with temperature, but the predicted values are not reliable.

23.1.3 Ignition Requirements

Ignition is usually considered to be a combustion reaction with the evolution of heat and emission of light. Most combustible-oxidant systems are not capable of sufficient self-reaction at ambient temperature to produce ignition. Sources of ignition can be considered for their spatial and temporal characteristics.

Fig. 23.4 shows a number of sources with their temporal and spatial characteristics [5].

At one extreme (electrical sparks), the source is very small but very high in temperature and the heating rate. Here the main concern is energy density and the ignition energy. On the other extreme (heated vessels), the source is large, heating rate is

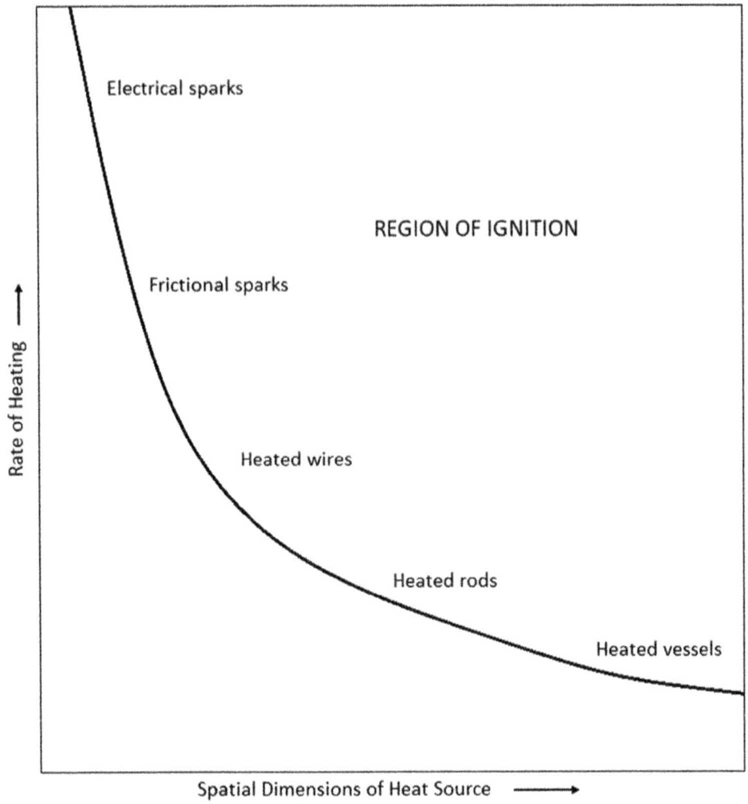

Figure 23.4 Temporal and spatial characterization of various ignition sources.

very low, and the temperature is also low. Here one should be more concerned about the ignition temperature.

23.1.3.1 Minimum Ignition Energy

Minimum ignition energy (MIE) for typical mine gases for both air and oxygen environments is shown in Table 23.3.

Generally speaking, all hydrocarbons have an MIE of 0.25 mJ. The MIE is lower by at least an order of magnitude for oxygen environments. Confining pressures greatly change MIE for oxygen environments.

23.1.3.2 Ignition Temperatures

Ignition temperature is a more important factor when a large vessel is the source of ignition. The minimum autoignition temperature (AIT) of gases is used for designing such vessels. Table 23.4 shows the AIT for typical mine gases in air and oxygen atmosphere.

Table 23.3 Minimum Ignition Energy for Mine Gases at 1 atm and 77°F

Combustibles	Minimum Energy, mJ	
	Air	**Oxygen**
Methane	0.30	0.003
Ethane	0.26	0.002
Propane	0.26	0.002
n-Butane	0.26	0.009
Acetylene	0.017	0.0002
Hydrogen	0.017	0.0012
Ethylene	0.07	0.01

23.1.4 Burning Velocities

Burning velocities of most mine gases are only a few cm/s, but it can increase to 40–50 cm/s at and around their stoichiometric concentrations. Assuming nonturbulent flames, the maximum flame speed can be roughly estimated at eight times the burning velocity, i.e., about 400 cm/s (13.1 ft/s). These numbers are well below the velocity of sound. Detonation velocities are supersonic and range from 2500 to 2700 m/s for typical mine gases. Only hydrogen has a detonating velocity that is close to 3000 m/s. In all cases, the maximum value is attained at a combustible concentration slightly higher than stoichiometric concentrations.

Table 23.4 Minimum Autoignition Temperature for Mine Gases

Combustible	Minimum AIT °C	
	Air	**Oxygen**
Methane	630	555
Ethane	515	505
Propane	450	NA
n-Butane	370	285
Acetylene	305	295
Ethylene	490	485
Carbon Monoxide	610	590
Hydrogen	520	400

23.1.5 Temperatures of Explosions

A general equation for a reliable estimate of flame temperatures for a combustible/explosive mixture can be obtained as follows. Assume the gas mixture to be 5% methane in air. For a lower limit mixture like this, the reaction for complete combustion can be written as

$$0.05\ CH_4 + 0.20\ O_2 + 0.75\ N_2 \rightarrow 0.05\ CO_2 + 0.10\ H_2O\ (gas) + 0.10\ O_2 + 0.75\ N_2 \tag{23.7}$$

The heat of combustion (reaction) ΔH is derived from standard heat of formation, H_f, at 298K.

$$\Delta H_{298} = \Sigma n \Delta H_{fn}(\text{products}) - \Sigma n \Delta H_f(\text{reactants})$$

Hence,

$$\Delta T(\text{products}) = \frac{\Delta H_{298}}{\Sigma n C_p(\text{products})} = \frac{\Delta H_{298}}{\overline{C}_p\ \text{mixture}} \tag{23.8}$$

where C_p is the specific heat at constant pressure.

$$\overline{C}_p, \text{mixture} = 0.05\ C_p(CO_2) + 0.10\ C_p(H_2O) + 0.10\ C_p(O_2) + 0.75\ C_p(N_2) \tag{23.9}$$

The upper limit, T_L, for flame temperature for a mixture being discussed can be calculated by assuming ΔH_{298} for most C–H–O–N combustibles to be 10–11 k cal per mole of mixture and $\overline{C}_p$ for their products at about 8×10^{-3} k cal/mol K.

The upper limit of T_L is about 1300°C, but significant variations from this value can occur with very high or very low reactivity [5].

The temperature rise for a constant volume combustion is, similarly, given by

$$\Delta T(\text{products}) = \frac{\Delta \Sigma_{298}}{\overline{C}_V(\text{mixture})} \tag{23.10}$$

where $\Delta\Sigma_{298}$ is constant volume heat release and $\overline{C}_V$ is average heat capacity for constant volume.

These temperature rises tend to be 20% higher than those for constant pressure as shown in Eq. (23.8).

23.1.6 Pressure Rise in Explosions

The pressure rise in an explosion of gas mixture is considered under two categories: (1) deflagration (subsonic propagation) and (2) detonation (supersonic propagation).

23.1.6.1 Deflagration

The explosion pressure (P_2) of deflagration is the highest in adiabatic combustion—at constant volume. In this case, Eq. (23.8) can be modified for pressure and is given below in Eq. (23.11).

$$\frac{P_2}{P_1} = \frac{n_2 T_2}{n_1 T_1} \tag{23.11}$$

Assuming $P_1 = 1$ atm at ambient temperature of 25°C (298K), the highest pressure developed is eight to nine times the atmospheric pressure [5]. n_1 and n_2 are the number of molecules in reactants and products of combustion, respectively.

23.1.6.2 Detonation

When a gas explosion is reflected by a confined space, it can build higher pressures causing supersonic propagation rates. Detonations are unique in that their combustion wave is coupled with a leading shock wave. The detonation pressure (P_2) for gaseous mixtures can be estimated by Eq. (23.12).

$$P_2 = 2P_V \tag{23.12}$$

where P_V is the maximum pressure P_2 in Eq. (23.12). Thus the maximum pressure in detonation of gases can be 18 times higher than the atmospheric pressure.

Fig. 23.5 shows a typical pressure history from a well-developed gaseous detonation [7].

These data are for a stoichiometric acetylene–air mixture. Three pressure levels are noticeable here.

1. "C," the initial spike or shock front which is of very short duration,
2. "D," the well-defined pressure spike corresponding to complete combustion, and
3. "E," the static pressure, which reflects the expansion of gases. This phase has the highest impulse ($\int$ Pdt). P_2 refers to this pressure.

Just like deflagration pressures, detonation pressures can also get reflected and increase to 2.5 times P_2 creating large-scale damages in the mines. Stoppings to seal off a mined out area must be designed to withstand this highest pressure (approximately 650 psi) in the detonation of gaseous mixtures.

23.2 Dust Explosions

Just like gases in coal mines, fine coal dust mixed with air can explode. In general, a deflagration or detonation of a methane–air mixture triggers a coal dust explosion. A blown-out explosive shot is another cause of coal dust explosion. A mixed explosion usually causes most fatalities and damage to mine structures.

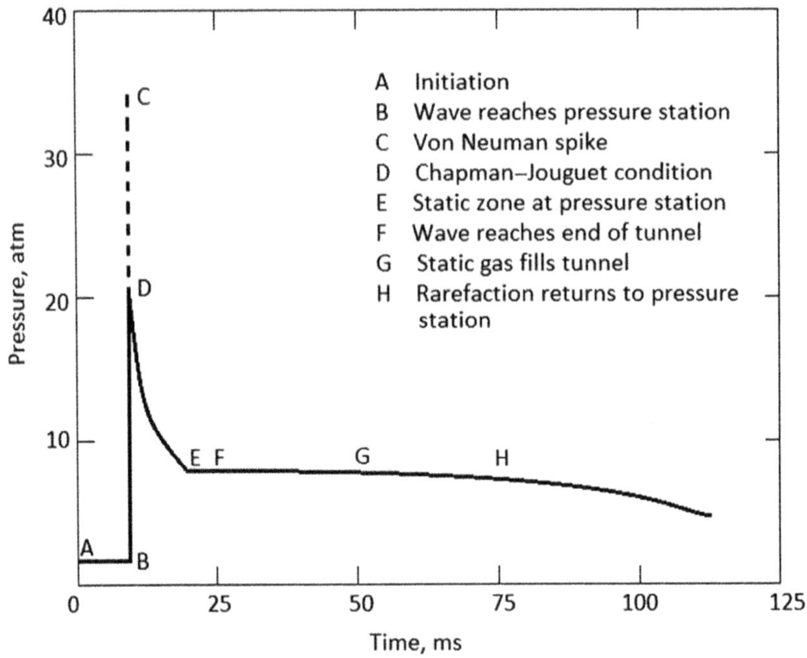

Figure 23.5 Pressure–time transient for detonation of a stoichiometric acetylene–air mixture in a 45.7 m (150 ft) long tunnel with initiation end closed. Instrument station is 18.3 m (60 ft) from initiator point.

23.2.1 Explosive Limits of Dust in Air

The range of explosive dust concentration in air is large and is largely dependent on particle size and the amount of volatile matter in coal. Tables 23.5 and 23.6 show pertinent data for Pittsburgh coal seam (HVB) and Pocahontas #3 coal seam (LVB) [8].

The minimum explosive concentration (MEC) decreases with particle size ranging from 130 to 85 g/m^3 for Pittsburgh seam that has a higher fraction of volatiles. The MEC is generally lower for a finer dust cloud as compared with a coarser dust cloud for any coal.

The best way to make the coal dust nonexplosive or inert is to mix a noncombustible dust, such as calcium carbonate, gypsum, or even shale dust, with it. The amount of noncombustible dust needed to inert coal dust increases with decreasing coal dust particle size reaching a value of 83%–87% when coal dust is 13–14 μm in size.

The upper limit of dust flammability is poorly defined—because it is difficult to carry out such experiments. In an experimental mine, the flame was quenched at a coal dust concentration of 5 kg/m^3 [9].

Table 23.5 Lower Explosive Limit for Pittsburgh Coal Seam

Characteristic	Sample Number					
	1	2	3	4	5	6
Average particle size, dw (mass basis, μm)	179	76	52	34	32	14
Percent—75 μm (200 mesh)	24	54	80	100	94	100
Minimum explosive conc. (g/m^3)	130	85	80	65	60	85
Maximum pressure (atm)	6.0	6.3	6.6	6.7	6.8	7.1
Amount of rock dust to inert (%)	53	68	74	79	83	87

23.2.2 Minimum Ignition Energy

Available MIE for combustible dust (coal) is not as precise as it is for mine gases. The MIE for dusts is generally 10–100 times higher than that for gas–air mixtures. It is also very dependent on particle size. Fig. 23.6 shows it for atomized aluminum [5].

MIEs for dust are usually the minimum at concentrations five to 10 times the MEC limit.

23.2.3 Ignition Temperature

Ignition temperature of coal and other dust is not affected by particle size. A big difference is made by the way coal dust comes in contact with a heat source. Thus, layered powder of coal can ignite at a much lower temperature (160°C approx.), whereas a cloud of coal dust will need 450–650°C to ignite. Table 23.7 shows some relevant data [5].

Table 23.6 Lower Explosive Limit for Pocahontas #3 Coal Seam

Characteristic	Sample Number					
	1	2	3	4	5	6
Average particle size, dw (mass basis, μm)	58	63	60	41	23	13
Percent—75 μm (200 mesh)	75	65	75	86	94	100
Minimum explosive conc. (g/m^3)	120	130	100	90	80	80
Maximum pressure (atm)	6.0	6.3	–	6.3	6.5	6.5
Amount of rock dust to inert (%)	60	64	76	78	82	83

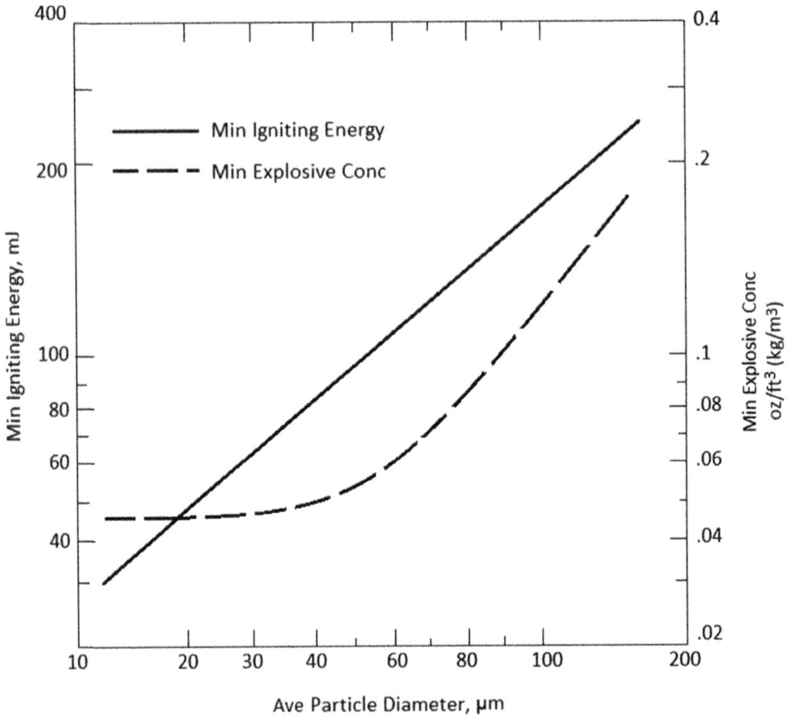

Figure 23.6 Variation of minimum ignition energy and minimum explosive concentration with particle size (atomized aluminum dust) at 25°C at 1 atm.

The thicker the layer of coal dust, the lower is the ignition temperature [2]. The minimum ignition temperature for coal dust layer is 155°C. It is related to the logarithm of the depth of layer, λ, between 2.9 and 59.0 mm.

Table 23.7 Ignition Temperatures for Coals

	Ignition Temperature (°C)		
Combustible Dust	**Layer**	**Cloud**	**MEC (g/m^3)**
Lignite, North Dakota	180	440	45
Coal, Wyoming (HVC)	180	575	40
Coal, Bituminous Colorado (HVC)	180	440	45
Coal, Bituminous Illinois (HVC)	160	600	40
Coal, Bituminous Pennsylvania (HVA)	170	610	55
Coal, low-volatile West Virginia	220	640	60

MEC, minimum explosive concentration.

$$t = 326.3 - 95.5 \log \lambda \tag{23.13}$$

where t is in centigrade and λ in mm.

The coal dust explosion temperature is difficult to calculate but measured values range from 2500 to 2600°C.

23.2.4 Dust Explosion Pressure

Maximum and near maximum explosion pressure and pressure rise rates for coal dust occur within a concentration range of 200–1000 g/m^3. Thus, there is a broad range of dust concentrations exhibiting the most damaging effects. Fig. 23.7(A) shows a typical pressure rise curve for Pittsburgh coal seam (HVA) in a 20 L chamber [8]. The maximum pressure of almost 7 atm is achieved at around 600 g/m^3, a concentration 10 times the MEC.

It is also clear from Fig. 23.7(B) that the rate of pressure rise, $\frac{dp}{dt}$, generally increases with dust concentration reaching a limit at concentration of 600 g/m^3. The rate of pressure rise also increases with a decrease in particle size [5].

23.3 Prevention of Gas Explosions

Both gas and dust explosions can be practically eliminated if proper actions are taken. Two major sources of ignition, namely, spontaneous combustion and frictional ignitions, are already discussed in Chapters 21 and 22, respectively. Some additional precautionary measures are discussed here.

23.3.1 Methane Drainage

As discussed in Chapters 13–16, draining methane from coal prior to mining and post-mining can be a big help in minimizing the risk of mine explosions. All coal seams with a gas content of above 100 ft^3/t must be degassed prior to mining. All longwall panels should have some vertical gob wells to prevent excessive gas pressures in the gobs. Coal seams underlying the working coal seam must be degassed to prevent methane outburst and consequent methane inundation.

23.3.2 Ventilation

The second line of defense against mine explosion is good ventilation. The air quantity and its velocity should be large enough to dilute methane below the statutory limit of 1% and eliminate the risk of gas layering near the roof or floor. The subject is fully discussed in Chapters 1–6 in this book.

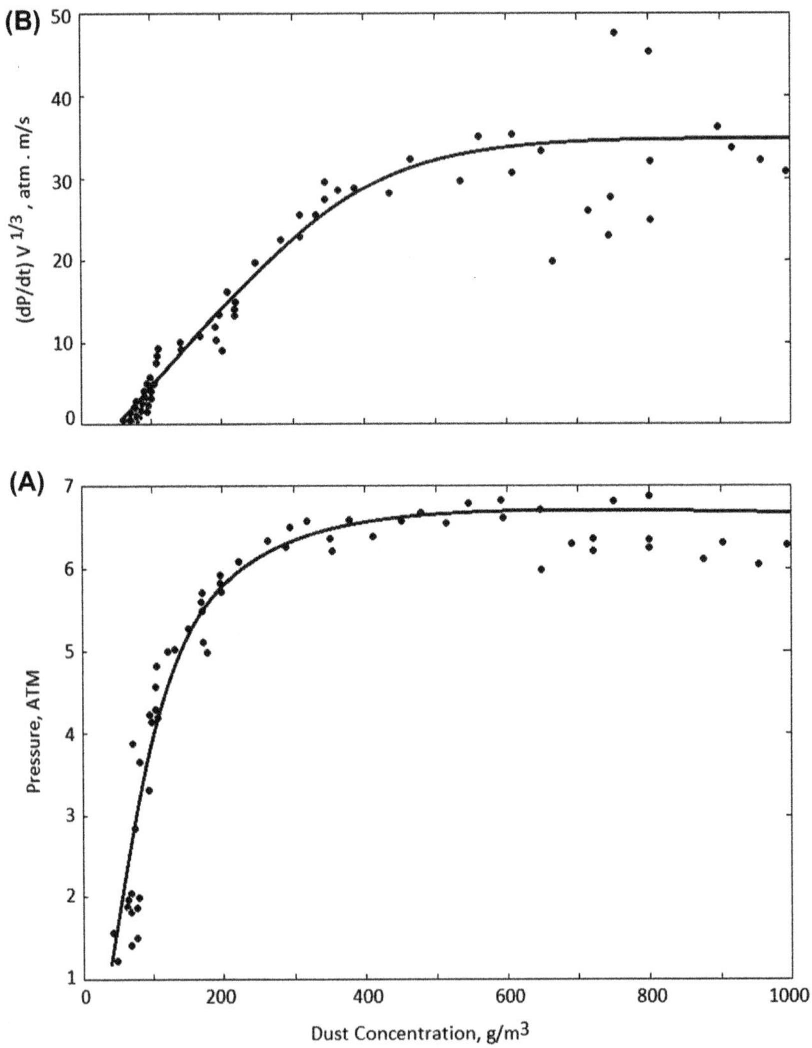

Figure 23.7 (A and B) Explosibility data for Pittsburgh (high-volatile bituminous) pulverized coal dust in the 20 L chamber. From Cashdollar KL: Coal Dust Explosibility, J of Loss Prevention, Processing and Industry; vol 9, No 1; 1996, pp. 65-76.

23.3.3 Preventing Ignition of Methane–Air Mixtures

Besides the potential sources of ignition, such as spontaneous combustion and frictional ignitions, the only other major source is a flame or an electrical spark. The following is a list of preventative measures.

1. No open lights or smoking should be allowed in underground coal mines.
2. Mine safety lamps or approved battery-powered cap lights should be used for lighting. Mine safety lamps are almost obsolete in the United States.

3. All equipment used at the working face (an area inby of the last open crosscut) should be either explosion proof or intrinsically safe.
4. A gas concentration check should be made prior to starting any mining equipment. Mining machines should have methane detectors that can sound an alarm at 1.5% of methane and cut off the power at 2%.
5. Continuous monitoring of areas where methane accumulation is likely should be required.
6. Routine and extensive inspection of all areas in a mine should be done to minimize methane ignition risks.
7. Mined out areas should be isolated by explosion-proof stoppings.

The US Federal Regulations (and State regulations) [10] provide a comprehensive list of all these precautions. Compliance with law is required. Citations and fines are issued for any and all violations.

23.3.4 Prevention of Dust Explosions

The first step in this direction is the same as the control of respirable dust discussed earlier in the book. Prevention of dust formation and suppression by water sprays are strongly recommended. Prevention of methane–air explosion is the second step in the prevention of dust explosions. The third and the last step is to make the coal dust inert by mixing some noncombustible dust with it. Gypsum powder, crushed shale, and limestone powder are the most commonly used inerting material. In the United States, limestone is mostly used because of its availability in the proximity of coal mines.

The percentage of inert dust in the mixture with coal dust that will stop the propagation of a coal dust explosion depends on a number of factors, such as

1 Volatile content of coal.
2 Moisture content.
3 Fineness of coal dust.
4 Strength of the initiator (explosive charge or a fixed volume methane explosion).
5 Concentration of any combustible gas, such as methane in air.

Excellent references on the subject, such as Cashdollar [8], Nagy [11], Cybulski [12], and Richmond [13], are available. A summary of their findings will be presented here.

23.3.4.1 Effect of Volatile Matter in Coal

Fig. 23.8 summarizes the data by Cashdollar [8], Nagy [11], Cybulski [12], and Richmond [13].

Percentage of inert dust increases with the weight percentage of volatiles in coal from a low of 40% to a high of 80%. The US law now requires 80% noncombustible in all airways. Legal requirements for inert dust are shown in Table 23.8 for some coal mining countries. The higher the "s" value, the better is the level of protection.

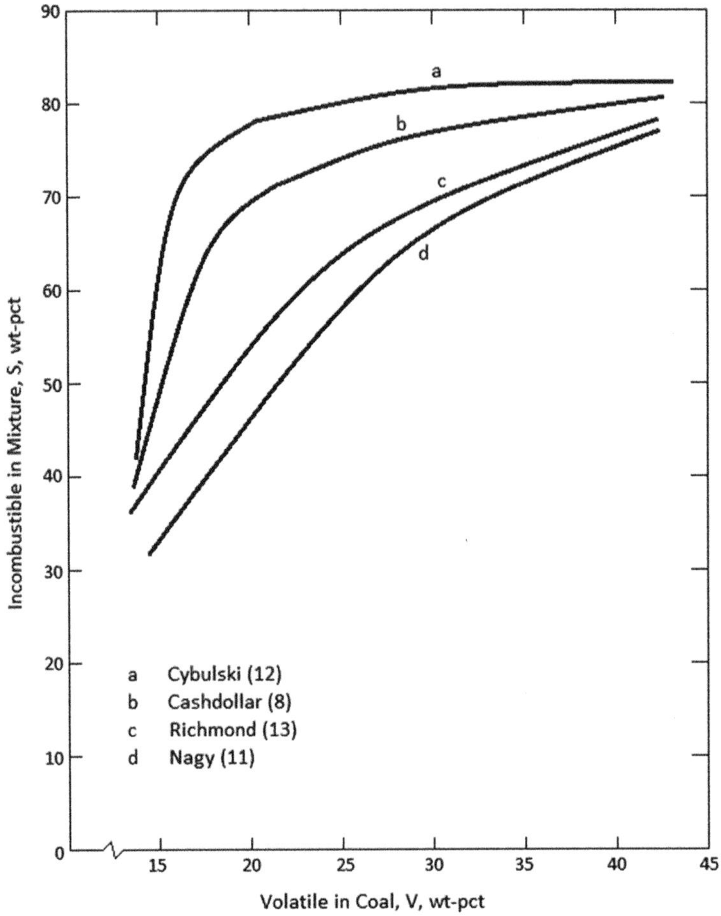

Figure 23.8 Incombustibles required to prevent propagation of coal dust explosions.

23.3.4.2 Effect of Moisture

Moisture acts as an inert material, but it is neither regulated nor is any credit given to its concentration in the coal mine dust. However, rock dusting can be dispensed if the coal is so wet that water comes out when the coal is squeezed in hand [10].

23.3.4.3 Effect of Methane in Air

Cashdollar [8] shows the impact of methane on the MEC of two coals in Fig. 23.9. Increasing methane concentrations decreases the MEC for both coals. Hence, the inert concentration is usually increased if there is methane present in mine airways. US laws require 1% increase in inert dust content for each 0.1% of increase in methane concentration.

Table 23.8 Minimum Incombustibles in Coal Mine Dust

Country	Year	Minimum Incombustible (S) Required in the Dust Mixture, pct	Minimum Ratio Incombustible to Coal $\frac{S}{100-S}$
Belgium	1965	60–78	1.5–3.5
Canada	1959	65	1.9
Czechoslovakia	1957	80	4.0
Germany	1965	80	4.0
Poland	1959	70–80	2.3–4.0
South Africa	1959	50	1.0
United Kingdom	1961	50–75	1.0–3.0
USA (Federal)	2014	80	4.0
USSR	1953	60–75	1.5–3.0

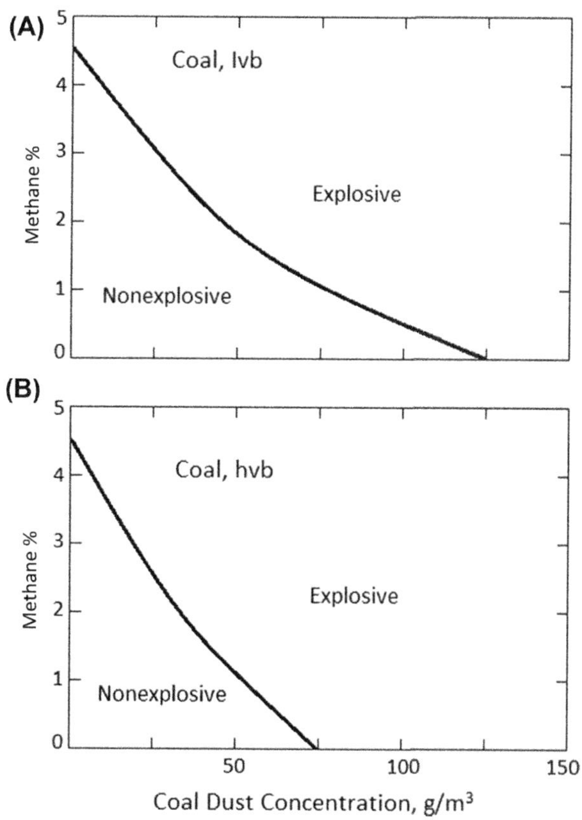

Figure 23.9 Minimum explosive concentrations of a mixture of coal dust and methane gas.

23.4 Stone Dust Barriers for Explosion Propagation Prevention

The most effective way of controlling coal dust explosion is to minimize them at the source—the initial ignition. But this is not always possible. The second step in this direction is to make the coal dust inert by mixing it with noncombustible dust discussed earlier. The third line of defense is the use of stone dust barriers. They consist of a number of planks suspended from the roof that are loaded with very fine stone dust, such as limestone or gypsum powder. The shock wave that leads the flame front in a detonation of gas or dust knocks the plank off its support. The resulting cloud acts as a barrier to the forward progress of the flame by quenching it on the spot or only a short distance away from it. The explosion is thus contained in the section and minimizes widespread destruction. Many countries require stone dust barriers by law in coal mines. Water can be used in place of dust to considerably cut down the cost. Such barriers are called passive barriers. Cybulski [12] did extensive work on the design and effectiveness of such barriers.

Fig. 23.10 shows the effectiveness of these barriers. They appear to work if they are kept close to the working face, where an explosion is likely to initiate. Thus, they must be moved inby as the development entries are driven inby. By contrast, there are some barriers that are activated by remote sensors on the mining machines or in the nearby areas.

Besides limestone and gypsum, marble, dolomite, and anhydrites have been used, but gypsum (calcium sulfate) and rock phosphates appear to do a better job with

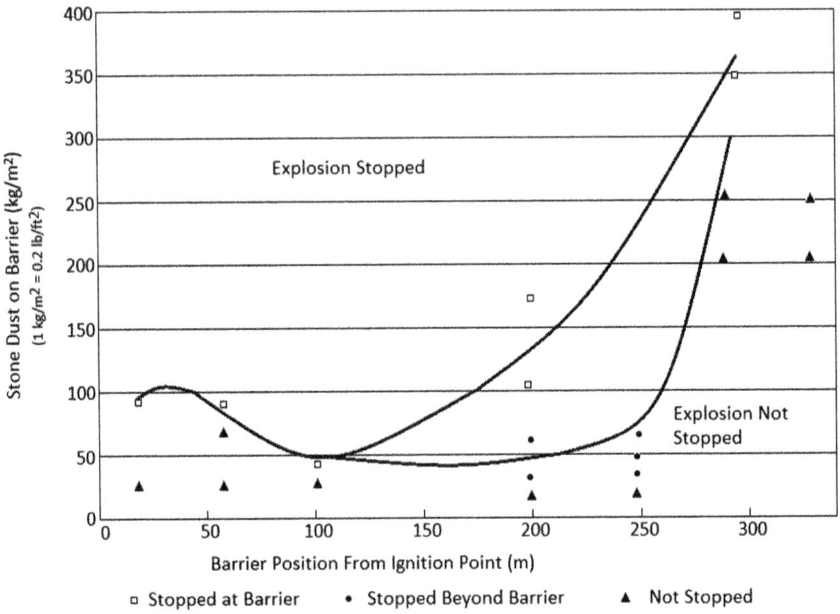

Figure 23.10 Effectiveness of a stone dust explosion barrier [12].

smaller loading of the dust. In heavy loading, the total weight on a plank is about 8–9 lbs per square feet of the airway cross section. The lightly loaded barriers may contain only half as much. The typical length of the barriers is 100–200 ft. The volumetric stone dust concentration to quench a gas/dust explosion is 7–12 kg/m^3. This is 150–200 times the lower explosion concentration limit for coal dust. A typical set of stone dust barriers may have a total stone dust load of 5–7 tons.

The fineness of the dust is the same as that of the rock dust used to inert the coal dust. In the United States, 100% of the dust must be smaller than 850 microns, 70% finer than 75 microns with less than 5% combustibles and less than 4% silica. Most rock dusts tested by the author were finer than 37 microns (400 mesh) and had no silica in it. Finer the dust, the better is its ability to disperse in air and higher its residence time in air. Some chemicals, such as stearic acid, are mixed with dust to prevent caking.

Stone dust barriers appear to work better in test galleries but do not do as well in actual mines. It would be, therefore, prudent to (1) degas the coal seam, (2) use only permissible equipment in the working face, and (3) inert the coal dust with noncombustible dust. US coal mine regulations do not require stone dust barriers.

Problems

The following gas analysis was received after a mine fire. Calculate the composition point and determine if the atmosphere is explosive.

Gas	Concentration (by Volume)
CO	2.5
CO_2	0.2
CH_4	2.3
H_2	2.0
O_2	10
N_2	(100 for all other gases)

References

[1] Byrer C, et al. Coalbed methane: a miner's curse and a valuable resource. In: Thakur P, et al., editors. Coalbed methane from prospect to pipeline. Elsevier Publishing; 2014. p. 1–6.

[2] Bodurtha FT. Industrial explosions prevention and protection. McGraw-Hill Book Company; 1980. p. 157.

[3] Coward HF, Jones GW. Limits of flammability of gases and vapors. USBM Bulletin #503. 1952.

[4] Zabetakis MG. et al. Determining the explosibility of mine atmospheres. USBM I.C. 7901 1959:11.
[5] Kuchta JM. Investigation of fire and explosion accidents in chemical, mining and fuel related industries – a manual. USBM Bulletin #686 1985:84.
[6] Le Chatelier H, Boudguard O. Limit of flammability of gaseous mixtures. Bulletin de la Société Chimique de France 1898;19:483–8.
[7] Burgess DS, Murphy JN, Hanna NE, Van Dolah RW. Large scale studies of gas detonations. USBM RI 7196 1968:53.
[8] Cashdollar KL. Coal dust explosibility. Journal of Loss Prevention, Processing and Industry 1996;9(1):65–76.
[9] Palmer KN. In: The quenching of flame by wire gauzes, the seventh symposium on Combustion, Buttersworth; 1959. p. 497–503.
[10] Code of Federal Regulations, Title 30, Chapter 74 & 75, Mine Safety and Health Administration, U.S. Department of Labor; 2007. (Published annually.).
[11] Nagy DL. The explosion hazard in mining, U.S. Mine Safety and Health Administration, IR 1119 1981.
[12] Cybulski W. Coal dust explosion and their suppression (in Polish), TT 73-54001, National Technical Information Service, U.S. Department of Commerce, Springfield, VA; 1975.
[13] Richmond JK, et al. Effect of rock dust on explosibility of coal. USBM RI 8077 1975:34.

Mine Sealing and Recovery

24

Chapter Outline

In the last three chapters, various methods of mine design, equipment design, and other preventative techniques to prevent a mine fire have been discussed. In summary they are as follows:

1. Design a mine to minimize spontaneous combustion,
2. Prevent frictional ignitions and use only permissible equipment in the working face,
3. Degas the coal seam, premining, and postmining to minimize methane ignitions,
4. Make coal dust noncombustible with adequate rock dusting,
5. Use stone dust barriers, where required,
6. Better housekeeping and keeping some firefighting equipment handy. These include dry powder fire extinguishers, water hydrants, and equipment to create high-expansion foam. Detailed descriptions of these items are available in the literature [1]. They are also often required by the Federal and State regulations [2].

In spite of all efforts, there are occasions when a mine must be sealed. The rule of thumb is that if the fire is not "controlled" within a few hours, planning for sealing must start [1]. "Controlled" does not mean extinguished: it means the fire has been

Advanced Mine Ventilation. https://doi.org/10.1016/B978-0-08-100457-9.00024-9

surrounded and manpower and materials gathered to prevent its propagation. Mine sealing is misconstrued to mean an "end of mining." On the contrary, it can be, sometimes, the quickest way to extinguish the fire and recover the mine, especially if the mine is gassy. The subject of mine sealing and recovery will be discussed under the following three headings:

- Mine sealing
- Sampling the sealed mine atmosphere and interpretations of data
- Mine recovery

24.1 Mine Sealing

The work of mine sealing can be broadly divided into two teams: (1) surface sealing team and (2) in-mine sealing team. Such teams are generally a part of normal mine planning.

24.1.1 Surface Sealing

All openings that can feed air to the underground mine must be closed. These include:

1. Shafts, drifts, and slopes.
2. All boreholes for rock dust, power cable, water, degasification, and diesel fuels. Some of these boreholes and shafts can be used to monitor the mine atmosphere.
3. Locations for new boreholes to surround the fire area should be selected and materials to extinguish fire should be gathered. This may be just plain water, nitrogen foam, silica gel solutions, or nitrogen generating equipment, such as a PSA (pressure swing adsorption) unit that can extract nitrogen from atmosphere. Most seals on these mine openings are made of steel plates that are covered with rigid urethane foam to make an airtight seal.

24.1.2 In-Mine Seals

All seals/stoppings (a term more popular in mining parlance) used in underground mines can be divided into three categories:

1. Temporary/ventilation stoppings,
2. Permanent stoppings, and
3. Explosion-proof stoppings.

24.1.2.1 Temporary Stoppings

Such stoppings are routinely made to separate intake airways from return airways. These need to be sealed and made airtight. Most commonly, they are made of concrete blocks, but brattice cloth and metal stoppings are also used. The latter can be recovered and reused. All such stoppings are made airtight by urethane foam sprays.

Temporary stoppings are also built in a mine fire emergency situation to protect people engaged in firefighting. They are made of a wooden frame with brattice cloth or belting on top [1]. Another variation of a temporary stopping is a parachute stopping developed by Kissell [3]. They look similar to the parachute used to jump off an airplane and are made of impermeable fabric. They can be installed in minutes and are mostly used during rescue and recovery operations.

24.1.2.2 Permanent Stoppings/Seals

Permanent seals are mostly used to isolate old/abandoned workings in a coal mine. They may not be explosion proof but are generally much stronger than temporary stoppings. They are typically made of masonry/concrete blocks with occasional steel reinforcing. US regulations [2] require them to be strong enough to withstand:

1. 50 psi overpressure when the atmosphere in the sealed area is monitored and maintained inert. Here inert means that the oxygen content is less than 10% and methane concentration is either less than 3% or greater than 20%. Each seal will have two sampling pipes in it. One pipe shall extend about 15 ft into the sealed area and another will extend to the center of the first connecting crosscut inbye the seal.
2. 120 psi overpressure if the sealed area atmosphere is not monitored and is not maintained inert but it is not explosive.
3. An overpressure greater than 120 psi if the sealed area is not monitored and is likely to contain an explosive/flammable gas mixture: methane between 4.5% and 17% and oxygen above 17% throughout the entire sealed area.

These stoppings to withstand 120 psi are typically masonry stoppings about 4 ft in thickness and are hitched in the roof and floor. No steel reinforcing is used.

Using the equation for failure in a pressurized cylindrical vessel, we can check if this is adequate.

$$\sigma = \frac{\mathrm{Pd}}{4\mathrm{t}} \tag{24.1}$$

where σ is the tensile strength of the vessel wall in psi, P is the inside pressure behind the seal, d is the entry/vessel diameter, and t equals thickness of the seal, in feet.

Typical safe tensile strength for:

Masonry = 100 psi
$P = 120$ psi;
d = 10 ft (75 ft^2 area of tunnel)

Hence, $\mathrm{t} = \frac{120\times 10}{100\times 4} = 3$ ft.
A 4 ft thick seal is, therefore, adequate to withstand 120 psi.

24.1.2.3 Explosion-Proof Stoppings

If a sealed area is likely to contain an explosive gas mixture and it is not monitored, the seals must be designed to withstand an internal pressure up to 640 psi (refer to

Table 24.1 Minimum Thickness for Explosion Proof Stoppings

Country	Material	Formula	Thickness for a 20 × 6 ft Entry (ft)
United Kingdom	Rock/Cement Grout	$t = \left(\frac{W+H}{2}\right) + 2$	15
Germany	Rock/Cement Grout	–	5–7
Poland	Rock/Cement Grout	–	7–12
United States [4]	Rock/Cement Grout	$\left(\frac{W+H}{2}\right)$	13
United States [4]	Sandbags	$\left(\frac{W \times H}{3}\right)$	40
United States [4]	Loose Rock or Sand	3 W	60

Chapter 23). They are also known as "bulkheads." They are typically made of two end walls (3–4 ft thick masonry) 10–50 ft apart depending on local law, materials, and method of construction. The gap between the walls is tightly filled with bulk materials and Class A cement grout. Concrete, gypsum and anhydrite powder, fly ash, and bentonite are also used in some cases. In the United States, many seals are made of concrete blocks, Omega 384 foam blocks, cementitious or polymer foam seals. Two sampling tubes with steel caps are also installed as in the permanent stopping for sampling the air in the sealed area.

Using Eq. (24.1) again and assuming that the minimum explosion pressure can be as high as 640 psi, the thickness of an explosion-proof stopping should be at least 16 ft.

Minimum thickness required in different countries are listed in Table 24.1.

In the author's experience, two masonry walls of 4 ft thickness with at least 10 ft of space between the walls filled with rock and pressurized cement grout provide the best protection against any explosion in a sealed area.

24.2 Inertization of the Sealed Area

To expedite the quenching of fire in a sealed area, two options are generally available:

Flooding with water or
Inertization with a nonreactive gas such as N_2 or CO_2.

24.2.1 *Flooding With Water*

Flooding with water is an excellent technique for the extinction of fire under suitable circumstances, such as inclined workings. It first cools the fire and secondly excludes all oxygen making sure the fire would be extinguished promptly. It also cools the roof and floor and fills them up with fines preventing all access to air/oxygen. No serious damage is done to mining equipment by waterflooding.

On the negative side, usually a vast amount of water is needed, and pumping costs for dewatering may be very high. In some cases, there is a slightly increased danger of rekindling of fire, especially if the coal is of low rank and more liable to spontaneous combustion.

24.2.2 *Inertization With Nonreactive Gas*

As discussed in Chapter 21, Spontaneous Combustion of Coal, inertization with N_2 and CO_2 is a very old, established technique to extinguish a sealed fire.

Nitrogen is available commercially, but it can be produced much more cheaply by a PSA type generator at the mine site. Liquid nitrogen is transported in cryogenic tanker trucks that convert the liquid N_2 (at $-196°C$) to a gas at temperatures higher than $0°C$. The volumetric change from liquid to gas creates a 700 times higher expansion and it drives the gas into the sealed area through steel pipes. The PSA type N_2 generators are becoming popular. It strips nitrogen out of the atmosphere using molecular sieves and can generate a flow of 500–1000 CFM at 98% purity. The air has to be compressed but the delivered nitrogen remains pressurized and can be easily pumped into the sealed area. A plant of this capacity is not expensive, costing about $500,000. It is designed to run almost unattended using solid-state electronics and Programmable Logic Control (PLC) controls for the process.

Advantages of using nitrogen include easy availability, nontoxic nature, and ease of transport. On the negative side, liquid nitrogen may present some transport and maintenance problems. PSA N_2 generators have no such limitations.

Carbon dioxide is widely used for extinguishing fire. It is available as liquid or solid form (dry ice). Just like N_2, it can be stored and transported with ease and it is very effective in quenching the fire. Disadvantages of using CO_2 are as follows:

1. It is a noxious gas and can be problematic in the recovery of the mine.
2. Liquid CO_2 presents fire hazards in the actual operation.
3. Dry ice can cause explosion when it comes in contact with fire.

For these reasons, N_2 is preferred for all inertization work. In most cases, N_2 is cheaper than CO_2.

24.3 Sampling the Sealed Mine Atmosphere and Interpretation of Data

Sampling the mine atmosphere before and after sealing can provide very useful data to manage a mine fire. The gas composition prior to sealing can provide a reliable baseline, such as the normal ambient CO concentration. Once the mine or a section of the mine is sealed, air samples should be withdrawn on a routine basis. It could be every 4 hours or at less frequent intervals. There are a number of indices derived from the gas composition that can predict the status of mine fire, including the final question: "Is the fire dead yet?"

24.3.1 Indices Used to Predict the Status of Fire Behind Seals

There are many indices that help decide if the fire is dead or not. The most commonly used are the following five:

1. CO Index
2. CO_2 Index
3. Jones–Trickett Ratio
4. Litton Ratio
5. Hydrocarbon ratio

The CO Index and Jones–Trickett Ratio are already discussed in Chapter 21, Spontaneous Combustion of Coal. They are normally used when a gas sample is collected from a ventilated airway but in the author's experience (about 20 mine fires), they are also useful when the sample comes from the sealed area. Besides these indices, a "trend analysis" for all gases at each sampling location is also very important. The concentration of gases, as well as the resulting indices listed above, are plotted against time to determine if the fire is dying or dead and if mine recovery should start.

24.3.2 Sampling and Gas Analysis Procedure

Where possible, real-time instruments are used to provide instantaneous results. These instruments are handheld stain tube chemical sensors, infrared sensors, or electrochemical sensors permanently installed in the mine airways.

The other and more reliable technique is to withdraw gas samples from boreholes drilled all around the fire area, mine shafts, vent holes, etc. A gas laboratory is immediately established at the mine to analyze gas samples.

A gas chromatograph (GC) is the most commonly used apparatus for gas analysis. Instantaneous data obtained with different instruments should be confirmed by GC analysis because the latter is more reliable.

There are two types of GCs based on the gas concentrations. The flame ionization detector-type GC is used for low gas concentrations. It is very sensitive and can detect very low (up to 1 ppm of CO and H_2) concentrations of many gases. It, however, destroys the sample. The thermal conductivity detector–type GC is used when gas concentrations are high. It is not very precise, but it does not destroy the gas sample.

The gas samples are collected by either pulling the gas into an air-evacuated cylinder or by forcing the gas sample into a container with an air pump. Air-evacuated cylinders are made of plastic or glass and have a volume of 10–22 cc. The air evacuation is usually 95%. They usually deteriorate with age and retain some CO. Their shelf life is 1–2 months [5]. Some operators use syringes that can be cleaned and reused. Even Mylar bags have been used to collect large volumes of samples.

Some good rules for proper gas sampling procedure are listed below:

1. Always make sure the sampling line is outgassing (not intaking) when an air sample is drawn.
2. Preferably sample each location at the same time every day at the same intervals (that is, every 4 or 8 h and so on).
3. Track total emissions of CO, CH_4, and other gases if the mine is not sealed yet.
4. Purge the sampling line by letting the air pump run for a while when it is used to collect a gas sample.
5. Increase sampling frequency if there is sudden change in the barometric pressure.
6. Trained personnel should be used for sample collection/gas analysis.

24.3.3 Interpretation of Air Analysis Data

The CO Index and CO_2 Index usually confirm if there is fire still burning. CO Index has been defined in Chapter 21, Spontaneous Combustion of Coal. CO_2 Index was also developed by Graham [6] and is given by Eq. (24.2).

$$CO_2\ \text{Index} = \frac{CO_2 - 0.03}{100 - O_2} \tag{24.2}$$

where CO_2 and O_2 are expressed as volume percents.

In a normal mine atmosphere, the CO_2 Index tends to be zero. Once ignition occurs, CO_2 Index increases and continues to increase with rising temperature. A persistently increasing CO_2 Index also confirms existence of an active fire just as the CO Index does.

The Jones–Trickett Ratio discussed in Chapter 21 can also be used for sealed areas to determine what is burning. Values above 0.5 indicate that coal is burning.

When the above indices go to their baseline value (prior to a fire), they indicate that the fire may be dead. Litton Ratio and Hydrocarbon Ratio are also used to confirm if the fire is dead.

24.3.3.1 Litton Ratio (R)

It was developed by Litton [7] to confirm if fire in a sealed area is dead and is safe to open up the sealed area. In order to derive an equation for R, we need to find three parameters.

1. Equivalent air
2. Methane plus other hydrocarbons
3. Residual gas; Rg

Equivalent air is the oxygen equivalent of air and is equal to 4.774 O_2 (% by volume). In coal mines besides methane, only ethane is present in measurable quantities.

Hence,

$$Rg(\%) = 100 - 4.774O_2 - (CH_4\% + C_2H_6\%) \tag{24.3}$$

The concentration of CO in Rg is defined as:

$$(\mathrm{CO})_{\mathrm{Rg}} = \frac{(\mathrm{CO})\mathrm{s}}{\mathrm{Rg}} \times 100 \tag{24.4}$$

where (CO)s is the concentration of CO in the original gas sample in percent.

Let us call (CO) Rg/*Rg*, R_1.

Hence,

$$R_1 = \frac{(\mathrm{CO})_{\mathrm{Rg}}}{\mathrm{Rg}} \times 100 = \frac{(\mathrm{CO})\mathrm{s}}{\mathrm{Rg}^2} \times 100 \tag{24.5}$$

Based on experimental data, R_1 appears to be a function of $\left(\frac{\mathrm{Rg}}{O_2}\right)$ only [7] and can be expressed as:

$$R_1 = 150\left(\frac{\mathrm{Rg}}{O_2}\right)^{-1/2} \tag{24.6}$$

The upper limit of Rg is

$$R_1 = 300\left(\frac{\mathrm{Rg}}{O_2}\right)^{-1/2} \tag{24.7}$$

Eq. (24.7) is mostly used for safety.

Taking a logarithm on both sides of Eq. (24.7):

$$\ell\mathrm{n}\ R_1 = \ell\mathrm{n}\ 300 - \frac{1}{2}\ell\mathrm{n}\left(\frac{\mathrm{Rg}}{O_2}\right) \tag{24.8}$$

A plot of Eq. (24.8) is shown in Fig. 24.1.

The Litton Ratio is defined as the ratio of R_1 (Eq. 24.5) derived from sealed area gas sample to the upper limit of R_1 shown in Eq. (24.7) or:

$$R = \frac{1}{3}\frac{(\mathrm{CO})\mathrm{s}}{\mathrm{Rg}^{3/2}\cdot {O_2}^{1/2}} \tag{24.9}$$

If this value is greater than 1.00, ambient temperatures have not been reached. Smoldering coal combustion at above-ambient temperature is likely to be in progress. This condition is unsafe for mine entry or reopening of seals. If the ratio is less than 1.00 and continues to decline, then it is safe to assume that the fire is dead and ambient equilibrium has been achieved. Table 24.2 shows some field data to illustrate the conclusion [5].

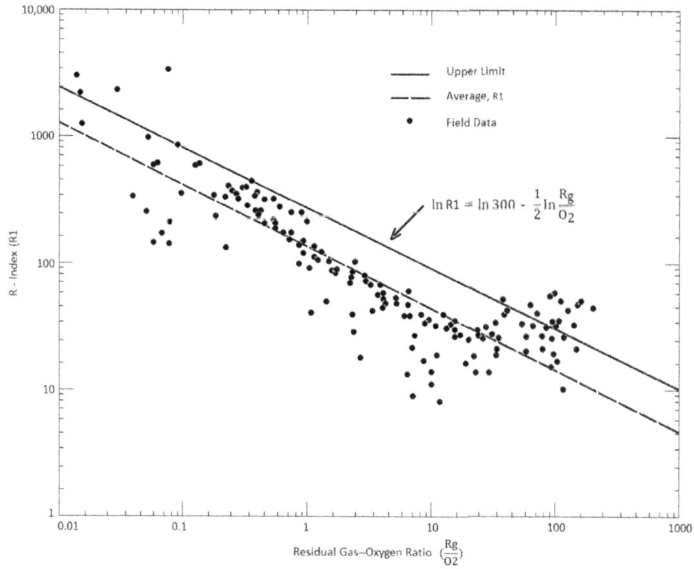

Figure 24.1 R-Index versus the dimensionless ratio of residual gas to oxygen (Rg/O_2).

Table 24.2 Litton Ratio for a Dying Mine Fire

Day	O_2%	CH_4%	C_2H_6 (ppm)	Ratio, R
1	13.89	0.38	740	4.45
10	12.69	2.61	700	3.68
30	9.45	3.93	660	1.19
60	5.81	5.23	440	0.16
80	2.25	5.42	410	0.04

Adapted from Timko RJ, Derick RL. Methods to determine the status of mine atmosphere – an overview. In: Paper presented at 2006 SME annual meeting – mine fire prevention; 2006. p. 9.

It is prudent to wait until R values stabilize at a point below 1.00 for 30 days or so before reopening the mine.

24.3.3.2 Hydrocarbon Ratio (HR)

Justin and Kim [8] developed the hydrocarbon ratio based on their finding that low molecular weight hydrocarbons desorbed from coal in direct proportion to increasing temperature. Methane is liberated first followed by CO_2, CO, H_2, C_2H_4, C_3H_6, and C_2H_2 in that sequence.

The HR is defined by Eq. (24.10) below:

$$HR = \left[\frac{1.01(THC) - CH_4}{THC + C}\right] \times 1000 \quad (24.10)$$

where THC is the sum of all hydrocarbons in ppm; CH_4 is the methane concentration in ppm; C is a constant = 0.01 ppm; HR = 0 when no hydrocarbons are present; HR = 10 when only methane is present; Maximum HR = 1010.

A limitation in Eq. (24.10) is that methane concentration must be above 20 ppm.

Interpretation of HR data is as follows:

When:	$0 < HR < 50$, normal conditions prevail.
	$50 < HR < 100$, alleviated temperatures exist.
	$HR > 100$, high temperature oxidation is in progress.

Table 24.3 shows field data from an actual fire.

The stable value at 9.9 (which is smaller than 50) indicates that the fire is dead.

It is advisable to plot all indices, that is, CO Index, CO_2 Index, Jones–Trickett Ratio, Litton Ratio, R and HR on a semi log paper with time on x-axis and logarithm of all indices on the "y" axis to draw the final conclusion.

Fig. 24.2 shows a typical plot.

24.4 Recovery of the Sealed Mine

When all air analysis data discussed in Section 24.3 indicate that the fire is dead and waiting period of at least 30 days have elapsed, plans are made to reopen the mine. A good rule to remember is that "one should never be sure that the fire is dead until one physically verifies it by walking through the area that was on fire."

A "command center" is established at the mine office where management and government (federal and state) agencies are present. Recovery operation is a continuous process and hence a sufficient number of rescue teams are gathered. They are properly

Table 24.3 Decline in HR With Time

Day	CH_4 (ppm)	C_2H_6	HR	Remarks
1	4000	1000	210	Active fire
10	30,000	600	29.6	Dead fire
30	40,000	400	9.9	Dead fire
60	60,000	200	9.9	Dead fire

Adapted from Timko RJ, Derick RL. Methods to determine the status of mine atmosphere – an overview. In: Paper presented at 2006 SME annual meeting – mine fire prevention; 2006. p. 9.

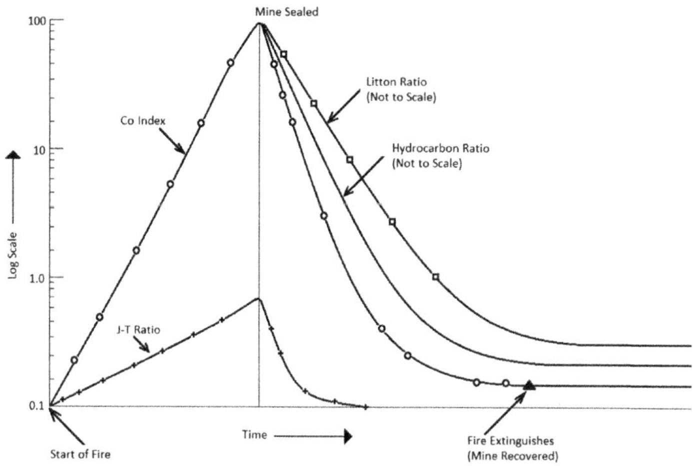

Figure 24.2 A hypothetical plot of major indices during fire and after sealing.

trained on the mine airways and escape airways in case the fire rekindles. Mitchell [1] lists a set of "laws" for recovery work that is quite comprehensive and is noted here in brief.

1. Rope off and guard all openings to the mine and surface buildings.
2. Restrict entry to the area.
3. Start the fans but maintain good communication with the command center.
4. Do not energize any electrical equipment in the mine except some pumps to dewater an area.
5. Create a resting place and provide plenty of food for all rescue-trained people.
6. Maintain a check-in and check-out list for all personnel engaged in the recovery work. Each team should also keep a separate log on their workers.
7. All agencies in the command center should develop together a plan for recovery and agree on it.

There are two ways to recover the sealed mine:

1. Reventilation of the mine.
2. Air locking (to recover in steps).

24.4.1 Reventilation

It is the preferred way if the fire is actually dead and there is no chance for rekindling. There is a definite length of time that the fans need to run to clean the airways. Fans should run at full capacity and mine airways should be monitored for gas composition and all indices that indicate that there is no rekindling should be calculated to verify it.

The mine rescue teams walk every roadway from shaft bottom to the working faces and back reporting on air composition, roof, and floor conditions. They may run into some hot spots (small amount of coal still smoldering) that can be quickly extinguished with water.

Regular mine people follow the rescue teams and do all that is necessary to resume normal mining.

24.4.2 Air Locking

This technique is used when one is not absolutely sure that the fire is actually dead. It was permitted in the past but may not be approved today by government agencies, such as Mine Safety and Health Administration of the United States.

The logic here is to recover the mine or a mine entry in steps. An air lock is built against the sealed area and the outbye area is fully recovered. The fresh air base is moved inbye and another air lock is built to recover the next section of the mine/airway. In repeated steps, thus, the whole area that was on fire is inspected and recovered. Step by step procedures are listed by Mitchell [1]. Reference can be made for details.

Problem

The following air analysis is received from a sealed area in a mine that was on fire. Calculate CO Index, Litton Ratio, and Hydrocarbon Ratio to conclude if the fire is dead.

Gas	Concentration
O_2	6% (60,000 ppm)
CH_4	6.5% (65,000 ppm)
CO	700 ppm
C_2H_6	500 ppm
C_3H_8	100 ppm
CO_2	8.5%
N_2	= (100 − all above gases)

References

[1] Mitchell DW. Mine fires: prevention, detection, fighting. Maclean Hunter Publishing Company; 1990. p. 167.
[2] Code of federal regulations, title 30, mineral resources 75.335 and 75.336. 2007. p. 752 [published annually].
[3] Kissell FN, Thimons ED. New stopping designs – an update. Mining Congress Journal July 1981.
[4] Mitchell DW. Personal communications.

[5] Timko RJ, Derick RL. Methods to determine the status of mine atmosphere – an overview. In: Paper presented at 2006 SME annual meeting – mine fire prevention; 2006. p. 9.
[6] Graham JI. Pyrites as the cause of spontaneous combustion in coal mines. Trans Institute of Mining Engineers 1924;67(2):100–13. London.
[7] Litton CD. Gas equilibrium in sealed coal mines. US Bureau of Mines RI 9031; 1986. p. 13.
[8] Justin TR, Kim AG. Mine fire diagnostics to locate and monitor abandoned mine fires. US Bureau of Mines IC 9184; 1988. p. 348–55.

Appendix A

Nature of Loss	Sketch	Formulas	Symbols	Source
Sharp bend	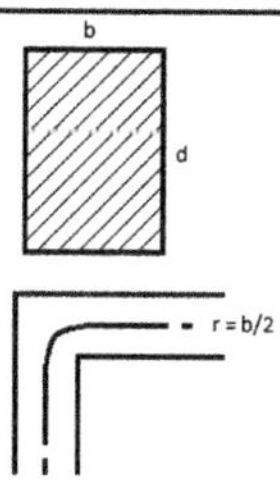	1. $X = \frac{0.60}{ma^{0.5}}(\frac{\theta}{90})^2$	d: airway height m: radius ratio = r/b a: aspect ratio = d/b θ: angle of bend	
Abrupt contraction	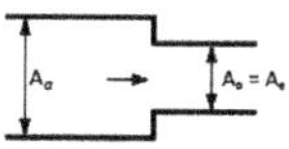	2. $X = (\frac{1}{C_c} - 1)^2$	C_c : coefficient of contraction = $\frac{A_a}{A_o}$ N_c : area ratio = $\frac{A_o}{A_a}$ N_e : area ratio = $\frac{A_o}{A_e}$	
Abrupt expansion	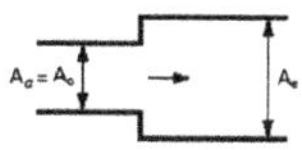	3. $X = (\frac{1}{N_e} - 1)^2$		
Regulator (or doorway)	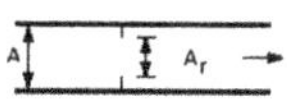	4. $X = (\frac{\frac{1}{Cc} - N}{N})^2$	A : area of airway = A_e A_r: area of regulator = A_o N : area ratio = $\frac{A_r}{A} = \frac{A_o}{A_e}$	

Appendix B: Ventilation Network Analyzer in Fortran IV

Variable Names with Definitions

Input Variables

DELQDG flow of diesel exhaust per unit length of the roadway—general
DELQDS flow of diesel exhaust per unit length of the roadway—special
E error considered acceptable for the quantity flow
FF friction factor for the branch
FX(I) fan pressure for data point no. I
HEIGHT(I) height of branch no. I
JJ junction number for the outside junction
KEY a control number to prevent repetitive calculations on identical fan curves
LENGTH(I) length of branch no. I
MAXIT maximum number of iterations
MAXJ maximum numerical designation for a junction
NB number of branches in the mine network
NBWNVP number of branches with natural ventilation pressure
NF number of fans in the mine
NFIXB number of fixed quantity flow branches
NJ number of junctions in the mine
NP number of data points for a fan curve
NQDBS number of branches with diesel exhaust generations
NQDJS number of junctions with diesel exhaust generations
NVP(I) natural ventilation pressure in branch no. I
QDBG quantity flow of diesel exhaust species in branches
QDBS quantity flow of diesel exhaust species in branches having diesel engines as a source
QDJG quantity flow of diesel exhaust species in junctions from sources other than diesel engines
QDJS quantity flow of diesel exhaust species in junctions having diesel engines as a source
TLV threshold limit value of concentration of diesel exhaust
WIDTH(I) width of branch no. I
X(I) quantity flow for fan data point no. I

Other Symbols for Quantity Flow Analysis Section

BRANCH(I) locates branch numbers in decreasing order of resistance
J1(I) inlet junction number of branch no. 1
J2(I) exit junction number of branch no. 1
JC(I) labels each tree at junction no. 1

ME(L) marks last subscript location in array NA for mesh no. L
NA(JK) contains the branch number in each mesh
NFBPF number of fixed flow branches plus the number of fans
NM number of meshes in the mine
OUT(I) marks the basic branch with a one and marks branches in the mesh or those not to be considered for a mesh with a minus one
Q(I) quantity of air flowing through branch I
R(I) resistance of branch no. I
RR(I) auxiliary storage (for sorting) for resistance

Other Symbols for Fan Curve Section

ALPHA(I) the constant αi, in polynomial Fi(Q)
BETA(I) the constant $\beta i-1$, in polynomial Fi(Q)
C(L, I) the constant Ci-1, of the conventional polynomial of fan no. L
DD(JJ) pre-iteration flow estimate in branch with fan or fixed quantity flow; both in branch no. JJ
ND(L) degree of polynomial for fan no. I
NI number of constants in the polynomial
P(I, J) the polynomial F_{i-1} (Q) evaluated at Q
S(I) the constant, S_{i-1}, of the orthogonal polynomial
SIGMA2(I) factor to test curve fit of polynomial of degree I
SMALL temporary storage location for the smallest value of σ
W(I) denominators used in calculation of ALPHA(I) and S(I)

Symbols for Iteration Procedure

DHF derivative of the fan curve
FQ(IM) fan pressure of fan no. IM
IM the fan number
SUMD the sum of all flow corrections
SUMDH denominator in the equation for calculation of flow corrections
SUMH numerator in the equation for calculation of flow corrections

Other Symbols for Diesel Exhaust Concentration Calculation Section

CDB(I) concentration of diesel exhaust in branch no. I
CDJ(JJ) concentration of diesel exhaust at junction counter no. JJ
CDWALL(I) bulk flow of diesel exhaust in branch no. I
J1(I) junction number of entrance of branch no. I
J2(I) junction number at exit of branch no. I
JCTN(JJ) junction number (JJ is junction counter no.)
JIN(JJ) highest value of LL (in NBIN) for junction counter no. JJ
JOUT(JJ) highest value of L (in NBOUT) for junction counter no. JJ
MB(I) branch marker array (I is the branch number)
MJ(I) junction marker array (I is the junction counter number)
NBIN(LL) branch number having flow toward junction (LL is a counter number)
NBOUT(L) branch number having flow away from junction (L is a counter number)

Q(I) quantity flow of air in branch no. I
QD quantity flow of diesel exhaust
QDJUNC(I) flow of diesel exhaust at junction no. I
SUMAIR summation of quantity flows of air through a junction
SUMMTN summation of quantity flow of diesel exhaust through a junction
TLVMXB(I) mixture TLV in Branch, I
TLVMXJ(I) mixture TLV at Junction, I

Program Listings

```
*********************************************************************
C
C ***A DIGITAL SIMULATOR FOR DIESEL EXHAUST CONTAMINATION OF MINE ***
C ******                                                     *****
                                                              ***
                                                              ***
***                                                           ***
*********************************************************************
C *** IT IS A GENERAL PURPOSE PROGRAM FOR THE STEADY STATE ***
C TURBULENT DISPERSION OF ANY GASEOUS POLLUTANT IN THE MINE ***
C *** AIRWAYS NETWORK. IN THIS CASE IT CALCULATES THE CONCENTRATION ***
C *** OF VARIOUS SPECIES OF DIESEL EXHAUST IN THE MINE NETWORK ***
C *** THE GEOMETRY OF THE ROADWAY NAMELY, THE LENGTH, THE WIDTH, THE ***
C *** HEIGHT AND THE FRICTION FACTOR, THE FAN CHARACTERISTIC, AND THE ***
C ***SOURCE STRENGTH OF THE POLLUTANTS AS WELL AS ITS LOCATIONS IN ***
C *** THE NETWORK ARE USED AS INPUT TO THE PROGRAM. FIRSTLY, THE ***
C ***PROGRAM CALCULATES THE QUANTITY OF AIR IN EACH BRANCH BY ***
C FORMING MESHES AROUND A TREE DETERMINED UNIQUELY AND APPLYING ***
C CORRECTION TO EACH MESH ACCORDING TO THE HARDY CROSS ITERATIVE ***
C TECHNIQUE AND KIRCHHOFF'S LAWS. THE FAN CHARACTERISTIC IS **
C APPROXIMATED BY ORTHOGONAL POLYNOMIALS AND ITS OPERATING ***
C *** POINT IS ALSO DETERMINED. FINALLY, IT COMPUTES CONCENTRATIONS ***
C *** OF DIESEL EXHAUST IN ALL THE BRANCHES AND AT ALL THE ***
C *** JUNCTIONS OF THE NETWORK. IT ALSO PRINTS OUT IF THE THRESHOLD ***
C ***VALUES OF THE COMPONENTS OF DIESEL EXHAUST OR THAT OF THE ***
C *** MIXTURE THEREOF ARE EXCEEDED ANYWHERE. THIS PROGRAM HAS BEEN ***
C *** SUCCESSFULLY RUN ON THE IBM 370/168 COMPUTER USING FORTRAN IV G ***
C *** LEVEL COMPILER. IT NEEDS 280 K STORAGE SPACE. THE TIME AND ***
C *** RECORDS DEPEND ON THE SIZE OF THE MINE NETWORK. ***
*********************************************************************
```

```
C
C
C
IMPLICIT REAL*8(A-H, O-Z)
REAL*8 NVP(550), LENGTH(550)
REAL*8 CDB(550), CDJ(400), QDWALL(550), QDJUNC(400)
REAL*8 TLVMXB(550), TLVMXJ(400)
INTEGER BRANCH
INTEGER*4 JCTN(400), NBOUT(550), JOUT(400), NBIN(550), JIN(400),
1MB(550), MJ(400)
C
DIMENSION NA(4200), JC(999), BRANCH(55 0), R(550), Q(550), RR(550)
1 OUT(550), J1(400), J2(400), ME(350), SUMNVP(350), ND(40), DD(80)
2 X(20), FX(20), W(6), Z(6), S(6), ALPHA(6), BETA(6), SIGMA2(6),
C(40, 6), 3P(7, 20), FQ(40), HH(550), TITLE(10), HEIGHT(550),
WIDTH(550)
DIMENSION SUBTIT(10)
C
EQUIVALENCE (NA, JC), (BRANCH, SUMNVP), (BRANCH(351), C), (Q, RR, OUT
1 NVP, X), Q(21), FX), CQ(41), W), (Q(51), Z), (Q(61), S), (Q(71),
SIGMA2)
2 (Q(81), ALPHA), (Q(91), BETA), (Q(101), P)
C
C READ IDENTIFICATION OF THE PROBLEM
10 READ 500, (TITLE(I), I=1, 10) R-1
C
C READ PARAMETERS
READ 501, NB, NJ, MAXJ, NF, MAXIT, NBWNVP, NFIXB, E R-2
IF(NB)11, 11, 12
11 STOP
12 PRINT 610, (TITLE(I), I=1, 10)
NM=NB-NJ+1
NFBPF=NFIXB+NF
PRINT 502, NB, NJ, MAXJ, NM, NF, MAXIT, NBWNVP, NFIXB, E P-1
C
C READ BRANCH CHARACTERISTICS
DO 15 I=1 , NB
READ 503, J1(I), J2(I), R(I) , FF, HEIGHT(I), WIDTH( I), LENGTH( I)
BRANCH(I)=I
IF(R(I)) 13, 13, 14
13 T=HEIGHT(I)*WIDTH(I)
R(I) = FF*(HEIGHT(I)+WIDTH(I))*LENGTH(I)/(2.6*T*T*T )
```

```
14 RR(I)=R(I)
IF(FF.GT.0.0.AND.HEIGHT(I).GT.0.0) PRINT 504
1, I, J1(I), J2(I), R(I), FF, HEIGHT(I), WIDTH(I), LENGTH(I)
IF(FF.LE.0.0.AND.HEIGHT(I).GT.0.0) PRINT 528, I, J1(I), 1J2(I), R(I),
HEIGHT(I), WIDTH(I), LENGTH(I)
IF(HEIGHT(I).LE.0.0)PRINT 504, I, J1(I), J2(I), R(I)
15 CONTINUE
C
C ARRANGE GENERAL BRANCHES IN DECREASING ORDER OF RESISTANCE
IS=NFBPF+1
IE=NB-1
JE=IE
L = 0
DO 18 I=IS, IE
DO 17 J=IS, JE
IF(RR(J+1)-RR(J)) 17, 17, 16
16 T=RR(J)
RR(J)=RR(J+1)
RR( J+1 )=T
T=BRANCH(J)
BRANCH(J) = BRANCH(J+1)
BRANCH(J+1)=T
L=1
17 CONTINUE
IF(L)19, 19, 18
18 JE=JE-1
C
C DETERMINE BASIC BRANCHES
19 DO 20 I=1, MAX J
20 JC(I)=0
I=NB+1
L=0
N =0
DO 31 IJ=IS, NB
I = I-1
OUT(I)=0.
K=BRANCH(I)
JA=J1(K)
JB=J2(K)
IF(JC(JA)-JC(JB))26, 28, 21
21 IF(JC(JB))22, 25, 22
22 JJ=JC(JB)
DO 24 J=1, MAXJ
IF(JC(J)- JJ)24, 23, 24
```

```
23 JC(J)=JC(JA)
24 CONTINUE
GO TO 31
25 JC(JB)=JC(JA)
GO TO 31
26 IF(JC(JA))22, 27, 22
27 JC(JA)=JC(JB)
GO TO 31
28 IF(JC(JA))29, 30, 29
29 OUT(I)=1.
N=N+1
GO TO 31
30 L=L+ 1
JC(JA)=L
JC(JB)=L
31 CONTINUE
IF(N+NFBPF-NM)32, 33, 32
32 PRINT 505, N
STOP 32
33 IF(NFBPF)36, 36, 34
34 DO 35 I=1, NFBPF
35 OUT(I)=1.
C
C FIND MESHES
36 JK=0
JE=0
L=0
DO 54 I=1, NB
IF(OUT(I))54, 54, 37
37 K=BRANCH(I)
L=L+1
JK=JK+1
NA(JK)=K
JA=J1(K)
JB=J2(K)
N=I+1
38 DO 45 J=N, NB
IF(OUT(J))45, 39, 45
39 K=BRANCH(J)
IF(JB-J1(K))41, 40, 41
40 JB=J2(K)
JK=JK+1
```

```
NA(JK)=K
GO TO 43
41 IF(JB-J2(K))45, 42, 45
42 JB=J1(K)
J K= J K+1
NA (JK)=-K
43 IF(JB-JA)44, 51, 44
44 OUT(J)= -1.
GO TO 50
45 CONTINUE
K=IABS(NA(JK))
IF(NA(JK))47, 46, 46
46 JB=J1(K)
GO TO 48
47 JB=J2(K)
48 JK=JK-1
IF(JK-JE)49, 49, 38
49 K-BRANCH(I)
PRINT 525, K
STOP 49
50 GO TO 38
51 DO 53 J=N, NB
IF(OUT(J))52, 53, 53
52 OUT(J)=0.
53 CONTINUE
ME(L)=JK
JE=JK
54 CONTINUE
DO 55 I=1, NB
55 NVP(I)=0.
IF(NBWNVP) 57, 57, 56
C
C READ AND PRINT NATURAL VENTILATION PRESSURE
56 READ 506, (I, NVP(I), J=1, NBWNVP)
PRINT 507
PRINT 508, (I, NVP(I), I=1, NB)
C
C PRINT MESH TABLE
57 PRINT 509
JE=0
DO 61 I=1, NM
JS=JE+1
```

```
JE=ME(I)
SUMNVP(I)=0.
K=IABS(NA(J))
DO 60 J=JS, JE
IF(NA(J))59, 58, 58
58 SUMNVP(I)=SUMNVP(I)+NVP(K)
GO TO 60
59 SUMNVP(I)=SUMNVP(I)-NVP(K)
60 CONTINUE
L=JE-JS+1
PRINT 510, I, L, SUMNVP(I)
61 PRINT 511, (NA(J), J=JS, JE)
PRINT 512, JE
IF(NFBPF)88, 88, 261
C
C APPROXIMATION OF FAN CHARACTERISTICS
261 IF(NF)287, 287, 62
62 PRINT 513
DO 87 L=1, NF
JJ=L+NFIXB
READ 501, NP, KEY
PRINT 514, L, J1(JJ), J2(JJ), KEY
IF(KEY)63, 63, 85
63 READ 515, (X(I), FX(I), I=1, NP)
PRINT 516, (I, X(I), FX(I), I=1, NP)
DD(JJ)=X(NP)
IF(NP-2)910, 910, 970
910 ND(L)=1
N1=2
IF(NP-1)920, 930, 940
920 STOP 920
930 C(L, 1)=FX(1)
C(L, 2)=0.
GO TO 840
940 IF(X(1)-X(2))960, 950, 960
950 STOP 950
960 C(L, 1)=(FX(1)*X(2)-FX(2)*X(1))/(X(2)-X(1))
C(L, 2) = (FX(2)-FX(1))/(X(2)-X(1))
GO TO 840
970 DSQ=0.
DO 64 J=1, NP
DSQ=DSQ+FX(J)*FX(J)
```

```
P(1, J)=0.
64 P(2, J)=1.
W(1)=NP
BETA(1)=0.
IF(NP-7)66, 65, 65
65 N1=6
GO TO 67
66 N1=NP-1
67 DO 73 1=1, N1
K=I+1
Z(I)=0.
DO 68 J=1, NP
68 Z(1)=Z(I)+FX(J)*P(K, J)
S(I)=Z(I)/W(I)
DSQ-DSQ-S(I)*S(I)*W(I)
SIGMA2(I)-DSQ/FLOAT(NP-I)
IF(SIGMA2(I)-1.D-6)78, 78, 69
69 IF(I-N1)70, 74, 74
70 ALPHA(I)=0.
DO 71 J=1, NP
71 ALPHA(I)=ALPHA(I) +X (J)*P(K, J)*P(K, J)
ALPHA(I)=ALPHA(I)/W(I)
W(K)=0.
DO 72 J=1, NP
P(K+1, J)=(X(J)-ALPHA(I))*P(K, J)-BETA(I)*P(I, J)
72 W(K)=W(K)+P(K+1, J)*P(K+1, J)
73 BETA(K)=W(K)/W(I)
74 SMALL=SIGMA2(1)
I = 1
DO 77 J=2, N1
IF(SMALL-SIGMA2(J))77, 77, 76
76 I=J
SMALL=SIGMA2(J)
77 CONTINUE
78 PRINT 517, SIGMA2(I)
N1 = I
N=N1-1
IF (N)780, 780, 79
780 ND(L) = 1
C(L, 1)= S(1)
C(L, 2)=0.
GO TO 840
79 ND(L)=N
```

```
DO 81 I=1, N1
DO 80 J=1, I
80 P(I, J)=0.
81 P(I, J+1)=1.
DO 82 J=1, N
82 P(1, J+2)=-P(1, J+1)*ALPHA(J)-P(1, J)*BETA(J)
DO 83 I=2, N
DO 83 J= I, N
83 P(I, J+2) = P(I-1, J+1)-P(I, J+1)*ALPHA(J)-P(I, J)*BETA(J)
DO 84 I=1, N1
C(L, I)=0.
DO 84 J=I, N1
84 C(L, I)=C(L, I)+P(I, J+1)*S(J)
840 PRINT 518, (C(L, I), I=1 , N1)
GO TO 87
85 ND(L)=ND(KEY)
N1=ND(KEY)+1
DD(JJ)=DD(KEY+NFIXB)
DO 86 J=1, N1
86 C(L, J)=C(KEY, J)
87 CONTINUE
287 IE = NFBPF
IF(NFIXB)89, 89, 288
288 READ 515, (DD(I), I=1, NFIXB)
GO TO 89
C
C SET INITIAL VALUES
88 DD(I)=100.
IE=1
89 DO 90 I=1, NB
Q(I) = 0.
90 R(I)=R(I)*1.D-04
JE=0
DO 93 I=1, IE
JS=JE+1
JE=ME(I)
DO 93 J=JS, JE
K=1ABS(NA(J))
IF(NA(J))92, 91, 91
91 Q(K) = Q(K)+DD(I)
GO TO 93
92 Q(K)=Q(K)-DD(I)
93 CONTINUE
```

```
C
C BEGIN ITERATION
IB=NFIXB+1
DO 108 IT=1, MAXIT
IF(NFIXB)820, 820, 810
810 JE=ME(NFIX8)
GO TO 830
820 JE=0
830 L=0
SUMD=0.
DO 107 I=IB, NM
IM=I-NFIXB
JS=JE+1
JE=ME(I)
SUMH= -SUMNVP(I)
SUMDH=0.
DNF=0.
IF(NFBPF-I)98, 94, 94
94 N=ND(IM)
FQ(IM)=C(IM, N+1)*Q(I)
IF(N-1)97, 97, 95
95 J=N
DO 96 IJ=2, N
FQ(IM)=(FQ(IM)+C(IM, J))*Q(I)
DHF=(DHF+FLOAT(J)*C(IM, J+1))*Q(I)
96 J=J-1
97 FQ(IM)=FQ(IM)+C(IM, 1)
SUMH=SUMH-FQ(IM)
DHF=DHF+C(IM, 2)
98 DO 101 J=JS, JE
K=IABS(NA(J))
DH=R(K)*DABS(Q(K))
H=DH*Q(K)
SUMDH=SUMDH+DH
IF(NA(J))100, 99, 99
99 SUMH=SUMH+H
GO TO 101
100 SUMH= SUMH-H
101 CONTINUE
SUMDH=SUMDH+SUMDH-DHF
IF(DABS(SUMDH)-1.D-20)106, 106, 102
102 D=-SUMH/SUMDH
DO 105 J=JS, JE
```

```
K=IABS(NA(J))
IF(NA(J))104, 103, 103
103 Q(K)=Q(K)+D
GO TO 105
104 Q(K) = Q(K)-D
105 CONTINUE
SUMD= SUMD+DABS (D)
IF(DABS(D)-E)107, 107, 106
106 L-1
107 CONTINUE
IF(L)109, 109, 108
108 CONTINUE
109 PRINT 519, IT, SUMD P-4
PRINT 520
GO TO 110
109 PRINT 519, IT, SUMD P-5
C
C COMPUTE HEAD LOSS
C PRINT SOLUTION
110 PRINT 521
DO 111 I=1, NB
HH(I)=R(I)*DABS(Q(I))*Q(I )
111 PRINT 522, I, J1(I ), J2(I), R(I), Q(I), HH(I) P-6
C
C COMPUTE THE PRESURE TO BE ADJUSTED IN A FIXED QUANTITY BRANCH
C -- R(I)*DABS(Q(I) )*Q(I ) IS ZERO IF R(I) OF A FIXED QUANTITY BRANCH IS
C -- NOT GIVEN
IF(NFIXB)880, 880, 850
850 PRINT 600
JE=0
DO 870 I=1, NFIXB
JS=JE+1
JE=ME(I)
SUMH=-SUMNVP(I)
DO 860 J=JS, JE
K=IABS(NA(J))
IF(NA(J)) 856, 855, 855
855 SUMH=SUMH+HH(K)
GO TO 860
586 SUMH=SUMH-HH(K)
860 CONTINUE
PRINT 522, I, J1(I), J2(I), R(I), Q(I), SUMH
C CALCULATE THE SIZE OF THE REGULATOR USING MURGUE'S FORMULA
```

```
C OR THE CAPACITY OF THE BOOSTER FAN AS THE CASE MAY BE
IF(SUMH) 865, 866, 866
865 SUMH=-SUMH
REGSIZ=0.389*Q(I)/DSQRT(SUMH)
PRINT 530, REGSIZ
GO TO 870
866 PRINT 540, SUMH, Q(I)
870 CONTINUE
C
C PRINT FAN PRESSURE
880 IF(NF) 885, 885, 890
890 PRINT 523
PRINT 524, (I, J1(I+NFIXB), J2(I+NFIXB), Q(I+NFIXB), FQ(I), I=1, NF) P-7
885 READ 501, K
IF(K.LE.0) GO TO 10
INDIC=0
C MAKE NECESSARY PREPARATIONS FOR THE CALCULATION
C OF POLLUTANT CONCENTRATIONS....
C
C MAKE ALL FLOWS POSITIVE
IF(INDIC.EQ.I) GO TO 1050
DO 1010 I=1, NB
Q(I)=Q(I)*1. E3
IF(Q(I).GT.0.0) GO TO 1010
L=J1(I)
J1(I)=J2(I)
J2(I)=L
Q(I)=-Q(I)
1010 CONTINUE
C
C TABULATE INFLOWS AND OUTFLOWS
L = 0
LL=0
JJ= 1
DO 1040 I=1, MAXJ
K=L
DO 1020 J=1, NB
IF(J1(J).NE.I) GO TO 1020
L = L+1
NBOUT(L)=J
1020 CONTINUE
JOUT(JJ)=L
KK=LL
DO 1030 J=1, NB
```

```
IF(J2(J).NE.I) GO TO 1030
LL=LL+1
NBIN(LL)=J
1030 CONTINUE
JIN(JJ)=LL
IF((KK.EQ.LL).AND.(K.EQ.L)) GO TO 1040
JCTN(JJ) = I
JJ=JJ+1
1040 CONTINUE
INDIC= 1
C
C SET MIXTURE T.L.V. TO ZERO...
DO 1043 I=1, NB
1043 TLVMXB(I)=0.0
DO 1045 I=1, NJ
1045 TLVMXJ(I)=0.0
C
C SET DIESEL EXHAUST CONCENTRATIONS TO ZERO
1050 DO 1060 I=1, NB
CDB(I)=0.0
1060 MB(I)=0
DO 1070 I =1, NJ
COJ(I)=0.0
1070 MJ(I) = 0
C
C IDENTIFY THE COMPONENT OF THE DIESEL EXHAUST
READ 500, (SUBTIT(I), I=1, 10)
PRINT 610, (SUBTIT(I), I=1, 10)
C
C READ OUTSIDE CONDITIONS
READ 651, JJ
C
C FIND JUNCTION COUNTER FOR OUTSIDE JUNCTION
DO 1080 J=1, NJ
IF(JCTN(J).NE.JJ) GO TO 1080
JJ=J
GG TO 1090
1080 CONTINUE
1090 CDJ(JJ)=0.0
MJ(JJ)=1
C
C READ GENERAL DIESEL EXHAUST GENERATION RATES
READ 652, QDBG, DEL, QDG, QDJG, TLV, NQDBS, NQDJS IF(DELQDG.EQ.0.0) GO TO 1110
DO 1100 1=1, NB
```

```
1100 QDWALL(I)=DELQDG*LENGTH(I)
GO TO 1140
1110 DO 1120 I=1, NB
1120 QDWALL(I)=QDBG
DO 1130 I=1, NJ
1130 QDJUNC(I)=QDJG
C
C READ SPECIAL DIESEL EXHAUST GENERATION RATES
1140 IF(NQDBS.EQ.0) GO TO 1160
DO 1150 J=1, NQDBS
READ 653, I, QDBS, DELQDS
IF(QDBS.NE.0.0) QDWALL(I)=QDBS
IF(DELQDS.NE.0.) QDWALL(I)=DELQDS*LENGTH(I)
1150 CONTINUE
1160 IF(NQDJS.EQ.0) GO TO 1200
DO 1180 J=1, NQDJS
READ 654, 11, QDJS
DO 1170 1=1, NJ
IF(JCTN(I).NE.II) GO TO 1170
QDJUNC(I)=QDJS
GO TO 1180
1170 CONTINUE
1180 CONTINUE
C
C ROADWAY CALCULATIONS
1200 JE=1
IF(JJ.NE.1) JE=JOUT(JJ-1)+ 1
JS=JOUT(JJ)
DO 1300 K=JE, JS
I=NBOUT(K)
QD=CDJ(JJ)*Q(I)+QDWALL(I)
COB(I)=QD/Q(I)
MB(I)=1
1300 CONTINUE
C
C JUNCTION CALCULATIONS
DO 1400 I=1, NJ
IF (MJ(I).NE.0) GO TO 1400
JE= 1
IF(I.NE.1) JE=JIN(I-1)+1
JS=JIN(I)
SUMMTN=0.0
SUMAIR=0.0
```

```
DO 1310 K=JE, JS
J=NBIN(K)
IF(MB(J).EQ.0) GO TO 1400
SUMMTN=SUMMTN+Q(J)*CDB(J)
SUMAIR=SUMAIR+Q(J)
1310 CONTINUE
JJ=I
SUMMTN=SUMMTN+QDJUNC(I)
IF(SUMNTN.LE.SUMAIR*1.E-70) GO TO 1320
CDJ(I)=SUMMTN/SUMAIR
1320 MJ(I)=1
GO TO 1200
1400 CONTINUE
C
C PRINT POLLUTANT CONCENTRATIONS
PRINT 660
DO 1410 I=1, NB
CDB(I)=CDB(I)*10.0**6
C MULTIPLICATION BY 10.0**6 CONVERTS THE CONCENTRATION INTO,
C PARTS PER MILLION.........
PRINT661, I, QDWALL(I), CDB(I) , TLV
C CHECK IF TLV IS EXCEEDED
IF(CDB(I).GT.TLV) PRINT 663, I
C CALCULATE THE T.L.V. OF THE MIXTURE IN BRANCHES...
TLVMXB(I)=TLVMXB(I)+CDB(I)/TLV
1410 CONTINUE
PRINT 662
DO 1420 I=1, NJ
CDJ(I)=CDJ(I)*10.0**6
C MULTIPLICATION BY 10.0**6 CONVERTS THE CONCENTRATION INTO
C PARTS PER MILLION.........
PRINT 661, JCTN(I), QDJUNC( I ), CDJ(I), TLV
C CHECK IF TLV IS EXCEEDED
IF(CDJ(I).GT.TLV)PRINT 664, 1
C CALCULATE THE T.L.V. OF THE MIXTURE AT JUNCTIONS......
TLVMXJ(I)=TLVMXJ(I)+CDJ(I)/TLV
1420 CONTINUE
READ 651.K
IF(K-1) 1425, 1050, 1050
1425 PRINT 665
PRINT 666
DO 1430 I=1, NB
PRINT 667, I, TLVMXB(I)
C CHECK IF THE MIXTURE TLV IS EXCEEDED...
```

```
IF(TLVMXB(I).GT.1.0) PRINT 668, I
1430 CONTINUE
PRINT 669
DO 1440 I=1, NJ
PRINT 667, I, TLVMXJ(I)
C CHECK IF THE MIXTURE TLV IS EXCEEDED....
IF(TLVMXJ(I).GT.1.0) PRINT 670, I
1440 CONTINUE
READ 651, K
IF(K-1) 10, 1500, 1500
1500 CONTINUE
C
C FORMAT STATEMENTS
500 FORMAT(10A8)
501 FORMATS (7I10, F10.4)
502 FORMAT(//1H0, 12X, 2HNB, 7X, 2HNJ, 5X, 4HMAXJ , 7X, 2HNM, 7X, 2HNF, 4X,
1 5HMAXIT, 3X, 6HNBWNVP, 4X, 5HNFIXB, 10X, 1HE/6X, 819, F11.4//
2 1H0, 8X, 6HBRANCH, 7X,
3 2HJ1, 4X, 2HJ2, 14X, 1HR, 13X, 2HFF, 4X6HHEIGHT, 5X, 5HWIOTH, 4X,
6HLENGTH/)
503 FORMAT(2I10, 5F10.2)
504 FORMAT(1X, I14, 3X, 216, F15.6, 5X, 4F10.1)
505 FORMAT(//27H0** NO. OF BASIC BRANCHES =, I4, 5H + NF)
506 FORMAT(4(I10, F10.3))
507 FORMAT(1H1*5(4X, 6HBRANCH, 11X, 4HNVP)/)
508 FORMAT(5(I11, F14.7))
509 FORMAT(1H1, 10X, 4HMESH, 7X, 6HNO. OF, 16X, 6HSUMNVP/22X,
8HBRANCHES/)
510 FORMAT(2I15, F20.7)
511 FORMAT (60X, 10I6)
512 FORMAT(/18X, 7HTOTAL =, 15)
513 FORMAT (1H1)
514 FORMAT(//12X, 3HFAN, 8X, 2HJ1, 4X, 2HJ2, 7X, 3HKEY, 15X, 5 HPOINT,
11X, 1HX, 1 8X, 2HFX//I15, 4X, 2I6, I10)
515 FORMAT (8F10. 3)
516 FORMAT(I61, 2X, 2F10.4)
517 FORMAT(/18X, 8HSIGMA2 =, 1PE14.7)
518 FORMAT(1H0, 17X, 12HCOEFFICIENTS//18X, 1P6E18.7)
519 FORMAT(1H1, 18, 11H ITERATIONS, 9X, 6HSUMD =, 1PE14.7)
520 FORMAT(10X, 25H** D STILL GREATER THAN E)
521 FORMAT(///14X, 6HBRANCH, 8X, 2HJ1, 3X, 2HJ2, 19X, 1HR, 19X, 1HQ,
19X, 1HH/)
522 FORMAT(120, 5X, 215, 4PF20.6, OPF20.4, OPF20.7)
523 FORMAT(1H1, 16X, 3HFAN, 8X, 2HJ1, 3X, 2HJ2, 19X, 1HQ, 18X, 2HFQ/)
```

```
524 FORMAT(I20, 5X, 2I5, F20.4, F20, 7)
525 FORMAT(//27HC** NO MESH FOUND AT BRANCH, 14)
528 FORMAT (1X, I 14, 3X, 216, F15.6, 15X, 3F10.1)
530 FORMAT('0%', 'THE SIZE OF THE REGULATOR IS' , F10.4,' SQ FT')
540 FORMAT ('0', 'THE CAPACITY OF THE BOOSTER FAN IS', F10.4, 1' INCHES
OF W.G. AT', 3PF10.4, 'C.F.M.')
600 FORMAT(1H1, 13X'HEAD LOSS TO BE ADJUSTED IN FIXED QUANTITY BRANCH'
1 ///14X, 6HBRANCH, 8X, 2HJ1, 3X, 2HJ2, 19X, 1HR, 19X, 1HQ, 16X,
4HSUMH/)
610 FORMAT(1H1, 4X, 10A8)
651 FORMAT(I10)
652 FORMAT (F10.6, F10.6, F10.6, F10.4, 2110)
653 FORMAT(I10, F10.6, F10.6)
654 FORMAT(I10, F10.6)
660 FORMAT(1H1, 24X, 36HDIESEL EXHAUST CONCENTRATION RESULTS///
133X, 13HBRANCH VALUES//35X, 14HDIESEL EXHAUST, 13X,
113HCONCENTRATION, 10X, 3HTLV/
213X, 13HBRANCH NUMBER, 8X, 10HFROM WALLS, 18X, 6HIN PPM)
661 FORMAT(18X, I3, 12X, F10, 4, 19X, F15.1, 5X, F10.1)
662 FORMAT(1H1, 30X, 15HJUNCTION VALUES//
135X, 14HDIESEL EXHAUST, 13X, 13HCONCENTRATION, 10X, 3HTLV/13X,
215HJUNCTION NUMBER, 4X, 13HFROM JUNCTION, 17X, 6HIN PPM)
663 FORMAT('0', '**THE DIESEL EXHAUST CONCENTRATION EXCEEDED,
1'ITS TLV IN BRANCH NO.', I3)
664 FORMAT('0', **THE DIESEL EXHAUST CONCENTRATION EXCEEDED, 1'ITS TLV
AT JUNCTION NO,', I3)
665 FORMAT(1H1, 'THE FOLLOWING ARE THE VALUES OF TLV OF 1 THE MIXTURE OF
DIESEL EXHAUST COMPONENTS')
666 FORMAT(1H1, 13X, 'BRANCH NUMBER', 5X, 'MIXTURE T.L.V.')
667 FORMAT('0', 18X, 13, 15X, F10.2)
668 FORMAT('0', '**THE MIXTURE OF DIESEL EXHAUST EXCEEDED ITS TLV
IIN BRANCH NO.', 13)
669 FORMAT(1H1, 13X, 'JUNCTION NUMBER', 5X, 'MIXTURE T.L.V.')
670 FORMAT('0', '** THE MIXTURE OF DIESEL EXHAUST COMPONENTS EXCEEDED
IITS T.L.V. AT JUNCTION NO.', 13)
STOP
END
/*
```

Brief Description of Program Listings

Lines 1–9 are declaration and dimension statements.

Lines 10–351 constitute the program developed for quantity flow analysis in a network of airways with minor changes. Excellent description of this part of the program is available elsewhere (16, 18) and, consequently, it is described here very briefly. Essentially, this part of the program reads in the branch characteristics, e.g., length, width, height, friction factor, et cetera, and forms meshes around a tree determined uniquely on the basis that the basic branch in a mesh shall have the highest resistance. Natural ventilating pressures, if any, are also read in. Next, the program approximates the fan characteristic using orthogonal polynomials. The solution for each mesh is obtained by the well-known Hardy Cross iterative technique in conjunction with Kirchhoff's laws for networks. The program prints out all the input parameters and data as well as the quantity and pressure loss in each branch of the network. The operating point on the fan characteristic is also determined and printed out. For branches where a fixed quantity of air is provided, the size of the regulator or the capacity of the booster fan as the case may be is printed out.

Line 352 initializes a computational switch INDIC.

Line 353 checks the value of INDIC which avoids repeated calculations when several runs are to be made on the same mine with different sets of data.

Lines 354–361 convert all the quantity flows, originally in thousands of cubic feet per minute into cfm and makes them positive wherever they are negative. In the latter case, it also interchanges the junction numbers of that branch accordingly.

Lines 362–383 gather information that make it possible to follow the flow of air and diesel exhaust from one junction to the next. There are five one-dimensional arrays here which store the following information:

Array JCTN(I) stores all the junction numbers in the network in a sequence, compacting them in the process and saving storage.

Array NBOUT(J) stores the branch number of roadways, having flow away from junction JCTN(I).

Array JOUT(JJ) stores the location of branch J in array NBOUT(J).

Array NBIN(J) stores the branch number of roadways having flow into the junction, JCTN(I).

Array JIN(JJ) stores the location of branch J in array NBIN(J).

Line 384 sets the value of INDIC equal to 1.

Lines 385–394 initialize the arrays of concentration of diesel exhaust components as well as that of the mixture TLV of these components and the branch and junction counter arrays, MB(I) and MJ(I), respectively.

Lines 395–396 identify the component of the diesel exhaust and print out the same.

Line 397 reads the outside junction, i.e., the junction representing the open atmosphere. Calculations of concentrations start here.

Lines 398–403 find out the junction counter number for the outside junction. This is labeled zero at line 403. CDJ(I) is the array which holds diesel exhaust concentration at all junctions.

Line 404: MJ(JJ) is set equal to one to indicate that the JJ junction, i.e., outside junction, has been processed.

Lines 405–413 read the general diesel exhaust emission rate in all branches and at all junctions. These are values of those components of diesel exhaust which are given out by sources other than diesel engines in all roadways, e.g., hydrocarbons. Normally, these quantities would be exceedingly small and can be set equal to zero. On the other hand, the number of branches and junctions with diesel exhaust sources are treated as branches and junctions with special diesel exhaust emission rates. The last two variables on line 405, viz., NQDBS and NQDJS, are the number of such branches and junctions, respectively, in the network. The general exhaust emission rates can be only read in as bulk flow in cubic feet per minute for junctions, but for a branch it can be read in either as bulk flow or incremental flow, e.g., cfm/ft. of the roadway. In any case, the bulk flow is always calculated and stored in the arrays QDWALL(I) and QDJUNC(I) for branches and junctions, respectively.

Lines 414–438 handles all the branches and junctions which are not covered by the general case. From line 416 to 419, special diesel exhaust emissions are read in for individual branches and stored in the array QDWALL(I). Those lines are skipped if NQDBS is zero. As with the previous case, the special diesel exhaust emission for a branch can be read in as a bulk flow or as incremental flow over the entire length of the roadway. From line 421 to 428, special diesel exhaust emissions at particular junctions are read in and stored in QDJUNC(I). Those lines are skipped if NQDJS is zero at line 420. Lines 423–424 determine the junction counter number corresponding to junction no. II. As mentioned before, QDJUNC(I) is the quantity of diesel exhaust flow available at junction number 1.

Lines 429–437 are labelled as roadway calculations. They add the diesel exhaust emitted in each roadway to the total diesel exhaust flowing into that roadway. They also contain operations which determine the branches having flow away from a given junction. When line 429 is entered for the first time, the starting junction is the outside junction counter number, JJ. The next four lines determine the branch number I, having flow away from the junction JJ. For the first pass through the "ROADWAY CALCULATION" section, these branches will be all the roadways leading into the mine from the surface. Line 434 adds diesel exhaust emission in the roadway (equation [3.55]) and line 435 determines the concentration (equation [5.8]). Line 436 marks these branches (in the array MB) with a 1, indicating that calculations for them are completed. Control is now passed to the section called junction calculations.

Lines 438–457determine the concentration of diesel exhaust at the junctions from the air and diesel exhaust flowing into that junction. In order to do this, a junction must be known where all these quantities are known. Starting on line 438, a search begins through all the junctions using the junction counter number. If the junction has already been processed, then that junction is bypassed (line 439).

Lines 440–447 determine the branch number J with inflow to the junction having counter no. I. If any of these in flowing roadways has not been processed, that is the diesel exhaust concentration in that roadway is not known (MB[J] is marked with a zero), then that junction is bypassed. If the diesel exhaust concentration for all the inflowing roadways are known, the total diesel exhaust flow at the junction (Eq. 5.9) and total air flow at the junction (Eq. 5.10) can be determined. This is accomplished on lines 448, 449, 452. On line 454, the diesel exhaust concentration at junction I is determined (Eq. 5.11) and this junction is marked with 1 on the next line, At this point, the control is passed back to line 429. The junction where diesel exhaust concentration was just found is now the new starting junction (line 451

JJ = I) from which the roadway calculations begin. The process is repeated again and again until concentrations for all the junctions are calculated.

Lines 458–471 prints the results after converting the concentrations into ppm and checking the values against their respective TLVs. Lines 459–462 do the above for branches in the network. At line 463, the mixture TLV is calculated as defined by the American Conference of Governmental Hygienists (1). Lines 466–471 do the same operations for all the junctions in the network.

Line 472 reads the value of another computational switch, K. A zero value of K indicates that all components of diesel exhaust have been processed and mixture TLV can now be calculated.

Line 473: If K is equal to one or larger, it indicates that more data have to be processed and control is transferred to line 389.

Lines 474–475 print titles for mixture TLV calculations.

Lines 476–484 check the value of mixture TLV. If the value of TLVMXB(I) or TLVMXJ(I) is greater than 1.0, a message is printed to that effect.

Lines 485–486 read another computation switch, K. A value of K equals to zero or less transfers the control to line 10 which eventually stops the program through a check on lines 12 and 13. The rest of the program is simply the format statements for input and output.

Appendix C: Ventilation Network Analyzer in C++ With Input and Output

HMineDeclarations_h

```
#include <iostream>
#include <string>
#include <iomanip>
#include <math.h>

using namespace std;

#define ARRAY_SIZE 100

string title, subtit;

double nvp[ARRAY_SIZE], sumnvp[ARRAY_SIZE], tlvmxb[ARRAY_SIZE],
qdwall[ARRAY_SIZE], length[ARRAY_SIZE];
double tlvmxj[ARRAY_SIZE], qdjunc[ARRAY_SIZE];
double qd, sumtn, sumair, summtn, dh, h, d, e, t, sumd, sumdh, sumh,
dhf, dsq, regsiz, small;
double fx[20], x[20], sigma2[6], q[50], w[6], dd[40], beta[6], z[40],
s[6], alpha[6], hh[40], fq[50];
double c[6][40], p[20][7];

unsigned int ja, jb;
unsigned int jj1[ARRAY_SIZE], j2[ARRAY_SIZE];

float qdbg, delqdg, qdjg, tlv, nqdbs, nqdjs, qdbs, delqds, qdjs, ff;
float r[ARRAY_SIZE], height[ARRAY_SIZE], width[ARRAY_SIZE],
ff_temp[ARRAY_SIZE], rr[ARRAY_SIZE], cdj[ARRAY_SIZE], mj[ARRAY_SIZE];
float cdb[ARRAY_SIZE], mb[ARRAY_SIZE], jout[ARRAY_SIZE],
jin[ARRAY_SIZE],
nbin[ARRAY_SIZE], nbout[ARRAY_SIZE];

int nb, nj, maxj, nf, maxit, nbwnvp, nfixb, cont_flag, nm, nfbpf, ib,
np, key, n1, im,
```

```
iterations, indic;
int jj, n, k, jk, je, js, is, tmp, ie, l, ll, kk, ii;
int nd[40], jc[ARRAY_SIZE], out[ARRAY_SIZE], na[ARRAY_SIZE],
jctn[ARRAY_SIZE],
branch[ARRAY_SIZE], me[ARRAY_SIZE];

void readBranchCharacteristics();
void printBranchCharacteristics();
void toFindBasicBranches();
void tofindMeshes();
void toReadPrintNVP();
void toPrintMeshTable();
void printc(int l);
void toApproximateFanCharacteristics();
void setInitialValues();
void toBeginIteration();
void printIterationandSum(int iterations,bool flag);
void computeAndPrintHeadLoss();
void adjustHLinFixedQuantityBranch();
void printFanPressure();
void setFlowsPositive();
void tabulateInAndOutFlows();
void setMixtureTLVZero();
void setDieselExhaustConcZero();
bool printPollutantConc();
void sortBranches();
void assignDEGenRates();
```

HMineDeclarations_sc

```
#include "HMineDeclarations.h"
   int main()
   {
     // Read Identification of the Problem
     cout << "\nInput title of the problem : " << endl;
     getline(cin, title);

       cout << "\nEnter nb, nj, maxj, nf, maxit, nbwnvp, nfixb values :
       " << endl;
       cin >> nb >> nj >> maxj >> nf >> maxit
       >> nbwnvp >> nfixb;

       cout << "\nEnter acceptable error value : " << endl ;
       cin >> e;

       // Number of branches should be a positive value
     if (nb <= 0)
          exit(0);

     cout << endl << title << endl;
       nm = nb - nj + 1;
       nfbpf = nfixb + nf;
     cout << "nb" << setw(10) << "nj" << setw(10) << "maxj" <<
setw(10) << "nm" << setw(10) << "nf" << setw(10) << "maxit" <<
setw(10) << "nbwnvp" << setw(10) << "nfixb" << setw(10) << "e" <<
endl;
     cout << nb << setw(10) << nj << setw(10) << maxj <<
setw(10) << nm << setw(10) << nf << setw(10) << maxit
<< setw(10) << nbwnvp << setw(10) << nfixb << setw(10) <<
e << endl ;

     // Read Branch Characteristics
     readBranchCharacteristics();

     printBranchCharacteristics();

     // Arrange General Branches in Decending Order of Resistance
     sortBranches();

     // Determine Branch Characteristics
     toFindBasicBranches();
```

```
    if((n+nfbpf-nm) != 0)
    {
      cout << "\n NUMBER OF BASIC BRANCHES = " << n << " AND
FANS = " << nf << " IN THE MINE" << endl;
      exit(0);
    }

    if(nfbpf > 0)
    {
      for(int i=1; i<=nfbpf; i++)
      {
        out[i] = 1;
      }
    }

    tofindMeshes();
    toReadPrintNVP();

    toPrintMeshTable();
    if(nfbpf > 0)
    {
      toApproximateFanCharacteristics();
      ie = nfbpf;

      if(nfixb > 0)
      {
        cout << "\n\nInput " << nfixb << " dd values : "<< endl;

        for(int i = 1; i<=nfixb ; i++)
        {
          cin >> dd[i];
        }
      }
    }
    else
    {
      dd[1] = 100;
      ie = 1;
    }

    setInitialValues();
    toBeginIteration();
    computeAndPrintHeadLoss();
```

```
/* Compute the pressure to be adjusted in a fixed quantity branch
 * r[i]*dabs(q[i])*q[i] is zero if r[i] of a fixed quantity
 * branch is not given
 */
if(nfixb > 0)
{
  adjustHLinFixedQuantityBranch();
}

printFanPressure();

cout << endl << "Input K value : " << endl;
cin >> k;

if(k <= 0)
{
  exit(0);
}
indic = 0;

// Make Necessary Preparations for the Calculation of Pollutant
   concentrations
// Make all the flows positive
if(indic !=1 )
{
   setFlowsPositive();
    // Tabulate Inflows and Outflows
    tabulateInAndOutFlows();

    indic = 1;
    // Set Mixture TLV to zero.....
    setMixtureTLVZero();
}
// Set Diesel Exhaust Concentrations to zero.......
label1050 : setDieselExhaustConcZero();

cin.clear();
cin.sync();

cout << "\nInput title of the component of diesel exhaust : \n";
getline(cin, subtit);
cin.clear();
cin.sync();
```

```
    //READ OUTSIDE CONDITIONS
    cout << "\nInput jj value" << endl;
    cin >> jj;

    // FIND JUNCTION COUNTER FOR OUTSIDE JUNCTION
    for(int j=1; j<=nj; j++)
    {
       if(jctn[j] == jj)
        {
          jj=j;
          break;
        }
    }
    cdj[jj] = 0.0;
    mj[jj] = 1;

    // READ GENERAL DIESEL EXHAUST GENERATION RATES
    cout << "\nInput qdbg, delqdg, qdjg, tlv and nqdjs values : " <<
endl;
    cin >> qdbg >> delqdg >> qdjg >> tlv >> nqdbs >> nqdjs;
    assignDEGenRates();

    //READ SPECIAL DIESEL EXHAUST GENERATION RATES
    if(nqdbs != 0)
    {
        int i;
        cout << "\nInput i, qdbs and delqds values : " << endl;
        for(int j=1; j<=nqdbs; j++)
        {
          cin >> i >> qdbs >> delqds ;
          if(qdbs != 0)
          {
            qdwall[i] = qdbs;
          }
          if(delqds != 0)
          {
            qdwall[i] = delqds*length[i];
          }
        }
    }
    if(nqdjs != 0)
    {
        cout << "\n Input ii and qdjs values : " << endl;
        for(int j=1; j<=nqdjs; j++)
        {
```

```
            cin >> ii >> qdjs;
            for(int i=1; i<=nj; i++)
            {
               if(jctn[i] == ii)
               {
                 qdjunc[i] = qdjs;
                 break;
               }
            }
         }
   }

   // ROADWAY CALCULATIONS

 label1200: je = 1;

   if(jj != 1)
      je = jout[jj-1]+1;
   js = jout[jj];

   for(int k=je; k<=js; k++)
   {
      int index = nbout[k];
        qd = cdj[jj] * q[index] + qdwall[index];
        cdb[index] = qd/q[index];
        mb[index] = 1;
   }

   // JUNCTION CALCULATIONS
   for(int i=1; i<=nj; i++)
   {
      if(mj[i] == 0)
       {
          je = 1;
          if(i != 1)
          {
              je = jin[i-1]+1;
          }
          js = jin[i];
          summtn = 0.0;
          sumair = 0.0;
          bool flag = false;
          for(int k=je; k<=js; k++)
          {
              int j = nbin[k];
              if(mb[j] == 0)
```

```
                {
                  flag = true;
                  break;
                }

                summtn = summtn + q[j] * cdb[j];
                sumair = sumair + q[j];
            }
            if (!flag)
            {
                jj = i;
                summtn = summtn + qdjunc[i];
                if(summtn > sumair * 1e-70)
                    cdj[i] = summtn/sumair;

                mj[i] = 1;
                goto label1200;
            }
        }
    }
    // PRINT POLLUTANT CONCENTRATIONS
    bool ret_val = printPollutantConc();
    if(!ret_val)
    {
       goto label1050;
    }

    return 0;
  }

  void sortBranches()
  {
    is = nfbpf + 1;
    ie = nb - 1;
    je = ie;
    l = 0;

    for (int i = is; i <= ie; i++)
    {
      for (int j = is; j <= je; j++)
      {
        if(rr[j+1] > rr[j])
        {
                t = rr[j];
                rr[j] = rr[j+1];
```

```
                rr[j+1] = t;
                t = branch[j];
                branch[j] = branch[j+1];
                branch[j+1] = t;
                l = 1;
            }
        }
        if (l > 0) je = je - 1;
        else break;
    }
}

void assignDEGenRates()
{
    if(delqdg == 0.0)
    {
        for(int i=1; i<=nb; i++)
        {
            qdwall[i] = qdbg;
        }
        for(int i=1; i<=nj; i++)
        {
            qdjunc[i] = qdjg;
        }
    }
    else
    {
        for(int i=1; i<=nb; i++)
        {
            qdwall[i] = delqdg * length[i];
        }
    }
}

bool printPollutantConc()
{
    cout << endl << endl << setw(10) << subtit << endl << endl;
    cout << "\n\t\tDIESEL EXHAUST CONCENTRATION RESULTS" << endl;
    cout << endl << setw(40) << "BRANCH VALUES\n" << endl ;
    cout << setw(30) << "DIESEL EXHAUST" << "\tCONCENTRATION" <<
endl;
    cout << " BRANCH NUMBER " << setw(15) << "FROM WALLS
"<< setw(10) << " IN PPM " << setw(10) << " TLV" << endl << endl;
```

```
    for(int i=1; i<=nb; i++)
    {
        cdb[i]=cdb[i] * pow(10.0,6);
         cout  <<  setw(10)   <<   i  <<  setw(15)   <<   qdwall[i]   <<
setw(15) << cdb[i] << setw(10) << tlv  <<  endl;
         // check if tlv is exceeded
         if(cdb[i]  >  tlv)
         {
           cout << "\n ** THE DIESEL EXHAUST CONCENTRATION EXCEEDED ITS
TLV IN BRANCH NO." << i  <<  endl;
         }
         //Calculate the TLV Of the Mixture in Branches.....
         tlvmxb[i]  =  tlvmxb[i]  +  cdb[i] / tlv;
    }

    cout  <<  endl <<  setw(40) << "JUNCTION VALUES" << endl <<  endl;
    cout << setw(26) << "DIESEL EXHAUST" << setw(20) << "CONCENTRATION "
<< endl;
    cout  <<  setw(10) << "JUNC NUM" << setw(15) << "FROM JUNCTION" <<
setw(15) << "IN PPM" << setw(10) << "TLV" << endl  <<  endl;

    for(int i=1; i<=nj; i++)
    {
         cdj[i]  =  cdj[i] * pow(10,6);
         cout  <<  setw(10)  <<  jctn[i]  <<  setw(15)  <<  qdjunc[i]  <<
setw(15) << cdj[i] << setw(10) << tlv  <<  endl;
         if(cdj[i]  >  tlv)
         {
             cout  <<  "\n  **  THE  DIESEL  EXHAUST  CONCENTRATION  EXCEEDED
ITS TLV IN JUNCTION NO." << i  <<  endl;
         }
         //Calculate the TLV Of the Mixture in Junctions.....
         tlvmxj[i]  =  tlvmxj[i]  +  cdj[i] / tlv;
    }

    cout << "\nInput k value" << endl;
    cin  >>  k ;
    if (k-1 >= 0)
    {
         return false;
    }
    cout  <<endl  <<  endl  <<  "THE  FOLLOWING  ARE  THE  VALUES  OF  TLV  OF
THE MIXTURE OF DIESEL EXHAUST COMPONENTS" << endl;
    cout  <<  endl  <<  endl  <<  setw(20)  <<  "BRANCH  NUMBER  "  <<
setw(20) << "MIXTURE TLV" << endl;
```

```
    for(int i=1; i<=nb; i++)
    {
         cout  <<  setw(15) << i  <<  setw(22) << tlvmxb[i] << endl;
         //Check if the Mixture TLV is Exceeded
        if(tlvmxb[i]  >  1.0)
        {
            cout << "\n *** THE MIXTURE OF DIESEL EXHAUST EXCEEDED IT'S
TLV IN BRANCH NO. " << i  <<  endl;
        }
    }

    cout  <<  endl  <<  endl  <<  setw(20)  <<  "JUNCTION  NUMBER"  <<
setw(20) << "MIXTURE TLV" << endl;

    for(int i=1; i<=nj; i++)
    {
        cout  <<  setw(15) << i  <<  setw(22) << tlvmxj[i] << endl;
        // Check if the Mixture TLV is Exceeded
        if(tlvmxj[i]  >  1.0)

        {
        cout << "\n *** THE MIXTURE OF DIESEL EXHAUST EXCEEDED IT'S TLV
IN JUNCTION NO. " << i  <<  endl;
        }
    }
    return true;
  }

  void setDieselExhaustConcZero()
  {
    for(int i=1; i<=nb; i++)
    {
        cdb[i]  =  0.0;
        mb[i]  =  0;
    }
    for(int i=1; i<=nj; i++)
    {
        cdj[i]  =  0.0;
        mj[i]  =  0;
    }
  }
```

```
void setMixtureTLVZero()
{
  for(int i=1; i<=nb; i++)
  {
      tlvmxb[i] = 0.0;
  }
  for(int i=1; i<=nj; i++)
  {
      tlvmxj[i] = 0.0;
  }
}

void tabulateInAndOutFlows()
{
  l = 0;
  ll = 0;
  jj = 1;

  for(int i=1; i<=maxj; i++)
  {
      k = l;
      for(int j=1; j<=nb; j++)
      {
        if(jj1[j] == i)
        {
            l = l+1;
            nbout[l] = j;
        }
      }
      jout[jj] = l;
      kk = ll;
      for(int j=1; j<=nb; j++)
      {
        if(j2[j] == i)
        {
          ll = ll + 1;
          nbin[ll] = j;
        }
      }

    jin[jj] = ll;
    if((kk == ll) && (k==l))
        continue;
```

```
        jctn[jj] = i;
        jj = jj+1;
      }
   }

   void setFlowsPositive()
   {
        for(int i=1; i<=nb; i++)
        {
            q[i] = q[i] * 1e3;
            if(q[i] <= 0)
            {
                l = jj1[i];
                jj1[i] = j2[i];
                j2[i] = l;
                q[i] = -q[i];
            }
        }
   }

   void toBeginIteration()
   {
       ib = nfixb+1;
       bool flag = false;

       for(int it=1; it<=maxit; it++)
       {
         if(nfixb <= 0)
         {
             je = 0;
         }
         else
         {
             je = me[nfixb];
         }
         l = 0;
         sumd = 0;

         for(int i=ib; i<=nm; i++)
         {
             im = i-nfixb;
             js = je+1;
             je = me[i];
             sumh = -sumnvp[i];
             sumdh = 0;
```

```
dhf = 0;

if(nfbpf-i >= 0)
{
    n = nd[im];
    fq[im] = c[im][n+1]*q[i];
    if(n-1 > 0)
    {
        int j = n;
        for(int ij=2; ij<=n; ij++)
        {
          fq[im] = (fq[im] + c[im][j]) * q[i];
          dhf = (dhf + float(j) * c[im][j+1]) * q[i];
          j = j-1;
        }
    }
    fq[im] = fq[im] + c[im][1];
    sumh = sumh - fq[im];
    dhf = dhf + c[im][2];
}
for(int j=js; j<=je; j++)
{
    k = abs(na[j]);
    dh = r[k] * fabs(q[k]);
    h = dh * q[k];
    sumdh = sumdh + dh;
    (na[j] >= 0) ? (sumh = sumh+h) : (sumh = sumh-h);
}

sumdh = sumdh + sumdh - dhf;
if(fabs(sumdh) - 1e-020 > 0)
{
    d = -sumh/sumdh;
    for(int j=js; j<=je; j++)
    {

       k = abs(na[j]);

  (na[j] >= 0) ? (q[k] = q[k] + d) : (q[k] = q[k] - d);
    }
    sumd = sumd+fabs(d);
    if(fabs(d) - e > 0)
        l = 1;
}
```

```
            else
            {
                l = 1;
            }
        }
        if(l > 0)
        {
          iterations = it;
          continue;
        }
        else
        {
          flag = true;
          iterations = it;
          break;
        }
    }

    printIterationandSum(iterations,flag);
}

void setInitialValues()
{
    for(int i=1; i<=nb; i++)
    {
        q[i] = 0;
        r[i] = r[i] * 1e-04;
    }
    je = 0;

    for(int i=1; i<=ie; i++)
    {
        js = je+1;
        je = me[i];
        for(int j=js; j<=je; j++)
        {
            k = abs(na[j]);
            if(na[j]<0)
            {
                q[k] = q[k]-dd[i];
            }
```

```
              else
              {
                  q[k] = q[k]+dd[i];
              }
          }
      }

  }


void toApproximateFanCharacteristics()
{
    bool flag = false;
    if(nf > 0)
    {
        for(int l=1; l<=nf; l++)
        {
        jj = l + nfixb;
        cout << "\nInput np and key values : " << endl;
        cin >> np >> key;
        cout << endl << setw(10) << "FAN" << setw(10) <<
"J1" << setw(10) << "J2" << setw(10) << "KEY" << endl;
        cout << setw(10) << l << setw(10) << jj1[jj] <<
setw(10) << j2[jj] << setw(10) << key << endl;

        if(key <= 0)
        {

          cout << "\nInput " << np << " values for x and fx : " << endl;

          for(int i=1; i<=np; i++)
          {
              cin >> x[i] >> fx[i];
          }

          cout << endl << setw(10) << "POINT" << setw(10) <<
"X" << setw(10) << "FX" << endl;

          for(int i=1; i<=np; i++)
          {
              cout << setw(10) << i << setw(10) << x[i] <<
setw(10) << fx[i] << endl;
          }
          dd[jj] = x[np];
```

```
        if(np-2 <= 0)
        {
            nd[1] = 1;
            n1 = 2;
            if(np-1 < 0)
            {
              exit(0);
            }
            else if (np-1 == 0)
            {
                c[1][1] = fx[1];
                c[1][2] = 0;
                printc(1);
            }
            else
            {
              if(x[1] == x[2])
              {
                  exit(0);
              }
              else
              {
                c[1][1] = (fx[1] * x[2] - fx[2] * x[1]) / (x[2] - x
[1]);

                c[1][2] = (fx[2] - fx[1]) / (x[2] - x[1]);
                printc(1);
              }
            }
          }
          else
          {
            dsq = 0;
            for(int j=1; j<= np; j++)
            {
              dsq = dsq + fx[j] * fx[j];
              p[1][j] = 0;
              p[2][j] = 1;
            }
            w[1] = np;
            beta[1] = 0;
            (np-7 >= 0) ? (n1 = 6) : (n1 = np-1);
```

```
            for(int i=1; i<=n1; i++)
            {
                k = i+1;
                z[i] = 0;

                for(int j=1; j<=np; j++)
                {
                    z[i] = z[i] + fx[j] * p[k][j];
                }
                s[i] = z[i]/w[i];
                dsq = dsq - s[i] * s[i] * w[i];
                sigma2[i] = dsq/(float)(np-i);

                if(sigma2[i] - 1e-6 > 0)
                {
                  if (i-n1 < 0)
                  {
                    alpha[i] = 0;

                    for(int j=1; j<=np; j++)
                    {
                      alpha[i] = alpha[i] + x[j] * p[k][j] * p
[k][j];
                    }
                    alpha[i] = alpha[i]/w[i];
                    w[k] = 0;

                    for(int j=1; j<=np; j++)
                    {
                        p[k+1][j] = (x[j] - alpha[i]) * p[k][j] -
beta[i] * p[i][j];
                        w[k] = w[k] + p[k+1][j] * p[k+1][j];
                    }
                    beta[k] = w[k]/w[i];
                }
                else
                {
                    break;
                }
            }
            else
            {
                goto label78;
            }
        }
```

```
        small = sigma2[1];
      tmp = 1;

        for(int j=2; j<=n1; j++)
        {
          if(small - sigma2[j] > 0)
          {
              tmp = j;
              small = sigma2[j];
          }
        }

    label78: cout << endl << setw(15) << "SIGMA2 = " << sigma2
[tmp] << endl;
        n1 = tmp;
        n = n1-1;

        if(n <= 0)
        {
            nd[1] = 1;
            c[1][1] = s[1];
            c[1][2] = 0;
            printc(1);
            continue;
        }
        else
        {
            nd[1] = n;

            for(int i=1; i<=n1; i++)
            {
              for(int j=1; j<=i; j++)
              {
                  p[i][j] = 0;
              }
              p[i][i+1] = 1;
        }

        for(int j=1; j<=n; j++)
        {
          p[1][j+2] = -p[1][j+1] * alpha[j] - p[1][j] * beta[j];
        }
```

```
        for(int i=2; i<=n; i++)
        {
            for(int j=i; j<=n; j++)
            {
                p[i][j+2] = p[i-1][j+1] - p[i][j+1] * alpha[j] - p[i]
[j] * beta[j];
            }
        }

        for(int i=1; i<=n1; i++)
        {
            c[l][i] = 0;

            for(int j=i; j<=n1; j++)
            {
                c[l][i] = c[l][i] + p[i][j+1] * s[j];
            }
          }
            printc(l);

        }
      }
    }
    else
    {
          nd[l] = nd[key];
          n1 = nd[key] + 1;
          dd[jj] = dd[key+nfixb];
          for(int j=1; j<=n1; j++)
          {
            c[l][j] = c[key][j];
          }
        }
      }
    }
  }

  void printc (int l)
  {
    cout << endl << setw(20) << "CO-EFFICIENTS " << endl;
    for (int i=1; i<=n1; i++)
    {
        cout << setw(10) << c[l][i] << setw(15);
    }
  }
```

```
void toPrintMeshTable()
{
      je = 0;
      cout << endl << setw(10) << "MESH" << setw(20) << "NO. OF
BRANCHES" << setw(20) << "SUMNVP" << endl;
      for(int i=1; i<=nm; i++)
      {
          js = je+1;
          je = me[i];
          sumnvp[i] = 0;
          k = abs(na[nb]);

          for (int j=js; j<=je; j++)
          {
              (na[j] < 0) ? (sumnvp[i] = sumnvp[i]-nvp[k]) :
(sumnvp[i] = sumnvp[i]+nvp[k]);

          }

          l = je - js + 1;
          cout << endl << setw(10) << i << setw(20) <<
l << setw(20) << sumnvp[i] << endl << setw(60) ;

          for (int j=js; j<=je; j++)
          {
            cout << na[j] << setw(5);
          }
    }
    cout << endl << setw(40) << "TOTAL = " << je << endl;
}

void toReadPrintNVP()
{
    for(int i=1; i<=nb; i++)
    {
      nvp[i]=0;
    }
    if(nbwnvp>0)
    {
      //Read and Print Natural Ventilation Pressure
      for(int j=1; j<=nbwnvp; j++)
      {
        cin >> j >> nvp[j] ;
      }
      cout << branch << endl << nvp << endl;
    }
}
```

```
void tofindMeshes()
{
   jk = 0;
   je = 0;
   l = 0;

   for(int i=1; i<=nb; i++)
   {
      if(out[i] <= 0)
           continue;

      k = branch[i];
      l = l+1;
      jk = jk+1;
      na[jk] = k;
      ja = jj1[k];
      jb = j2[k];
      n = i+1;
      cont_flag = 0;

loopj : for(int j=n; j<=nb; j++)
   {
      if(out[j] != 0)
           continue;
      k = branch[j];

      if(jb == jj1[k])
      {
         jb = j2[k];
         jk = jk+1;
         na[jk] = k;
      }
      else
      {
         if(jb == j2[k])
         {
            jb = jj1[k];
            jk = jk+1;
            na[jk] = -k;
         }
         else
            continue;
      }
```

```
      if(jb != ja)
      {
        out[j] = -1;
        goto loopj;
      }

      for (int j = n; j<=nb; j++)
      {
        if(out[j] < 0)
        out[j] = 0;
      }
      me[1] = jk;
      je = jk;
      cont_flag = 1;
      break;
    }

      if (cont_flag == 1)
          continue;

      k = abs(na[jk]);
      (na[jk] >= 0) ? (jb = jj1[k]): (jb = j2[k]);
      jk = jk-1;
      if(jk-je <= 0)
      {
          k = branch[i];
          cout << "\n\nNo mesh found at branch " << k << endl;
          exit (0);
      }
      goto loopj;
      for (int j = n; j<=nb; j++)
      {
          if(out[j] < 0)
          out[j] = 0;
      }
      me[1] = jk;
      je = jk;
    }
  }

  void toFindBasicBranches()
  {
      for(int i=1; i<=maxj; i++)
        jc[i] = 0;
```

```
    int i  =  nb+1;
    l  =  0;
    n  =  0;

    for(int ij=is; ij<=nb; ij++)
    {
      i  =  i-1;
      out[i]  =  0;
      k  =  branch[i];
      ja  =  jj1[k];
      jb  =  j2[k];

      if(jc[ja]  >  jc[jb])
      {
        if(jc[jb] != 0)
        {
            jj  =  jc[jb];
            for(int j=1; j<=maxj; j++)
            {
                if(jc[j] == jj)
                {
                    jc[j]  =  jc[ja];
                }
            }
        }
        else
        {
          jc[jb]  =  jc[ja];
        }
  }

  else if(jc[ja]  <  jc[jb])
  {
    if(jc[ja] != 0)
    {
      jj  =  jc[jb];
      for(int j=1; j<=maxj; j++)
      {
        if(jc[j] == jj)
        {
          jc[j]  =  jc[ja];
        }
      }
  }
```

```
      else
        {
          jc[ja] = jc[jb];
        }
      }

      else
      {
        if(jc[ja] != 0)
        {
          out[i] = 1;
          n = n+1;
        }
      else
      {
        l = l+1;
        jc[ja] = l;
        jc[jb] = l;
      }
    }
  }
}

   void readBranchCharacteristics()
   {
         for (int i=1; i<=nb; i++)
      {
         cout << "\nEnter J1, J2, FF, Height, Width, length for branch"
 << i  << endl;
         cin >> jj1[i] >> j2[i] >> ff >> height[i] >> width[i] >>
length[i] ;
         r[i] = 0;
         ff_temp[i] = ff;
         branch[i] = i;
         float t;

         if(r[i]<=0)
         {
                t = height[i] * width[i];
                r[i] = ff_temp[i]*(height[i] + width[i])*length[i]/
               (2.6*t*t*t);
         }
         rr[i] = r[i];
      }
   }
```

```
  void printBranchCharacteristics()
  {
    cout << "\n\n\n" << setw(10) << "BRANCH" << setw(10) << "J1" <<
setw(10) << "J2" << setw(15) << "R" << setw(10) << "FF" << setw(10) <<
"HEIGHT" << setw(10) << "WIDTH" << setw(10) << "LENGTH" << endl;
    for (int i=1; i<=nb; i++)
    {
      if(ff_temp[i] > 0 && height[i] > 0)
      {
        cout << setw(10) << i << setw(10) << jj1[i] <<
setw(10) << j2[i] << setw(15) << r[i] << setw(10) << ff_temp[i] <<
setw(10) << height[i] << setw(10) << width[i] << setw(10) << length
[i] << endl;
      }

      else if(ff_temp[i] <= 0 && height[i] > 0)
      {
        cout << setw(10) << i << setw(10) << jj1[i] <<
setw(10) << j2[i] << setw(15) << r[i] << setw(10) << height[i] <<
setw(10) << width[i] << setw(10) << length[i] << endl;
      }

      else if(height[i] <= 0)
      {
              cout << setw(10)<< i << jj1[i] << j2[i] << r[i] <<
endl;
      }
    }
  }

  void printIterationandSum(int iterations, bool flag)
  {
      cout << endl << setw(10) << iterations << " ITERATIONS " <<
setw(10) << "SUMD = " << sumd << endl << endl;
      if(!flag)
        cout << "**** d is still greater than e" << endl;
  }

  void computeAndPrintHeadLoss()
  {
      cout << setw(10) << "BRANCH " << setw(10) << "J1 " <<
setw(10) << "J2" << setw(10) << "R" << setw(15) << "Q" <<
setw(15) << "H" << endl;
```

```
	for(int i=1; i<=nb; i++)
	{
		hh[i] = r[i] * fabs(q[i]) * q[i];
		r[i] = r[i] / 1e-04;
		cout << setw(10) << i << setw(10) << jj1[i] <<
		setw(10) << j2[i]<< setw(10) << r[i] << setw(15) << q
		[i] << setw(15) << hh[i] << endl;
	}

}

void adjustHLinFixedQuantityBranch()
{
	cout << endl << setw(10) << "\tHEAD LOSS TO BE ADJUSTED IN
FIXED QUANTITY BRANCH\n" << endl;
	cout << setw(10) << "BRANCH" << setw(10) << "J1" <<
setw(10) << "J2" << setw(10) << "R" << setw(10) << "Q" <<
setw(15) << "SUMH" << endl;
	je = 0;

	for(int i=1; i<=nfixb; i++)
	{
		js = je+1;
		je = me[i];
	  sumh = -sumnvp[i];
	  for(int j=js; j<=je; j++)
	  {
	    k = abs(na[j]);
	    (na[j] >= 0) ? (sumh = sumh + hh[k]) : (sumh = sumh - hh
[k]);
	  }
	  cout << setw(10) << i << setw(10) << jj1[i] <<
setw(10) << j2[i] << setw(10) << r[i] << setw(10) << q[i] <<
setw(15) << sumh << endl;

	  // Calculate the size of regulator using Murgue's formula or the
capacity of the booster fan as the case may be
	  if(sumh < 0)
	  {
	    sumh = -sumh;
	    regsiz = 0.389 * q[i]/sqrt(sumh);
	    cout << endl << setw(10) << "\tTHE SIZE OF THE REGULATOR
IS "<< regsiz << "SQ FT" << endl;
	  }
```

```
        else
          {
            cout << endl << setw(10) <<"\tTHE CAPACITY OF THE
BOOSTER FAN IS "<< sumh << "INCHES OF W.G AT" << q[i] <<"C.F.M" <<
endl;
          }
        }
    }

    void printFanPressure()
    {
        if(nf > 0)
        {
          cout << endl << setw(10) << "FAN" << setw(10) << "J1" <<
setw(10) << "J2" << setw(10) << "Q" << setw(10) << "FQ" << endl;
        for(int i=1; i<=nf; i++)
        {
          cout << setw(10) << i << setw(10) << jj1[i+nfixb] <<
setw(10) << j2[i+nfixb] << setw(10) << q[i+nfixb] << setw(10) << fq
[i] << endl;
        }
      }
    }
```

Input

SIMULATION OF DIESEL EXHAUST CONTAMINATION OF A HYPOTHETICAL MINE

```
13
9
9
1
300
0
1
0.1
4
7
100
6.5
16
3000
9
1
100
6.5
16
1000
1
2
90
6.5
16
1000
2
3
90
6.5
16
5000
2
5
90
6.5
16
3000
3
4
```

```
90
6.5
16
2000
4
5
90
6.5
16
2000
3
6
100
6.5
16
3000
5
8
100
6.5
16
3000
6
7
100
6.5
16
2000
6
9
110
6.5
16
3000
7
9
110
6.5
16
3000
8
9
110
6.5
16
```

```
3000
3
0
80
6.6
100
5.6
120
4
30
1
THE FOLLOWING RESULTS ARE FOR CARBON MONO-OXIDE
1
0.0
0.0
0.0
50.0
2
1
4
0.0
0.0002
6
0.0
0.0002
4
0.3
1
THE FOLLOWING RESULTS ARE FOR NITRIC OXIDE
1
0.0
0.0
0.0
25
2
1
4
0
0.00015
6
0
0.00015
4
0.25
1
```

THE FOLLOWING RESULTS ARE FOR NITRIC DIOXIDE
1
0.0
0.0
0.0
5.0
2
1
4
0.05
0.0
6
0
0.00025
4
0.0001
1
THE FOLLOWING RESULTS ARE FOR HYDRO CARBONS
1
0.0
0.0
0.0
15
2
1
4
0
0.00013
6
0
0.001
4
0.25
0

Results

Input title of the problem :
SIMULATION OF DIESEL EXHAUST CONTAMINATION OF A HYPOTHETICAL MINE

Enter nb, nj, maxj, nf, maxit, nbwnvp, nfixb values :
13
9
9
1
300
0
1

Enter acceptable error value :
0.1

SIMULATION OF DIESEL EXHAUST CONTAMINATION OF A HYPOTHETICAL MINE

nb	nj	maxj	nm	nf	maxit	nbwnvp	nfixb	e
13	9	9	5	1	300	0	1	0.1

Enter J1, J2, FF, Height, Width, length for branch 1
4
7
100
6.5
16
3000

Enter J1, J2, FF, Height, Width, length for branch 2
9
1
100
6.5
16
1000

Enter J1, J2, FF, Height, Width, length for branch 3
1
2
90

6.5
16
1000

Enter J1, J2, FF, Height, Width, length for branch 4
2
3
90
6.5
16
5000

Enter J1, J2, FF, Height, Width, length for branch 5
2
5
90
6.5
16
3000

Enter J1, J2, FF, Height, Width, length for branch 6
3
4
90
6.5
16
2000

Enter J1, J2, FF, Height, Width, length for branch 7
4
5
90
6.5
16
2000

Enter J1, J2, FF, Height, Width, length for branch 8
3
6
100
6.5
16
3000

Enter J1, J2, FF, Height, Width, length for branch 9
5
8
100
6.5
16
3000

Enter J1, J2, FF, Height, Width, length for branch 10
6
7
100
6.5
16
2000

Enter J1, J2, FF, Height, Width, length for branch 11
6
9
110
6.5
16
3000

Enter J1, J2, FF, Height, Width, length for branch 12
7
9
110
6.5
16
3000

Enter J1, J2, FF, Height, Width, length for branch 13
8
9
110
6.5
16
3000

```
BRANCH J1 J2 R          FF   HEIGHT WIDTH LENGTH
 1      4  7  2.30797   100  6.5    16    3000
 2      9  1  0.769324  100  6.5    16    1000
 3      1  2  0.692391   90  6.5    16    1000
 4      2  3  3.46196    90  6.5    16    5000
 5      2  5  2.07717    90  6.5    16    3000
 6      3  4  1.38478    90  6.5    16    2000
 7      4  5  1.38478    90  6.5    16    2000
 8      3  6  2.30797   100  6.5    16    3000
 9      5  8  2.30797   100  6.5    16    3000
10      6  7  1.53865   100  6.5    16    2000
11      6  9  2.53877   110  6.5    16    3000
12      7  9  2.53877   110  6.5    16    3000
13      8  9  2.53877   110  6.5    16    3000


MESH NO.OF BRANCHES SUM NVP


 1    4               0    1-10-8 6
 2    5               0    2 3 5 9 13
 3    4               0    4 6 7 -5
 4    6               0    11 -13 -9 -7 -6 8
 5    7               0    12 -13 -9 -7 -6 8 10
TOTAL  =  26


Input np and key values :
3
0


FAN J1 J2 KEY
1   9 1  0


Input 3 values for x and fx :
80
6.6
100
5.6
120
4
```

```
POINT X     FX
1     80    6.6
2     100   5.6
3     120   4


SIGMA2 = 0.06


CO-EFFICIENTS
11.9 -0.065


Input 1 dd values :
30


12 ITERATIONS SUMD = 0.121272
```

BRANCH	J1	J2	R	Q	H
1	4	7	2.30797	30	0.207717
2	9	1	0.769324	121.078	1.12783
3	1	2	0.692391	121.078	1.01504
4	2	3	3.46196	53.8665	1.00452
5	2	5	2.07717	67.2118	0.938348
6	3	4	1.38478	7.05502	0.00689253
7	4	5	1.38478	-22.945	-0.0729049
8	3	6	2.30797	46.8115	0.50575
9	5	8	2.30797	44.2668	0.452259
10	6	7	1.53865	8.07051	0.0100217
11	6	9	2.53877	38.741	0.381035
12	7	9	2.53877	38.0705	0.36796
13	8	9	2.53877	44.2668	0.497485

HEAD LOSS TO BE ADJUSTED IN FIXED QUANTITY BRANCH

BRANCH	J1	J2	R	Q	SUMH
1	4	7	2.30797	30	-0.301161

THE SIZE OF THE REGULATOR IS 21.2653 SQ FT

```
FAN    J1    J2    Q          FQ
1      9     1     121.078    4.02924
```

```
Input K value :
1
```

Input title of the component of diesel exhaust :
THE FOLLOWING RESULTS ARE FOR CARBON MONO-OXIDE

Input jj value
1

Input qdbg, delqdg, qdjg, tlv and nqdjs values :
0.0
0.0
0.0
50.0
2
1

Input i, qdbs and delqds values :
4
0.0
0.0002
6
0.0
0.0002

Input ii and qdjs values :
4
0.3

THE FOLLOWING RESULTS ARE FOR CARBON MONO-OXIDE

DIESEL EXHAUST CONCENTRATION RESULTS

BRANCH VALUES

DIESEL EXHAUST CONCENTRATION

```
BRANCH NUMBER  FROM WALLS   IN PPM   TLV


    1            0            27.6991  50
    2            0            14.0405  50
    3            0  0                  50
    4            1            18.5644  50
    5            0            0        50
    6            0.4          75.2616  50

** THE DIESEL EXHAUST CONCENTRATION EXCEEDED ITS TLV IN BRANCH NO.6
    7        0      0         50
    8        0      18.5644   50
    9        0      0         50
    10       0      18.5644   50
    11       0      18.5644   50
    12       0      25.7626   50
    13       0      0         50
```

JUNCTION VALUES

```
  DIESEL EXHAUST CONCENTRATION
JUNC NUM   FROM JUNCTION   IN PPM     TLV

    1        0               0          50
    2        0               0          50
    3        0               18.5644    50
    4        0.3             27.6991    50
    5        0               0  50
    6        0               18.5644    50
    7        0               25.7626    50
    8        0               0          50
    9        0               14.0405    50
```

Input k value
1

Input title of the component of diesel exhaust :
THE FOLLOWING RESULTS ARE FOR NITRIC OXIDE

Input jj value
1

Input qdbg, delqdg, qdjg, tlv and nqdjs values :
0.0
0.0
0.0
25
2
1

Input i, qdbs and delqds values :
4
0
0.00015
6
0
0.00015

Input ii and qdjs values :
4
0.25

THE FOLLOWING RESULTS ARE FOR NITRIC OXIDE

DIESEL EXHAUST CONCENTRATION RESULTS

BRANCH VALUES

DIESEL EXHAUST CONCENTRATION

BRANCH NUMBER	FROM WALLS	IN PPM	TLV
1	0	21.6076	25
2	0	10.7369	25
3	0	0	25
4	0.75	13.9233	25
5	0	0	25
6	0.3	56.4462	25

```
** THE DIESEL EXHAUST CONCENTRATION EXCEEDED ITS TLV IN BRANCH NO.6
  7          0  0          25
  8          0  13.9233    25
  9          0  0          25
  10         0  13.9233    25
  11         0  13.9233    25
  12         0  19.9787    25
  13         0  0          25
```

JUNCTION VALUES

DIESEL EXHAUST CONCENTRATION

JUNC NUM	FROM JUNCTION	IN PPM	TLV
1	0	0	25
2	0	0	25
3	0	13.9233	25
4	0.25	21.6076	25
5	0	0	25
6	0	13.9233	25
7	0	19.9787	25
8	0	0	25
9	0	10.7369	25

Input k value
1

Input title of the component of diesel exhaust :
THE FOLLOWING RESULTS ARE FOR NITRIC DIOXIDE

Input jj value
1

Input qdbg, delqdg, qdjg, tlv and nqdjs values :
0.0
0.0
0.0
5.0
2
1

Input i, qdbs and delqds values :
4
0.05
0.0
6
0
0.00025

Input ii and qdjs values :
4
0.0001

THE FOLLOWING RESULTS ARE FOR NITRIC DIOXIDE

DIESEL EXHAUST CONCENTRATION RESULTS

BRANCH VALUES

DIESEL EXHAUST CONCENTRATION

```
BRANCH NUMBER   FROM WALLS  IN PPM  TLV

  1              0            16.8883 5

** THE DIESEL EXHAUST CONCENTRATION EXCEEDED ITS TLV IN BRANCH NO.1
     2     0      4.54334  5
     3     0      0        5
     4     0.05   0.92822  5
     5     0      0        5
     6     0.5    71.7997  5

** THE DIESEL EXHAUST CONCENTRATION EXCEEDED ITS TLV IN BRANCH NO.6
   7     0      0          5
   8     0      0.92822    5
   9     0      0          5
   10    0      0.92822    5
   11    0      0.92822    5
   12    0      13.5049    5
```

```
** THE DIESEL EXHAUST CONCENTRATION EXCEEDED ITS TLV IN BRANCH NO.12
  13  0  0  5
```

JUNCTION VALUES

DIESEL EXHAUST CONCENTRATION

```
JUNC NUM   FROM JUNCTION   IN PPM   TLV

  1        0               0        5
  2        0               0        5
  3        0               0.92822  5
  4        0.0001          16.8883  5

** THE DIESEL EXHAUST CONCENTRATION EXCEEDED ITS TLV IN JUNCTION NO.4
  5  0  0        5
  6  0  0.92822  5
  7  0  13.5049  5

** THE DIESEL EXHAUST CONCENTRATION EXCEEDED ITS TLV IN JUNCTION NO.7
  8  0  0        5
  9  0  4.54334  5
```

Input k value
1

Input title of the component of diesel exhaust :
THE FOLLOWING RESULTS ARE FOR HYDRO CARBONS

Input jj value
1

Input qdbg, delqdg, qdjg, tlv and nqdjs values :
0.0
0.0
0.0
15
2
1

Input i, qdbs and delqds values :
4
0
0.00013
6
0
0.001

Input ii and qdjs values :
4
0.25

THE FOLLOWING RESULTS ARE FOR HYDRO CARBONS

DIESEL EXHAUST CONCENTRATION RESULTS

BRANCH VALUES

DIESEL EXHAUST CONCENTRATION

```
BRANCH NUMBER  FROM WALLS  IN PPM   TLV

  1            0           77.8377  15

** THE DIESEL EXHAUST CONCENTRATION EXCEEDED ITS TLV IN BRANCH NO.1
  2   0   23.9514   15

** THE DIESEL EXHAUST CONCENTRATION EXCEEDED ITS TLV IN BRANCH NO.2
  3  0       0         15
  4  0.65    12.0669   15
  5  0       0         15
  6  2       295.553   15

** THE DIESEL EXHAUST CONCENTRATION EXCEEDED ITS TLV IN BRANCH NO.6
  7      0      0          15
  8      0      12.0669    15
  9      0      0          15
  10     0      12.0669    15
  11     0      12.0669    15
  12     0      63.8951    15
```

```
** THE DIESEL EXHAUST CONCENTRATION EXCEEDED ITS TLV IN BRANCH NO.12
  13   0  0  15
```

JUNCTION VALUES

DIESEL EXHAUST CONCENTRATION

```
JUNC NUM  FROM JUNCTION  IN PPM   TLV

  1        0              0        15
  2        0              0        15
  3        0              12.0669  15
  4        0.25           77.8377  15

** THE DIESEL EXHAUST CONCENTRATION EXCEEDED ITS TLV IN JUNCTION NO.4
  5   0   0            15
  6   0   12.0669      15
  7   0   63.8951      15

** THE DIESEL EXHAUST CONCENTRATION EXCEEDED ITS TLV IN JUNCTION NO.7
  8   0  0            15
  9   0  23.9514      15
```

** THE DIESEL EXHAUST CONCENTRATION EXCEEDED ITS TLV IN JUNCTION NO.9

Input k value
0

THE FOLLOWING ARE THE VALUES OF TLV OF THE MIXTURE OF DIESEL EXHAUST COMPONENTS

```
  BRANCH NUMBER  MIXTURE TLV
  1              9.98513

*** THE MIXTURE OF DIESEL EXHAUST EXCEEDED IT'S TLV IN BRANCH NO. 1
        2  3.21571

*** THE MIXTURE OF DIESEL EXHAUST EXCEEDED IT'S TLV IN BRANCH NO. 2
        3  0
  4   1.91832
```

```
*** THE MIXTURE OF DIESEL EXHAUST EXCEEDED IT'S TLV IN BRANCH NO. 4
  5  0
  6   37.8265

*** THE MIXTURE OF DIESEL EXHAUST EXCEEDED IT'S TLV IN BRANCH NO. 6
  7  0
  8  1.91832

*** THE MIXTURE OF DIESEL EXHAUST EXCEEDED IT'S TLV IN BRANCH NO. 8
        9   0
       10   1.91832

*** THE MIXTURE OF DIESEL EXHAUST EXCEEDED IT'S TLV IN BRANCH NO. 10
       11  1.91832

*** THE MIXTURE OF DIESEL EXHAUST EXCEEDED IT'S TLV IN BRANCH NO. 11
       12  8.27506

*** THE MIXTURE OF DIESEL EXHAUST EXCEEDED IT'S TLV IN BRANCH NO. 12
       13   0

JUNCTION NUMBER MIXTURE TLV
  1             0
  2             0
  3             1.91832

*** THE MIXTURE OF DIESEL EXHAUST EXCEEDED IT'S TLV IN JUNCTION NO. 3
  4   9.98513

*** THE MIXTURE OF DIESEL EXHAUST EXCEEDED IT'S TLV IN JUNCTION NO. 4
  5  0
  6  1.91832

*** THE MIXTURE OF DIESEL EXHAUST EXCEEDED IT'S TLV IN JUNCTION NO. 6
  7  8.27506

*** THE MIXTURE OF DIESEL EXHAUST EXCEEDED IT'S TLV IN JUNCTION NO. 7
       8   0
       9   3.21571

*** THE MIXTURE OF DIESEL EXHAUST EXCEEDED IT'S TLV IN JUNCTION NO. 9
```

Appendix D: The Input and Output Data for the Hypothetical Mine

Description of the Input

The input card deck consists of the following:

Card 1: The title of the problem can be punched here in 1–80 columns.

Card 2: Here, the following parameters are punched, giving each parameter a field width of 10: NB, JN, MAXJ, NF, MAXIT, NBWNVP, NFIXB, and E. Here, all variables are integer type, except E which is a floating-point–type variable. Integer-type variables are always punched right justified.

Card 3: Is a fixed quantity branch card. Only one branch (4-7) is labeled with fixed quantity here, but in general, there could be more. The following variables are punched here: J1 (I), J2 (I), RR, FF, HEIGHT, WIDTH, and LENGTH with a field width of 10 each. Here, J1 and J2 are integer-type variables, but the rest are floating-point variables.

Card 4: Contains the same information as above for a branch with fan. In this case, branch (9-1) has a fan. In general, there could be several fans in a network.

Card 5–15: Are airway cards that contain the same information about other branches in the network as card 3 does for the fixed quantity branch. There are 11 cards in this case for the remaining 11 branches.

Card 16: Is a key card for the fan. In the first 10 columns, it contains the number of data points on the fan characteristic to be read in. In the next 10 columns, it contains information if two similar fans are in use or not. A zero here makes sure that fan 2 will not use fan number 1's characteristic. In this case, this question does not arise because there is only one fan. Both data are integer-type variables.

Card 17: Contains the data for points on fan characteristic, e.g., quantity in 1000 cfm and corresponding pressure in inches of H_2O are punched in sequence for each point. In this case, three pairs of points have been taken from the fan characteristic. These are floating-point variables with a field width of 10 for each.

Card 18: Contains the air quantity in 1000 cfm for the fixed quantity branch. It is a floating-point variable with a field width of 10.

Card 19: Contains an integer equal to one or more in the first 10 columns if the calculation of concentrations is to be done. By making this number zero, the program can be forced to calculate the quantity of air only and skip the rest of the program. It is desirable to do it first for large networks where the branch number of roadways with special emission of diesel may not be known in advance.

Card 20: Title of first component of diesel exhaust, e.g., carbon monoxide.

Card 21: The outside junction number. It is 1 here.

Card 22: Contains values of QDGB, DELQDG, QDJG, TLV, NQDBS, and NQDJS for carbon monoxide. These are punched in the above sequence with a field width of 10 for each. QDGB, DELQDG, and QDJG are zero in this case.

Card 23–24: Read in special emissions of carbon monoxide with branch number in the first 10 columns. QDBS is punched in columns (11–20) or DELQDS in columns (21–30). Here, branches 4 and 6 are assumed to have special emissions. In this case, arbitrary values of these emission rates have been used, but in real cases, these will be calculated using the emission model described in , Chapter 3 (equation [3.51]) and shown in , Chapter 5.

Card 25: Contains the same information for junctions. Here, junction 4 is assigned a special make of carbon monoxide. For real cases, this has to be computed using emission model described in , Chapter 3 (equation [3.51]) and shown in , Chapter 5.

Card 26: Contains in the first 10 columns the value of the computational switch K. $K \geq 1$ means that more sets of data are to be processed.

Card 27–32: Carry the same information for nitric oxide as cards 20–25 do for carbon monoxide.

Card 33–39: Carry the same information as above for nitrogen dioxide. Card 33 is the same as card 26 above.

Card 40–46: Carry the same information as above for hydrocarbons.

Card 47: Because there are no more components of diesel exhaust to be processed, this is a blank card.

Card 48–50: Are also blank cards needed to stop the program through a computational sequence.

Card 51: Is a slash asterisk card.

Input Data Listings

```
SIMULATION OF DIESEL EXHAUST CONTAMINATION OF A HYPOTHETICAL MINE.                                Card 1
          13        9      9      1           300        0                     1  0.1             Card 2
           4        7        100.0       6.5        16.0          3000.0                          Card 3
           9        1        100.0       6.5        16.0          1000.0                          Card 4
           1        2         90.0       6.5        16.0          1000.0                          Card 5
           2        3         90.0       6.5        16.0          5000.0                          Card 6
           2        5         90.0       6.5        16.0          3000.0                          Card 7
           3        4         90.0       6.5        16.0          2000.0                          Card 8
           4        5         90.0       6.5        16.0          2000.0                          Card 9
           3        6        100.0       6.5        16.0          3000.0                          Card 10
           5        8        100.0       6.5        16.0          3000.0                          Card 11
           6        7        100.0       6.5        16.0          2000.0                          Card 12
           6        9        110.0       6.5        16.0          3000.0                          Card 13
           7        9        110.0       6.5        16.0          3000.0                          Card 14
           8        9        110.0       6.5        16.0          3000.0                          Card 15
           3        0                                                                             Card 16
80.0        6.6      100.0       5.6       120.0      4.0                                         Card 17
30.0                                                                                              Card 18
           1                                                                                      Card 19
THE FOLLOWING RESULTS ARE FOR CARBON MONOXIDE                                                     Card 20
           1                                                                                      Card 21
                  0.0         50.0             2          1                                       Card 22
           4     0.0002                                                                           Card 23
           6     0.0002                                                                           Card 24
           4  0.3                                                                                 Card 25
           1                                                                                      Card 26
THE FOLLOWING RESULTS ARE FOR NITRIC OXIDE                                                        Card 27
           1                                                                                      Card 28
                              25.0             2          1                                       Card 29
           4     0.00015                                                                          Card 30
           6     0.00015                                                                          Card 31
           4  0.25                                                                                Card 32
           1                                                                                      Card 33
THE FOLLOWING RESULTS ARE FOR NITROGEN DIOXIDE                                                    Card 34
           1                                                                                      Card 35
                               5.0             2          1                                       Card 36
           4          .05                                                                         Card 37
           6              0.00025                                                                 Card 38
           4  0.0001                                                                              Card 39
           1                                                                                      Card 40
THE FOLLOWING RESULTS ARE FOR HYDROCARBONS                                                        Card 41
           1                                                                                      Card 42
                              15.0             2          1                                       Card 43
           4              0.00013                                                                 Card 44
           6              0.001                                                                   Card 45
           4  0.25                                                                                Card 46
           0                                                                                      Card 47
                                                                                                  Card 48
                                                                                                  Card 49
                                                                                                  Card 50
/*                                                                                                Card 51
```

Output Results

The above input data were used in Program III. Output results are listed below.

simulation of diesel exhaust contamination of a hypothetical mine.

nb	nj	maxj	nm	nf	maxit	nbwnvp	nfixb	e
13	9	9	5	1	300	0	1	0.1000

branch	j1	j2	r	ff	height	width	length
1	4	7	2.307972	100.0	6.5	16.0	3000.0
2	9	1	0.769324	100.0	6.5	16.0	1000.0
3	1	2	0.692392	90.0	6.5	16.0	1000.0
4	2	3	3.461958	90.0	6.5	16.0	5000.0
5	2	5	2.077175	90.0	6.5	16.0	3000.0
6	3	4	1.384783	90.0	6.5	16.0	2000.0
7	4	5	1.384783	90.0	6.5	16.0	2000.0
8	3	6	2.307972	100.0	6.5	16.0	3000.0
9	5	8	2.307972	100.0	6.5	16.0	3000.0
10	6	7	1.538648	100.0	6.5	16.0	2000.0
11	6	9	2.538769	110.0	6.5	16.0	3000.0
12	7	9	2.538769	110.0	6.5	16.0	3000.0
13	8	9	2.538769	110.0	6.5	16.0	3000.0

MESH	NO. OF BRANCHES	SUMNVP							
1	4	0.0	1	-10	-8	6			
2	5	0.0	2	3	5	9	13		
3	4	0.0	4	6	7	-5			
4	6	0.0	11	-13	-9	-7	-6	8	
5	7	0.0	12	-13	-9	-7	-6	8	10

TOTAL = 26

FAN	J1	J2	KEY
1	9	1	0

POINT	X	DX
1	80.0000	6.6000
2	100.0000	5.6000
3	120.0000	4.0000

SIGMA2 = 6.0000000D-02
COEFFICIENTS

1.19000000 01 -6.5000000D-02

12 ITERATIONS SUMD = 1.2,127233D-01.

BRANCH	J1	J2	R	Q	H
1	4	7	2.307972	30.0000	0.2077175
2	9	1	0.769324	121.0783	1.1278255
3	1	2	0.692392	121.0783	1.0150430
4	2	3	3.461958	53.8665	1.0045223
5	2	5	2.077175	67.2118	0.9383479
6	3	4	1.384783	7.0550	0.0068926
7	4	5	1.384783	-22.9450	-0.0729049
8	3	6	2.307972	46.8115	0.5057496
9	5	8	2.307972	44.2668	0.4522587
10	6	7	1.538648	8.0705	0.0100217
11	6	9	2.538769	38.7410	0.3810347
12	7	9	2.538769	38.0705	0.3679600
13	8	9	2.538769	44.2668	0.4974846

HEAD LOSS TO BE ADJUSTED IN FIXED QUANTITY BRANCH

BRANCH	J1	J2	R	Q	SUMH
1	4	7	2.307972	30.0000	-0.3011612

THE SIZE OF THE REGULATOR IS 21.2653 SQ FT

FAN	J1	J2	Q	BQ
1	9	1	121.0783	4.0292401

The Following Results Are for Carbon Monoxide

Diesel Exhaust Concentration Results

Branch Values

BRANCH NUMBER	DIESEL EXHAUST FROM WALLS	CONCENTRATION IN PPM	TLV
1	0.0	27.7	50.0
2	0.0	14.0	50.0
3	0.0	0.0	50.0
4	1.0000	18.6	50.0
5	0.0	0.0	50.0
6	0.4000	75.3	50.0
** THE DIESEL EXHAUST CONCENTRATION EXCEEDED ITS TLV IN BRANCH NO. 6			
7	0.0	0.0	50.0
8	0.0	18.6	50.0
9	0.0	0.0	50.0
10	0.0	18.6	50.0
11	0.0	18.6	50.0
12	0.0	25.8	50.0
13	0.0	0.0	50.0

Junction Values

JUNCTION NUMBER	DIESEL EXHAUST FROM JUNCTION	CONCENTRATION IN PPM	TLV
1	0.0	0.0	50.0
2	0.0	0.0	50.0
3	0.0	18.6	50.0
4	0.3000	27.7	50.0
5	0.0	0.0	50.0
6	0.0	18.6	50.0
7	0.0	25.8	50.0
8	0.0	0.0	50.0
9	0.0	14.0	50.0

The Following Results Are for Nitric Oxide

Diesel Exhaust Concentration Results

Branch Values

BRANCH NUMBER	DIESEL EXHAUST FROM WALLS	CONCENTRATION IN PPM	TLV
1	0.0	21.6	25.0
2	0.0	10.7	25.0
3	0.0	0.0	25.0
4	0.7500	13.9	25.0
5	0.0	0.0	25.0
6	0.3000	56.4	25.0
** THE DIESEL EXHAUST CONCENTRATION EXCEEDED ITS TLV IN BRANCH NO. 6			
7	0.0	0.0	25.0
8	0.0	13.9	25.0
9	0.0	0.0	25.0
10	0.0	13.9	25.0
11	0.0	13.9	25.0
12	0.0	20.0	25.0
13	0.0	0.0	25.0

Junction Values

JUNCTION NUMBER	DIESEL EXHAUST FROM JUNCTION	CONCENTRATION IN PPM	TLV
1	0.0	0.0	25.0
2	0.0	0.0	25.0
3	0.0	13.9	25.0
4	0.2500	21.6	25.0
5	0.0	0.0	25.0
6	0.0	13.9	25.0
7	0.0	20.0	25.0
8	0.0	0.0	25.0
9	0.0	10.7	25.0

The Following Results Are for Nitrogen Dioxide

Diesel Exhaust Concentration Results

Branch Values

BRANCH NUMBER	DIESEL EXHAUST FROM WALLS	CONCENTRATION IN PPM	TLV
1	0.0	16.9	5.0
** THE DIESEL EXHAUST CONCENTRATION EXCEEDED ITS TLV IN BRANCH NO. 1			
2	0.0	4.5	5.0
3	0.0	0.0	5.0
4	0.0500	0.9	5.0
5	0.0	0.0	5.0
6	0.5000	71.8	5.0
** THE DIESEL EXHAUST CONCENTRATION EXCEEDED ITS TLV IN BRANCH NO. 6			
7	0.0	0.0	5.0
8	0.0	0.9	5.0
9	0.0	0.0	5.0
10	0.0	0.9	5.0
11	0.0	0.9	5.0
12	0.0	13.5	5.0
** THE DIESEL EXHAUST CONCENTRATION EXCEEDED ITS TLV IN BRANCH NO. 12			
13	0.0	0.0	5.0

Junction Values

JUNCTION NUMBER	DIESEL EXHAUST FROM JUNCTION	CONCENTRATION IN PPM	TLV
1	0.0	0.0	5.0
2	0.0	0.0	5.0
3	0.0	0.9	5.0
4	0.0001	16.9	5.0
** THE DIESEL EXHAUST CONCENTRATION EXCEEDED ITS TLV AT JUNCTION NO. 4			
5	0.0	0.0	5.0
6	0.0	0.9	5.0
7	0.0	13.5	5.0
** THE DIESEL EXHAUST CONCENTRATION EXCEEDED ITS TLV AT JUNCTION NO. 7			
8	0.0	0.0	5.0
9	0.0	4.5	5.0

The Following Results Are for Hydrocarbons

Diesel Exhaust Concentration Results

Branch Values

BRANCH NUMBER	DIESEL EXHAUST FROM WALLS	CONCENTRATION IN PPM	TLV
1	0.0	77.8	15.0
** THE DIESEL EXHAUST CONCENTRATION EXCEEDED ITS TLV IN BRANCH NO. 1			
2	0.0	24.0	15.0
** THE DIESEL EXHAUST CONCENTRATION EXCEEDED ITS TLV IN BRANCH NO. 2			
3	0.0	0.0	15.0
4	0.6500	12.1	15.0
5	0.0	0.0	15.0
6	2.0000	295.6	15.0
** THE DIESEL EXHAUST CONCENTRATION EXCEEDED ITS TLV IN BRANCH NO. 6			
7	0.0	0.0	15.0
8	0.0	12.1	15.0
9	0.0	0.0	15.0
10	0.0	12.1	15.0
11	0.0	12.1	15.0
12	0.0	63.9	15.0
** THE DIESEL EXHAUST CONCENTRATION EXCEEDED ITS TLV IN BRANCH NO. 12			
13	0.0	0.0	15.0

Junction Values

JUNCTION NUMBER	DIESEL EXHAUST FROM JUNCTION	CONCENTRATION IN PPM	TLV
1	0.0	0.0	15.0
2	0.0	0.0	15.0
3	0.0	12.1	15.0
4	0.2500	77.8	15.0
** THE DIESEL EXHAUST CONCENTRATION EXCEEDED ITS TLV AT JUNCTION NO. 4			
5	0.0	0.0	15.0
6	0.0	12.1	15.0
7	0.0	63.9	15.0
** THE DIESEL EXHAUST CONCENTRATION EXCEEDED ITS TLV AT JUNCTION NO. 7			
8	0.0	0.0	15.0
9	0.0	24.0	15.0

** THE DIESEL EXHAUST CONCENTRATION EXCEEDED ITS TLV AT JUNCTION NO. 9

The Following Are The Values of Tlv of The Mixture of Diesel Exhaust Components

BRANCH NUMBER MIXTURE T.L.V.

1 9.99
** THE MIXTURE OF DIESEL EXHAUST EXCEEDED ITS TLV IN BRANCH NO. 1
2 3.22
** THE MIXTURE OF DIESEL EXHAUST EXCEEDED ITS TLV IN BRANCH NO. 2
3 0.0
4 1.92
** THE MIXTURE OF DIESEL EXHAUST EXCEEDED ITS TLV IN BRANCH NO. 4
5 0.0
6 37.83
** THE MIXTURE OF DIESEL EXHAUST EXCEEDED ITS TLV IN BRANCH NO. 6
7 0.0
8 1.92
** THE MIXTURE OF DIESEL EXHAUST EXCEEDED ITS TLV IN BRANCH NO. 8
9 0.0
10 1.92
** THE MIXTURE OF DIESEL EXHAUST EXCEEDED ITS TLV IN BRANCH NO. 10
11 1.92
** THE MIXTURE OF DIESEL EXHAUST EXCEEDED ITS TLV IN BRANCH NO. 11
12 8.28
** THE MIXTURE OF DIESEL EXHAUST EXCEEDED ITS TLV IN BRANCH NO. 12
13 0.0

JUNCTION NUMBER MIXTURE T.L.V.

1 0.0
2 0.0
3 1.92
** THE MIXTURE OF DIESEL EXHAUST EXCEEDED ITS TLV IN JUNCTION NO. 3

4 9.99
** THE MIXTURE OF DIESEL EXHAUST EXCEEDED ITS TLV IN JUNCTION NO. 4

5 0.0
6 1.92
** THE MIXTURE OF DIESEL EXHAUST EXCEEDED ITS TLV IN JUNCTION NO. 6

7 8.28
** THE MIXTURE OF DIESEL EXHAUST EXCEEDED ITS TLV IN JUNCTION NO. 7

8 0.0
9 3.22
** THE MIXTURE OF DIESEL EXHAUST EXCEEDED ITS TLV IN JUNCTION NO. 9

Index

'*Note*: Page numbers followed by "f" indicate figures, "t" indicate tables.'

O

P

W

Printed and bound by CPI Group (UK) Ltd, Croydon, CR0 4YY
28/08/2024
01029756-0005